GCSE Mathematics

Course Notes, Examples and Exercises

R.C. Solomon M.A., Ph.D.

Bob Solomon has spent many years teaching and writing for mathematics at this level. In addition to this book, he has also written a series for Key Stage 4, and texts for GCSE Mathematics Higher and Advanced Level.

DP Publications Ltd
Aldine House, Aldine Place
142/144 Uxbridge Road,
London W12 8AW
1993

Acknowledgements

My thanks are due to the following:

To Hoverspeed for permission to reproduce their fare schedule.

To British Railways Board at Euston House, 24 Eversholt Street London NW1 for permission to reproduce their timetables.

A CIP catalogue record for this book is available from the British Library

Copyright RC Solomon © 1993

ISBN 1 85805 033 2

Printed by Loader Jackson Printers,
Old Oak Close,
Arlesey, Bedfordshire

Preface

Aim

The aim of this book is to provide all the support needed for a course in GCSE Mathematics. It is assumed that the principles of the subject have been taught, and the book contains all the extra material needed by students, such as course notes, examples, exercises, examination questions and answers.

Need

The need was seen for a book which did not duplicate the teaching of the lecturer. In the author's experience students understand the principles as explained by the lecturer, but have problems in remembering key points and in applying their knowledge.

This book supports any and all levels of the GCSE by providing the lecturer and student with what they want – thus making it easier to follow whilst minimising the cost.

Approach

The book sets out to achieve its aim in a number of ways:

(1) **Graded examples and exercises**. The many worked examples and exercises (with answers) are graded in line with the three tiers of the examination and provide all the practical self-assessment needed. This helps the student identify the level reached and enables progression to the next.

(2) **GCSE style examination questions**. There are twelve test papers modelled on the papers of the various GCSE boards. The answers are provided with marking guides for student self-assessment.

(3) **Common errors**. In every topic of Mathematics there are mistakes which occur frequently. Each chapter of the text contains a list of the common errors relevant to that chapter. This enables students to spot their mistakes and to correct them.

(4) **Summaries of principles**. All the principles, as taught by the lecturer, are summarised at each stage of the text (as appropriate) for student reference.

Preface to third edition

From 1994 the GCSE examination will form the Key Stage 4 assessment of the National Curriculum. This book has been thoroughly revised to take account of the changes in syllabus. The topics are labelled by the appropriate level of the National Curriculum. The content of the book has been changed – for example, the topics of Linear Programming and Critical Paths have been included, and the topic of Sets has been removed.

The marking guide for the exam papers allocates marks to the different Attainment Targets of the National Curriculum. Some of the boards assess Attainment Target 1 (using and applying Mathematics) by means of terminal examination. Appropriate questions are included in the papers.

RC Solomon
July 1993

Contents

Contents

Study and examination hints

1. Study hints

This book has been written to help you to pass the GCSE examination in Mathematics. Of course it must be used in an efficient way. The following hints will help you make good use of your study time.

A good rule to follow is: Little but often. An hour a day for seven days is more useful than seven hours of work on a single day.

Make sure you get a full 60 minutes out of every hour. Do your study in as quiet a room as possible, with all your equipment to hand, and with as few distractions as possible. Do not try to combine study with watching television or listening to the radio.

Write as well as read. There is little to be gained from simply reading Mathematics. Read the explanation and the worked examples, and then try the exercises. Write down the solutions. Do not try to convince yourself that you can do them in your head.

Do not give in too soon. There are answers at the back of the book, but use them for checking that you are right, rather than for finding out how to do a problem.

At the end of the book are twelve exams and three mental tests, for the various levels of the GCSE. The exams should be done under test conditions, with no reference books and a time limit. You are allowed a calculator for these exams. Make sure you are confident in its use. For the mental tests, get someone to read out the questions to you, repeating them once, and taking 20 minutes over each test. (For the first mental test you need to look at the illustrations.) You are not allowed a calculator for these tests.

2. Examination hints

How well you do in an exam depends mainly on how well you have prepared for it. But many candidates do poorly because of bad examination technique. The following hints will help you make the most of your examination time.

Read the question carefully. Very many marks are lost by candidates who have misunderstood a question. Not only do they lose the marks for that question, but they also lose time because the question is made more complicated by their mistake.

Do your best questions first. Try to amass as many marks as possible before tackling the harder questions.

Do not spend too much time on any one question. There is no point in spending a third of the time on a question which carries a twentieth of the marks. Leave that question, and come back to it later if you have time at the end of the exam.

Show your working. If you obtain the wrong answer just because of a slight slip, then you will not lose many marks provided that you show all your working. If you put down the wrong answer without any explanation then you will get no marks at all.

If a question has several parts, then usually each part leads on to the next. Part (a) helps with part (b), part (b) with part (c) and so on.

Make sure that your answer is sensible. If your answer gives eggs weighing 5 kilograms each, or mothers who are younger than their children, then almost certainly it is you who have made a mistake rather than that the question is wrongly set.

Go through your working, but do not spend too much time checking a question which is only worth a couple of marks.

Try to make your solutions neat. Use enough paper to ensure that your answer can be read easily. But do not spend too much time on this – there is no point in spending five minutes on a beautiful diagram for a question which is only worth two marks.

Recognise what each question is about. Many candidates find that they can answer a question when they are doing an exercise consisting of several similar problems, but could not answer the same question when it occurred in a test paper surrounded by different questions. Read each question carefully, and identify the topic or topics of the question. The examiners are not trying to 'catch you out'. They are trying to test that you have studied each part of the syllabus and are able to answer questions on it.

Recognise combinations of topics. Many questions start with a problem in one area of mathematics, and then require you to apply your answer to another, related, area. Make sure you appreciate what the examiners are looking for.

Above all: Make good use of your time! During the months that you are preparing yourself for the examination make sure that each hour of learning or revision is well spent. During the actual examination make the best use of the short time available.

Chapter 1

Numbers

1.1 The Four Operations

Add two numbers to get the *sum*. The sum of 3 and 5 is 8.

Subtract two numbers to get the *difference*. The difference of 9 and 4 is 5.

Multiply two numbers to get the *product*. The product of 2 and 3 is 6.

Divide one number by another to get the *quotient*. The amount left over is the remainder. When 13 is divided by 5, the quotient is 2 and the remainder is 3.

When a number is the product of two other numbers, those numbers are *factors*. 2 and 3 are factors of 6.

Numbers which are divisible by 2 are called *even*: numbers not divisible by 2 are *odd*.

A *prime* number has no factors except 1 and itself.

□ 1.1.1 Examples. *Foundation, levels 4, 5*

1) The postage for an overseas letter is 36p. How can I make this amount up out of 10p and 16p stamps?

   Solution   If I use one 16p stamp, that leaves 20p. This can be paid for by two 10p stamps.

   One 16p and two 10p stamps.

2) Jason needs £235 to buy a bike. He has saved up £57. How much more does he need?

   Solution   He must get the difference between £57 and £235.

   He needs £235 – £57 = £178.

3) I have 20 eggs to be shared fairly among 6 children. How many does each get, and how many eggs are left over?

   Solution   Divide 20 by 6. 6 goes into 20 3 times, with remainder 2. Hence:

   Each child gets 3 eggs, and 2 are left over.

4) Write down the factors of (a) 35 (b) 13.

   Solution   a)   35 can be divided by 5, with quotient 7.

   $$35 = 5 \times 7$$

   b)   13 can only be divided by 1 and 13.

   13 is a prime number

□ 1.1.2 Exercises. *Foundation, levels 4, 5*

1) You have 14p stamps and 19p stamps. How can you use them to make up the following amounts:

   a)   33p   b)   42p   c)   47p   d)   80p?

2) Your bus fare costs 39p, and the bus driver does not give change. How can you pay your fare out of three 10p, three 5p, and four 2p coins? Is there more than one way of paying?

3) You have two jugs which you can fill from the tap. Jug A holds 5 pints, Jug B holds 3 pints. How can you use them to pour out a quantity of:

   a) 9 pints   b)   11 pints   c)   17 pints.

4) A dress costs £23. How much change will there be from three £10 notes?

5) A 25 year mortgage will be paid off in 2000. When was it taken out?

6) A dealer buys a table for £230 and sells it for £355. How much was her profit?

7) A woman was born in 1938. How old will she be in 1996?

8) A cricket team scores 153. Their opponents have scored 34. How many more runs do they need (a) to draw (b) to win?

9) Find the quotient and remainder when:

   a)   40 is divided by 3   b)   72 is divided by 11   c)   100 is divided by 7.

10) 30 marbles are shared fairly between 4 boys. How many does each boy get, and how many are left over?

11) Jane has 83 p to spend on chocolate bars. If each bar costs 12 p, how many can she buy and how much will be left over?

12) Suppose that the calendar is to be decimalised so that a week is 10 days long. How many weeks will there be in a year of 365 days, and how many days will be left over at the end?

13) The hero of a fairy story wins a pair of 7 league boots, in which each stride takes him 7 leagues. (1 league = 3 miles). If he is to travel 300 miles, how many strides should he take? How many miles will be left to travel after he has taken the boots off?

14) Write down the following numbers in figures:

   a)   Four thousand three hundred and sixty three.

   b)   Two hundred and twenty seven thousand, four hundred and twelve.

   c)   Five hundred thousand and sixty two.

15) What is the next whole number after:

   a) 2901   b) 3980   c) 4849   d) 1999?

16) What is the last whole number before:

   a) 213   b) 4210   c) 2000   d) 9,000,000?

17) Fill in the boxes so that the following equations are correct:

   a) $\square + 4 = 13$ b) $\square \times 5 = 35$ c) $28 \div \square = 7$ d) $16 - \square = 4$

18) Paul wrote out some addition and subtraction problems. Ink was spilt on his paper, covering some of the numbers. Complete the sums:

   a)
   ```
     2 3
   + 9 7
   ─────
   □ □ 0
   ```
   b)
   ```
     4 8
   - 3 □
   ─────
     1 3
   ```
   c)
   ```
     2 4 1
   + □ 6 2
   ───────
     4 0 □
   ```
   d)
   ```
     7 4 6
   - □ 2 4
   ───────
     4 2 □
   ```

19) Which of the following numbers are prime? Write down factors of those numbers which are not prime.

   a) 11   b) 6   c) 23   d) 25   e) 29   f) 35   g) 41.

20) Anne has 12 square tiles, and wishes to glue them to the wall in the shape of a rectangle. One of the ways is shown on the right. What are the other possible rectangles she could make?

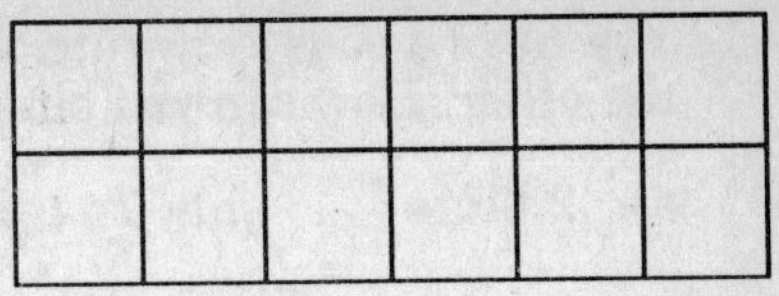

Fig 1.1

HCF and LCM

The greatest number which is a factor of two numbers is the *highest common factor (HCF)* or the *greatest common divisor (GCD)* of the numbers.

The smallest number which is a multiple of two numbers is the *least common multiple (LCM)* of the numbers.

☐ 1.1.3 Examples. *Intermediate, level 7*

1) Find the HCF and the LCM of 12 and 15.

Solution The factors of 12 are 2, 2 and 3, and the factors of 15 are 3 and 5. The greatest common factor is 3.

The HCF of 12 and 15 is 3

If both 12 and 15 are factors of a number, that number must be divisible by 2, 2, 3 and 5.

The LCM of 12 and 15 is $2 \times 2 \times 3 \times 5 = 60$

2) On the left side of a road the lamp-posts are 5 metres apart, and on the right side they are 6 metres apart. At the bottom of the road the lamp-posts are opposite each other. How far is it before two lamp-posts are again opposite each other?

Solution There will be a lamp-post on the left side after every multiple of 5 metres, and a lamp-post on the right after every multiple of 6 metres. There will be a lamp-post on both sides after the LCM of 6 and 5 metres.

Two lamp-posts are opposite after 30 metres.

☐ 1.1.4 Exercises. *Intermediate, level 7*

1) Find the HCF of the following pairs of numbers:

a) 6 and 10   b) 4 and 14   c) 9 and 15   d) 12 and 30   e) 14 and 15.

2) Find the LCM of the pairs of numbers in Question 1.

3) For each pair of numbers in Question 1 multiply together the HCF and the LCM. What do you notice?

4) You have several 2 p coins and several 5 p coins. What is the least sum of money that can be made up entirely from 2 p coins or entirely from 5 p coins?

5) You have a 3 pint jug and a 5 pint jug. What is the least volume which can be filled using either the 3 pint jug only or the 5 pint jug only?

6) Two hands move round a dial: the faster moves round in 6 seconds, and the slower in 8 seconds. If the hands start together at the top, when are they next together at the top?

7) Ben and Fred are playing with a pair of kitchen scales. Ben puts marbles which weigh 6 grams each into the left side, and Fred puts ball-bearings which weigh 15 grams into the right side. What is the smallest weight for which the sides balance?

8) A rectangular area of 150 cm by 120 cm is to be covered with square tiles. We want to use the largest tiles we can, and we don't want to break any. How big should the tiles be?

9) You have a source of water and a 3 pint jug and a 5 pint jug. You are allowed to pour from one jug to the other. How can you obtain volumes of:

a) 2 pints   b) 7 pints   c) 1 pint ?

1.2 Number Patterns

A number pattern is a sequence of numbers following a rule.

☐ **1.2.1 Examples.** *Foundation, levels 4, 5, 6*

1) Find the next 3 terms of the sequence: 1, 4, 7, 10, ...

     Solution   Notice that the terms are going up in steps of 3. Hence the next terms are:

13, 16, 19

2) Fig 1.2 shows a pattern of triangles. Write down the number of dots for each triangle. Continue the pattern of numbers for two more triangles.

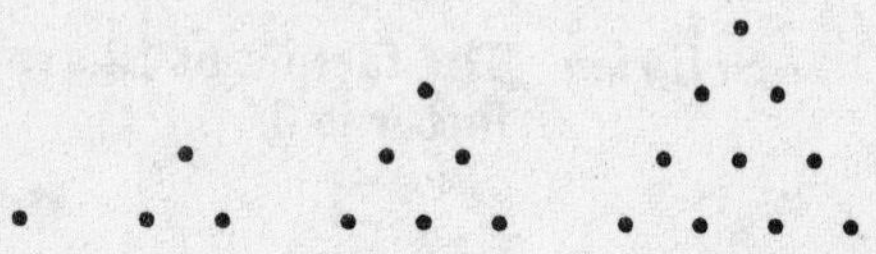

Fig 1.2

     Solution   The numbers of dots in the triangles are 1, 3, 6, 10. The fifth triangle will have one more row than the fourth. This extra row will have 5 dots in it. The next number is 10 + 5 = 15.

     The sixth triangle will have 6 more dots than the fifth.

The next two numbers in the pattern are 15, 21.

☐ **1.2.2 Exercises.** *Foundation, levels 4, 5, 6*

1) Find the next 2 numbers in the following patterns:

   a)   1, 3, 5, 7, 9, ...   b)   1, 6, 11, 16, 21, ...   c)   2, 6, 10, 14, 18, ...

   d)   1, 2, 4, 8, 16, ...   e)   1, 3, 9, 27, ...   f)   3, 6, 12, 24, ...

   g)   128, 64, 32, 16, ...   h)   162, 54, 18, ...   i)   1, 4, 9, 16, 25, ...

   j)   1, 8, 27, 64, ...   k)   2, 8, 18, 32, ...   l)   2, 5, 10, 17, 26, ...

2) For the numbers in Question 1 part (b) write down an odd number and a prime number.

3) For the numbers in question 1 part (l) write down an even number and a prime number.

4) Add together three of the numbers in question 1 part (b) to obtain a total of 28.

5) Add together three of the numbers in question 1 part (l) to obtain a total of 41.

6) Write down the next two lines in the following sets of equations:

   a)
$$2 \times 3 + 1 = 7$$
$$3 \times 4 + 1 = 13$$
$$4 \times 5 + 1 = 21$$

   b)
$$1 \times 3 - 1 = 2$$
$$2 \times 4 - 1 = 7$$
$$3 \times 5 - 1 = 14$$

   c)
$$2 \times 2 = 4$$
$$2 \times 2 \times 2 = 8$$
$$2 \times 2 \times 2 \times 2 = 16$$

   d)
$$1 \times 2 = 2$$
$$1 \times 2 \times 3 = 6$$
$$1 \times 2 \times 3 \times 4 = 24$$

7) In each of the following sequences of figures, write down the numbers of dots in each figure. Find the next two terms in each of the number patterns.

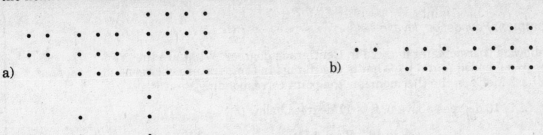

a)

b)

c)

Fig 1.3

8) Each of the following figures is made out of matches. Write down the number of matches for each figure. Find the next two terms the number patterns.

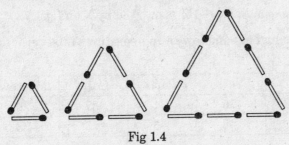

Fig 1.4

9) Repeat Question 8 for the following sequences of figures:

a)

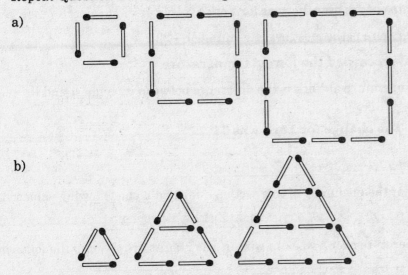

b)

Fig 1.5

10) The Fibonacci numbers are defined so that each number is the sum of the previous two numbers. The first two terms of the sequence are 1 and 1. The next terms are:

$$1 + 1 = 2 : 1 + 2 = 3 : 2 + 3 = 5 : 3 + 5 = 8.$$

Continue the pattern for 3 more numbers.

1.3 Directed Numbers

A line with numbers marked on it is called a *number line*.

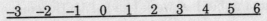

$$-3 \quad -2 \quad -1 \quad 0 \quad 1 \quad 2 \quad 3 \quad 4 \quad 5 \quad 6$$

Fig 1.6

The numbers to the right of 0 are *positive*, or *plus* numbers. The numbers to the left of 0 are *negative*, or *minus* numbers.

☐ 1.3.1 Examples. *Foundation, level 5*

1) Fig 1.7 shows a thermometer marked in Centigrade degrees. What are the temperatures labelled A and B? What is the change in temperature between these values? Mark on the thermometer the point corresponding to –15°.

Solution   A is 16 degrees above 0. B is 11 degrees below 0.

A is 16°, B is –11°

The difference in temperature is the sum of 11 and 16.

The change in temperature is 27°

The point corresponding to –15° is marked by X on fig 1.7.

Fig 1.7

2) The table below gives the highest and lowest temperatures at several towns over the year.

Town	York	Paris	Cairo	Moscow	Sydney
Highest temp.	26	31	37	29	34
Lowest temp.	–8	–5	5	–23	4

a)   Which town recorded the lowest temperature?

b)   What was the change in temperature over the year for York?

Solution   a)   The lowest temperature is the most negative number, –23°.

Moscow recorded the lowest temperature

b)   The change in temperature at York is the difference between its highest and its lowest value.

The change for York was 34°

☐ 1.3.2 Exercises. *Foundation, level 5*

1) Make a copy of the number line at the beginning of this section. Mark on it the following values:

   a) 3   b) 2   c) 0   d) –2   e) $1\frac{1}{2}$ f) $-1\frac{1}{2}$ g) $-\frac{1}{4}$ h) $-2\frac{3}{4}$.

2) John keeps a record of the temperature over a week in winter. His results are given in the following table:

Day	Mon	Tue	Wed	Thu	Fri	Sat	Sun
Highest Temp.	5	4	4	3	6	5	5
Lowest Temp.	0	–2	–1	–5	–3	–1	1

a)   What was the lowest temperature recorded?

b)   Which day had the greatest change in temperature?

3) A ship sails from 20 miles South of the equator to 130 miles North of the equator. How far North has it sailed?

4) I have £150 in my bank account, but last week I had –£55. How much has gone into my account over the week?

5) In Greece the clock is 2 hours ahead of London time, and in California it is 8 hours behind. If you fly from California to Greece, by how much do you alter your watch?

6) Jane stands on the edge of a cliff and throws a stone up in the air. After 1 second it is 3 m. high. After 5 seconds it has fallen to the bottom of the cliff 50 m. below. What distance has it travelled through between these times?

7) The altitudes in metres above sea-level of various places are given in the following table:

Place	London	Nairobi	Amsterdam	Dead Sea
Altitude	30	1800	−2	−400

   a) Which is the lowest place?

   b) How far do you climb when travelling from London to Nairobi?

   c) How far do you climb when travelling from the Dead Sea to Amsterdam?

   d) How far do you descend when travelling from Nairobi to Amsterdam?

8) The mileometer of a car has 5 digits, so that it only goes up to 99,999 miles. It now registers 98,567. What will it register after 3,945 more miles?

9) The level of a river is measured in inches above its average level. During a drought the level of the river is measured each week as follows:

Week	1	2	3	4	5	6
Level	−3	−7	−12	−15	−3	1

   a) What was the highest level? What was the lowest level?

   b) What was the greatest weekly change?

   c) By how much did the river rise from its lowest point to its highest?

10) How many years are there between 150 BC and 370 AD? What was the date 1250 years before the Battle of Hastings? (1066 AD)

11) A man was born in 27 BC, and lived for 62 years. What was the date when he died?

12) A woman was born in 158 BC, and lived for 84 years. What was the date when she died?

13) A man died aged 53 in 21 AD. What was his birth date?

14) The French revolutionaries reformed the calendar by measuring dates after the fall of the Bastille. (1789 AD).

   a) What date corresponds to year 8? What corresponds to year −15?

   b) What date would they give to 1800 AD? What to 1746 AD? What to 213 BC?

1.4 Fractions and Decimals

Fractions express quantities which are not whole numbers.

If a quantity is divided into 8 equal parts, then each of those parts is an *eighth*, or $\frac{1}{8}$, of the whole.

Decimals are another way of writing fractions. The digit one place to the right of the decimal point is the number of tenths, the digit two places to the right is the number of hundredths, and so on.

A fraction is unchanged if the top and the bottom are multiplied or divided by the same number.

❏ **1.4.1 Examples.** *Foundation, levels 4, 5, 6*

1) a) Convert $\frac{1}{4}$ to a decimal.

   b) Express 0.3 as a fraction.

   c) Simplify $\frac{5}{15}$.

Solution a) Divide 4 into 1.

$$\frac{1}{4} = 0.25$$

   b) 0.3 represents three tenths.

$$0.3 = \frac{3}{10}$$

   c) Divide top and bottom of this fraction by 5.

$$\frac{5}{15} = \frac{1}{3}$$

2) A man can walk at 6 km. per hour. How far does he walk in 5 minutes? Express your answer both as a fraction and as a decimal.

Solution 5 minutes is $\frac{1}{12}$ of an hour. The distance travelled in this time is therefore:

$$\frac{6}{12} = \frac{1}{2} \text{ km.} = 0.5 \text{ km.}$$

❏ **1.4.2 Exercises.** *Foundation, levels 4, 5, 6*

1) Express as fractions: a half, a third, a quarter, three quarters, five eighths.

2) Express in words: $\frac{1}{5}, \frac{1}{9}, \frac{5}{6}, \frac{7}{8}$.

3) Convert to decimal form: $\frac{1}{2}, \frac{1}{4}, \frac{3}{8}, \frac{7}{10}$.

4) Express as fractions: 0.5, 0.25, 0.7, 0.35.

5) Simplify as far as possible the following fractions:

$$\frac{2}{4}, \frac{3}{9}, \frac{4}{12}, \frac{7}{14}, \frac{5}{20}.$$

6) Jasmine was attempting to simplify fractions. She had to leave some of them unfinished. Complete them for her:

   a) $\frac{2}{\square} = \frac{1}{2}$ b) $\frac{\square}{6} = \frac{2}{3}$ c) $\frac{15}{20} = \frac{\square}{4}$ d) $\frac{4}{28} = \frac{1}{\square}$

7) 2 cakes are divided fairly between 10 children. What fraction of a cake does each child get?

8) Which is the greatest of $\frac{1}{3}$, 0.3, $\frac{3}{8}$? Which is the least?

9) Fig 1.8 shows several shapes in which part has been shaded. In each case find the fraction which has been shaded.

a)   b) 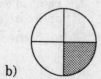   c)   d) 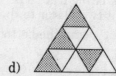   Fig 1.8

10) A supermarket sells rice in 500 g bags for 50 p, or in 200 g bags for 25 p. How much does a gram of rice cost in each bag?

11) The petrol tank of my car holds 20 gallons. How much petrol is there left when the needle is in the position shown?

Mark on the diagram the position of the needle when there are 15 gallons left.

Fig 1.9

12) a) How many minutes are there in half an hour? in three quarters of an hour?

    b) What fraction of an hour is 10 minutes? 5 minutes?

13) Out of 1,000 voters questioned, 450 said they voted Conservative. What fraction voted Conservative?

14) A school has 1,500 pupils, of which one fifth are in the sixth form. How many are not in the sixth form?

Mixed Numbers, Improper Fractions

The top part of a fraction is the *numerator*. The bottom part is the *denominator*.

A fraction such as $\frac{3}{5}$, in which the denominator is greater than the numerator, is called *proper*.

If the denominator is less than the numerator, as in $\frac{7}{5}$, it is called *improper*.

A number such as $5\frac{3}{8}$, containing both fractions and whole numbers, is a *mixed number*.

Positive whole numbers are called *natural numbers*.

Whole numbers, whether positive or negative, are called *integers*.

☐ **1.4.3 Examples.** *Foundation, level 6*

1) a) Express $\frac{15}{8}$ as a mixed number.

    b) Express $5\frac{3}{4}$ as an improper fraction.

    Solution   a) Divide 15 by 8. The result is 1 with remainder 7.

$$\frac{15}{8} = 1\frac{7}{8}$$

    b) Multiply 5 by 4. 5 is equal to $\frac{20}{4}$

$$5\frac{3}{4} = \frac{20}{4} + \frac{3}{4} = \frac{23}{4}$$

2) Three quarters of a school are under 16. 200 are 16 or over. How many are there in the school?

    Solution   One quarter of the school are 16 or over. There are 200 in this quarter.

The school has 4 × 200 = 800 pupils

☐ **1.4.4 Exercises.** *Foundation, level 6*

1) Express the following as mixed numbers:

    a) $\frac{5}{4}$ b) $\frac{7}{3}$ c) $\frac{20}{3}$ d) $\frac{12}{8}$ e) $\frac{27}{6}$.

2) Express the following as improper fractions:

    a) $1\frac{1}{2}$ b) $2\frac{3}{4}$ c) $3\frac{3}{8}$ d) $3\frac{4}{5}$ e) $2\frac{2}{7}$ f) $5\frac{3}{10}$

3) Write the following fractions so that their denominators are 100. Hence convert them to decimals.

    a) $\frac{1}{50}$ b) $\frac{2}{25}$ c) $\frac{7}{10}$ d) $\frac{3}{4}$

4) Convert the following decimals into improper fractions and into mixed numbers:

    a)   1.5   b)   2.3   c)   3.25   d)   1.125.

5) Two thirds of a class pass a French exam. 8 fail the exam. How many are there in the class?

6) Five eighths of a tennis club are women. There are 33 men in the club; how many women are there?

7) A man leaves a quarter of his money to his wife, and the remainder is divided equally between his two sons. What fraction does each son get? If each son gets £3,000, how much does the wife get?

Rational and Irrational Numbers

A fraction is sometimes called a *rational* number, because it is the ratio of two whole numbers.

Numbers which cannot be expressed in this way are *irrational*. There are infinitely many irrational numbers. One example is π, another is $\sqrt{2}$. If a whole number is not a perfect square, then its square root is irrational. Whole numbers are rational numbers. For example 3 can be written as $\frac{3}{1}$. In general, any number whose decimal terminates is rational. 1.75, for example, is $\frac{175}{100} = \frac{7}{4}$.

A decimal where the digits are repeated is called recurring. The digits which repeat are indicated by dots. For example:

$$0.\dot{3} = 0.333... \quad \text{or} \quad 1.7\dot{1}\dot{8} = 1.7181818....$$

Any number whose decimal is recurring is rational.

Numbers whose decimal is neither recurring nor terminating are irrational.

☐ 1.4.5 Example. *Higher, level 9*

Let x be the number 0.181818.... By writing x as a fraction show that it is rational.

    Solution   Shift the decimal point up two places by multiplying by 100.

$$100x = 18.181818...$$

    When we subtract x we obtain a whole number.

$$99x = 18$$

    Divide by 99 and a rational number is obtained.

$$x = \frac{18}{99} = \frac{2}{11}. \quad x \text{ is rational}$$

☐ 1.4.6 Exercises. *Higher, level 9*

1) Show that the following recurring decimals are rational, by converting them to fractions in simplest form.

    a)   0.1111...   b)   0.54545454...   c)   3.4444...

    d)   0.2333...   e)   0.135135...   f)   2.3454545...

2) Which of the following numbers are rational and which irrational? Write the rational numbers in the form $\frac{a}{b}$, where a and b are whole numbers.

a) $\sqrt{3}$ b) $\sqrt{2} \times \sqrt{2}$ c) 2π d) $\sqrt{121}$ e) $\sqrt{114}$ f) $\sqrt{0.25}$

g) $\sqrt{2} + 1$ h) $\sqrt{2} \times \sqrt{8}$ i) $\sqrt{2} \div \sqrt{3}$ j) $\frac{4}{9}$ k) $\sqrt{\frac{4}{9}}$ l) $0.\dot{3}$

3) Find two different irrational numbers whose product is rational.

4) Find two different irrational numbers whose ratio is rational.

5) Find two different irrational numbers whose sum is rational.

6) x and y are non-zero rational numbers: i.e. $x = \frac{a}{b}$ and $y = \frac{c}{d}$, where a, b, c, d are non-zero integers. Show that the following are also rational:

a) $x + y$ b) xy c) $\frac{x}{y}$

7) Find a rational number between 4 and 5.

8) Find an irrational number betweeen 4 and 5.

9) Show that there are infinitely many rational numbers between 0 and 1.

10) Show that there are infinitely many irrational numbers between 0 and 1.

Common errors

1) **The four operations**

a) Be sure that you do the correct operation.

If 5 people each have £300, then the total is found by *multiplying* £300 by 5.

If £600 is shared between 4 people, then each share is found by *dividing* £600 by 4.

b) The LCM of two numbers is not necessarily the product of the two numbers. Certainly the product is a common multiple, but it may not be the *least* common multiple.

Similarly, when finding the HCF of two numbers, be sure that you get the *greatest* common factor.

2) **Patterns**

Do not be too hasty when continuing a sequence. Look at all the terms you are given before deciding how they should go on. There are many possible patterns which begin 1, 2, 4, ... You must know the fourth term at least, before you can continue the pattern.

3) **Directed numbers**

a) The lowest of a set of negative numbers is the one which is most negative. So for example –17 is less than –3. Do not get this the wrong way round.

b) Be careful when finding the difference between directed numbers. The difference between –4 and 9 is 13, not 5.

4) **Fractions and decimals**

a) Notice carefully where the decimal point is. 0.05 is not the same as 0.5.

b) Do not confuse, for example, $\frac{1}{3}$ and 0.3.

Chapter 2

Calculation and Arithmetic

2.1 Mental Arithmetic. *Foundation, Levels 4, 5*

Arithmetic involving single digits can be done mentally. For example, you should know by heart the multiplication table up to 10 by 10. If the digits have 0's after them, as in 600 or 20, then when we multiply them we add the 0's. When we divide them we subtract the 0's.

❐ **2.1.1 Example.** *Foundation, Level 5*

Evaluate mentally: a) 600×20 b) $600 \div 20$.

    Solution a) Multiply the digits, obtaining 12. Follow this by three 0's.

$$600 \times 20 = 12,000$$

    b) Divide the digits, obtaining 3. Follow this by one 0.

$$600 \div 20 = 30$$

❐ **2.1.2 Exercises.** *Foundation, Levels 4 and 5*

Evaluate the following mentally:

1) a) $2 + 3 + 5$ b) $4 + 3 + 8$ c) $2 + 9 + 1$

     d) $7 + 5 - 2$ e) $4 + 9 - 3$ f) $4 + 6 - 7$

     g) $9 - 3 - 1$ h) $8 - 3 - 3$ i) $7 - 2 - 5$

2) a) 5×7 b) 6×8 c) 4×9

     d) 6×4 e) 3×9 f) 7×8

3) a) $28 \div 4$ b) $18 \div 3$ c) $45 \div 9$

     d) $42 \div 7$ e) $56 \div 8$ f) $63 \div 9$

4) a) 40×20 b) 30×200 c) $200 \times 4,000$

     d) 60×20 e) 400×900 f) $700 \times 8,000$

5) a) $800 \div 20$ b) $9,000 \div 30$ c) $8,000 \div 400$

     d) $120 \div 30$ e) $3,200 \div 80$ f) $63,000 \div 900$

Dividing and Multiplying by Decimals

When we multiply by 0.2, we multiply by 2 and *divide* by 10.

When we divide by 0.2, we divide by 2 and *multiply* by 10.

❐ **Example 2.1.3.** *Intermediate, Level 7*

Evaluate mentally a) 600×0.3 b) $8,000 \div 0.04$.

Solution a) Multiply the digits, obtaining 18. Then remove one of the 0's.

$$600 \times 0.3 = 180$$

b) Divide the digits, obtaining 2. Then add on two 0's.

$$8,000 \div 0.04 = 200,000$$

❐ **Exercises 2.1.4.** *Intermediate, Level 7*

Evaluate the following mentally

1) a) 400×0.2 b) $5,000 \times 0.3$ c) $7,000 \times 0.5$

   d) $60,000 \times 0.02$ e) 400×0.01 f) $600,000 \times 0.0007$

2) a) $12,000 \div 0.6$ b) $4,000 \div 0.2$ c) $25,000 \div 0.5$

   d) $270 \div 0.03$ e) $360,000 \div 0.09$ f) $1,000 \div 0.0005$

2.2 Arithmetic on Paper. *Foundation, Levels 4 and 5*

If a calculation is too complicated to be done mentally, then it can be done on paper.

❐ **2.2.1 Example.** *Foundation, Levels 4 and 5*

Evaluate without a calculator: a) $135 + 418$ b) $936 - 217$
c) 325×23 d) $432 \div 24$.

Solution a) Place the numbers under each other and add, making sure to "carry" 1 when 5 and 8 are added.

$$\begin{array}{r} 135 \\ 418 \\ \hline 553 \end{array}$$

$$135 + 418 = 553$$

b) Place the numbers under each other and subtract, making sure to "borrow" from 3 when subtracting 7 from 6.

$$\begin{array}{r} 936 \\ 217 \\ \hline 719 \end{array}$$

$$936 - 217 = 719$$

c) Multiply 325 by 3 and by 20, and add the results.

$$\begin{array}{r} 325 \\ 23 \\ \hline 975 \\ 650 \\ \hline 7475 \end{array}$$

$$325 \times 23 = 7,475$$

d)   Divide 24 into 43, find the remainder and bring down the 2. Then divide again by 24.

$$18$$

$$24)\overline{432}$$

$$\underline{24}$$

$$192$$

$$\underline{192}$$

432 ÷ 24 = 18

❑ **2.2.2 Exercises.** *Foundation, Levels 4 and 5*

Evaluate the following without a calculator:

1)	a)	352 + 542	b)	634 + 352	c)	381 + 558
	d)	492 + 745	e)	749 + 298	f)	846 + 859
2)	a)	745 − 395	b)	846 − 419	c)	782 − 764
	d)	824 − 641	e)	549 − 286	f)	742 − 488
3)	a)	234 × 4	b)	537 × 3	c)	639 × 7
	d)	385 × 6	e)	639 × 9	f)	375 × 8
4)	a)	408 ÷ 4	b)	624 ÷ 3	c)	492 ÷ 2
	d)	815 ÷ 5	e)	392 ÷ 7	f)	520 ÷ 8
5)	a)	311 × 14	b)	525 × 16	c)	672 × 26
	d)	492 × 45	e)	829 × 79	f)	999 × 99
6)	a)	812 ÷ 14	b)	966 ÷ 21	c)	864 ÷ 36
	d)	840 ÷ 56	e)	738 ÷ 82	f)	968 ÷ 11

2.3 Use of Calculator

Modern calculators follow the natural order of operations. To find out what – *Five plus six equals* – press the buttons in that order.

$$\boxed{5} \; \boxed{+} \; \boxed{6} \; \boxed{=}$$

The answer 11 appears.

❑ **2.3.1 Example.** *Foundation, level 4*

Evaluate 4.3 + 5.2 + 11.9.

   Solution   Press the buttons in the order of the sum.

$$\boxed{4} \; \boxed{.} \; \boxed{3} \; \boxed{+} \; \boxed{5} \; \boxed{.} \; \boxed{2} \; \boxed{+} \; \boxed{1} \; \boxed{1} \; \boxed{.} \; \boxed{9} \; \boxed{=}$$

4.3 + 5.2 + 11.9 = 21.4

☐ **2.3.2 Exercises.** *Foundation, level 4*

1) Use your calculator to evaluate the following.

   a) $2.7 + 5.6 + 1.1$ b) $6.3 + 6.6 + 9.8$ c) $12.43 + 43.65 + 23.84$

   d) $9.5 - 5.6$ e) $12.43 - 8.65$ f) $2.6 + 4.8 - 5.9$

   g) 2.65×4.65 h) 54.91×35.20 i) $2.5 \times 5.4 \times 8.2$

   j) $10.34 \div 2.2$ k) $10.85 \div 3.1$ l) $109.48 \div 9.52$

   m) $4.6 \div 0.23$

2) Find the total price of a meal costing £5.64 and drinks costing £3.17.

3) Find the total weight of coffee weighing 456 grams and its tin weighing 23 grams.

4) David is 12.3 km North of his home, and then walks 7.8 km further North. How far is he from home now?

5) I pay for a £7.89 pair of shoes with a £10 note. How much change should I get?

6) Patricia buys 10 metres of flex, and cuts off 3.65 metres. How much is there left?

7) The total weight of a full bottle is 1,019 grams. If the liquid weighs 923 grams, how much does the glass weigh?

8) The total cost of a shirt and a tie was £11.34. If the tie cost £3.72 find the cost of the shirt.

9) Find the cost of 3 kg of meat at £3.88 per kg.

10) Find the cost of 6 gallons of petrol at £2.15 per gallon.

11) Find the cost of 3.42 metres of flex at 50 p per metre.

12) Cinema tickets cost £3.50 per adult and £2.60 per child. How much does it cost for a family of two adults and three children to go to the cinema?

13) 5 pounds of apples cost £1.85. How much does each pound cost?

14) Seven yards of curtain material cost £10.01. How much does each yard cost?

15) £2,001 is shared equally between three sisters. How much does each sister get?

16) How many cakes will Joe be able to buy for £1.68, if each cake costs 24 p?

17) Complete the bill of fig 2.1.

    How much change will there be from a £20 note?

4 galls. petrol	@ £1.75 =	a
2 litre oil	@ £1.31 =	£2.62
	Total	b

Fig 2.1

18) The Jackson family stayed for a weekend at a hotel. Fig 2.2 shows their bill. Complete the bill.

6 nights at £4.50 each = £27

6 meals at £3.10 each = a

Total = b

Fig 2.2

19) Mr Singh receives his electricity bill as shown in fig 2.3. Fill in all the numbers.

Units used	Unit Price	Amount
580	4.75	a
	Standing charge	£9.73
	Total	b

Fig 2.3

20) A certain book has 432 pages, with an average of 392 words per page. Find the total number of words in the book.

When the book comes out in paperback there will be an average of 576 words per page. How many pages will there be?

Use of Brackets.

Multiplication and division are done before addition and subtraction. If you want to do the addition or subtraction first then you must use *brackets*.

$$2 \times 3 + 4 = 6 + 4 = 10 \qquad\qquad 2 \times (3 + 4) = 2 \times 7 = 14$$

A scientific calculator can work out calculations involving brackets. It is also possible to use the memory to store the result of a calculation.

☐ 2.3.3 Examples. *Intermediate, Level 7*

1) Nancy and Peggy go on holiday to France. Nancy takes £25, Peggy takes £34. How much do they have in French money, if there are 10.6 French Francs to the £?

    Solution Their total sum of money is £(25 + 34). In FF this becomes

    $$(25 + 34) \times 10.6$$

    Because the addition sum is in brackets, it must be done before the multiplication.

    $$25 + 34 = 59$$

    Now multiply by 10.6:

    They have 625.4 FF.

2) Evaluate $3 \times 2.4 + 5 \times 4.7 + 7 \times 9.3$.

    Solution If you have a scientific calculator, then you can press the keys in the same order as the problem.

    $$\boxed{3}\ \boxed{\times}\ \boxed{2.4}\ \boxed{+}\ \boxed{5}\ \boxed{\times}\ \boxed{4.7}\ \boxed{+}\ \boxed{7}\ \boxed{\times}\ \boxed{9.3}\ \boxed{=}$$

    If you have an elementary calculator, it may not be able to work out the problem all in one go. In this case you must perform all the multiplications and then add the results.

    $$3 \times 2.4 = 7.2 : 5 \times 4.7 = 23.5 : 7 \times 9.3 = 65.1$$

    $7.2 + 23.5 + 65.1 = 95.8$

3) Evaluate $\dfrac{13.3 \times 4.17}{2.4 - 2.21}$ without writing anything down.

    Solution This calculation can be done on a calculator using either brackets or the memory.

    One way is to put the bottom line in brackets. The sequence of operations is:

    $$\boxed{13.3}\ \boxed{\times}\ \boxed{4.17}\ \boxed{\div}\ \boxed{(}\ \boxed{2.4}\ \boxed{-}\ \boxed{2.21}\ \boxed{)}\ \boxed{=}$$

    Alternatively, put the bottom line into the memory and divide it into the top line. The sequence is:

    $$\boxed{2.4}\ \boxed{-}\ \boxed{2.21}\ \boxed{=}\ \boxed{Min}\ \boxed{13.3}\ \boxed{\times}\ \boxed{4.17}\ \boxed{\div}\ \boxed{MR}\ \boxed{=}$$

    The answer should be the same whichever method is used.

    $$\frac{13.3 \times 4.17}{2.4 - 2.21} = 291.9$$

❏ **2.3.4 Exercises.** *Intermediate, level 7*

1) Evaluate the following:

a) $5 \times 2 + 3$ b) $5 \times (2 + 3)$

c) $3 \times 7 - 5$ d) $3 \times (7 - 5)$

e) $5 + 8 \div 4$ f) $(5 + 8) \div 4$

g) $15 - 9 \div 3$ h) $(15 - 9) \div 3$

2) Insert brackets into the following equations to make them correct.

a) $3 \times 4 + 6 = 30$ b) $12 - 5 \times 2 = 14$

c) $7 - 1 \div 2 = 3$ d) $6 \div 17 - 14 = 2$

3) Evaluate the following:

a) $2 \times 9.3 + 3 \times 3.4 + 7 \times 2.9$ b) $11 \times 3.3 + 21 \times 6.8 - 42 \times 1.9$

c) $1.43 \times 5.73 + 4.43 \times 9.21 + 8.67 \times 2.21$ d) $34.2 \times 0.3 + 45.7 \div 0.2$

4) Evaluate the following:

a) $3.7 \times (3.8 + 2.8)$ b) $98.34 \times (5.32 - 5.29)$

c) $(34.83 + 57.39) \times (45.43 - 39.87)$ d) $\dfrac{4.95 - 2.99}{6.39 + 2.36}$

5) A second-hand book dealer is buying books: he buys 43 at 25 p, 87 at 35 p and 85 at 55 p. How much does he pay in all?

6) Muhammed is thinking of decorating his flat. He reckons that he will have to buy:

6 litres of emulsion paint at £1.75 per litre.
2 litres of gloss paint at £3.77 per litre.
2 brushes at £2.55 each.
1 litre of white spirit at £1.05.

How much will he have to spend in all?

7) A full bottle of whisky weighs 823 grams, of which 748 grams is the weight of the whisky. How many bottles can be made from 1,200 grams of glass?

2.4 Fractions and Negative Numbers.

When fractions are added or subtracted, they must first be put over the same denominator.

When fractions are multiplied, the tops are multiplied together and the bottoms are multiplied together.

To *divide* by a fraction a/b, *multiply* by b/a.

The product of two negative numbers is positive.

$$-3 \times -2 = 6.$$

The product of a negative number and a positive number is negative.

$$3 \times -2 = -6.$$

❒ **2.4.1 Examples.** *Intermediate, Level 8*

1) Evaluate $\frac{2}{5} + \frac{1}{4}$, leaving your answer as a fraction.

Solution Write both the fractions with a denominator of 20.

$$\frac{2}{5} + \frac{1}{4} = \frac{8}{20} + \frac{5}{20} = \frac{13}{20}$$

2) Evaluate   a) $\frac{3}{4} \times \frac{2}{5}$ b) $\frac{3}{4} \div \frac{2}{5}$.

Solution a) Multiply the tops together and the bottoms together.

$$\frac{3}{4} \times \frac{2}{5} = \frac{6}{20} = \frac{3}{10}$$

b) To divide by $\frac{2}{5}$ is to multiply by $\frac{5}{2}$.

$$\frac{3}{4} \div \frac{2}{5} = \frac{3}{4} \times \frac{5}{2} = \frac{15}{8} = 1\frac{7}{8}$$

3) Evaluate $(3.2 - 5.8) \times (2.7 - 4.1)$

Solution The subtractions in brackets must be evaluated first.

$$(3.2 - 5.8) \times (2.7 - 4.1) = (-2.6) \times (-1.4)$$

This is now the product of two negative numbers.

$$-2.6 \times -1.4 = 3.64$$

❒ **2.4.2 Exercises.** *Intermediate, Level 8*

1) Evaluate the following, leaving your answers as fractions.

a) $\frac{2}{3} + \frac{1}{4}$ b) $\frac{3}{5} + \frac{1}{6}$ c) $\frac{1}{7} + \frac{2}{3}$

d) $\frac{3}{4} - \frac{2}{5}$ e) $\frac{1}{4} + \frac{1}{2}$ f) $\frac{1}{10} + \frac{3}{5}$

g) $\frac{5}{7} - \frac{1}{3}$ h) $\frac{7}{8} - \frac{1}{5}$

2) Evaluate the following, leaving your answers as fractions.

a) $\frac{2}{3} \times \frac{4}{7}$ b) $\frac{5}{6} \times \frac{2}{15}$ c) $\frac{3}{2} \times \frac{3}{5}$

d) $\frac{2}{7} \times \frac{5}{3}$ e) $\frac{1}{3} \times 9$ f) $1\frac{1}{3} + 2\frac{1}{4}$

g) $4\frac{1}{5} - 3\frac{2}{3}$ h) $2\frac{7}{8} - 3\frac{1}{4}$ i) $\frac{1}{10} - 4\frac{1}{4}$

j) $\frac{2}{3} \div \frac{4}{7}$ k) $\frac{1}{2} \div \frac{3}{5}$ l) $\frac{5}{9} \div 3$

m) $\frac{11}{20} \div \frac{1}{40}$

3) Evaluate the following:

a) -3.2×-4 b) $-12 \div -6$ c) $7 - (3 - 5)$

d) $\dfrac{(-1.71)}{(3.5 - 7.3)}$ e) $-2 \times -3 \times 4$ f) $-4 \times -0.5 \times -6$

g) $(-3)^2$ h) $(-2)^5$.

4) In a horse race there was a very close finish. Amanda was $\frac{1}{3}$ of a length ahead of Brenda, and Brenda was $\frac{1}{2}$ a length ahead of Celine. By what fraction of a length did Amanda beat Celine?

5) George buys $1\frac{3}{4}$ yards of wire, and cuts off $\frac{1}{3}$ of a yard. How much does he have left?

6) $2\frac{1}{4}$ pints are poured from a 15 pint container. How much is left?

7) A man has three children: in his will he leaves half his estate to his wife, a quarter to his eldest child and a fifth to the middle child. What fraction does the youngest child get?

8) How many $\frac{1}{2}$ pint jugs can be filled from a 20 pint container?

9) I buy a batch of 32 apples, of which a quarter are bad. How many are good?

10) A ship sails at $12\frac{1}{2}$ km.p.h. How long will it take to cover 75 km.?

11) Michael walks one quarter of the distance to school, and then takes the bus. If the bus ride is $1\frac{1}{2}$ miles, how far does he walk?

12) Ailsa and Beatrice share a sum of money so that Ailsa gets two thirds of it. If Beatrice receives £70, how much was the sum?

13) A swimming bath is 50 metres long. Sharon can swim $2\frac{1}{2}$ lengths. How far can she swim?

14) Gabriel takes $1\frac{1}{2}$ minutes to peel an apple. How many apples can he peel in 9 minutes?

15) The profit of a firm is divided equally between its three partners. Each partner then has to pay three-tenths of his share in tax. What fraction of the original profit does each partner keep?

16) A cook uses eleven twelfths of a pound of flour to make 20 cakes. How much flour does each cake contain?

Common errors

1) **Use of calculator**

a) Don't forget to press the = button after a calculation. And don't press it *twice* – if you do so the results are unpredictable.

b) Be sure that you use brackets correctly. Do not confuse $2 \times 3 + 4$ and $2 \times (3 + 4)$.

$2 \times 3 + 4$ means: *Multiply 2 and 3 and then add 4.* The result is 10.

$2 \times (3 + 4)$ means: *Add 3 and 4 and then multiply by 2.* The result is 14.

2) **Fractions**

a) When adding fractions, do not add the numerators together and the denominators together.

$$\frac{2}{3} + \frac{4}{5} \neq \frac{6}{8}.$$

b) Do not invert fractions.

$$\frac{1}{3} + \frac{1}{6} = \frac{1}{2}. \text{ But } 3 + 6 \neq 2.$$

3) **Negative numbers**

Be careful when multiplying or dividing negative numbers. It is easy to forget the rules:

plus times plus is plus

plus times minus is minus

minus times minus is plus.

Chapter 3

Approximation

3.1 Rounding

When a number is *rounded* to a whole number, it is written as the nearest whole number. When a number is rounded to the *nearest 100*, it is written as the nearest multiple of 100.

☐ **3.1.1 Examples.** *Foundation, levels 3, 4*

1) Round to a whole number:

   a) 1.35   b) $7\frac{5}{8}$ c) $-2.8°$.

   Solution   a)   1.35 is nearer 1 than 2.

   Round to 1

   b) $7\frac{5}{8}$ is nearer 8 than 7.

   Round to 8

   c) $-2.8°$ is nearer $-3°$ than $-2°$.

   Round to $-3°$

2) Round 234,635   a)   to the nearest 100   b)   to the nearest 1,000   c)   to the nearest 10.

   Solution   a)   35 is nearer 0 than 100.

   Round to 234,600

   b)   635 is nearer 1,000 than 0.

   Round to 235,000

   c)   5 is halfway between 0 and 10. It could be rounded up or down. By convention it is usually rounded up.

   Round to 234,640

3) A train travels for 7 hours at 46 m.p.h. How far has it travelled, to the nearest 10 miles?

   Solution   The distance is the product of the speed and the time.

   $$7 \times 46 = 322 \text{ miles.}$$

   320 miles to the nearest 10 miles.

☐ **3.1.2 Exercises.** *Foundation, levels 3, 4, 5*

1) Round the following numbers to the nearest whole number:

   a) 3.56   b) 4.267   c) 129.67   d) 987.46

   e) 5.5   f) -8.45 g) -76.76 h) 0.67

i) 0.43   j) −0.87   k) −0.34   l) $9\frac{1}{4}$

m) $76\frac{1}{3}$ n) $\frac{6}{7}$

2) Round 1,097,382 to the nearest a) 1,000   b) 100   c) 10.

3) Round 463 to the nearest   a) 10   b) 100   c) 1,000.

4) Round −67.4 to the nearest a) whole number   b) 10   c) 100.

5) Work out the following, giving your answer to the nearest whole number:

a) 34.7 + 54.86   b) 2.9 − 1.7

c) 3.5 + 2.9 + 6.4   d) 10.6 + 2.3 − 8

6) Work out the following, giving your answer to the nearest whole number.

a) 3.6×4.7 b) $2.8 \times 3.2 \times 9.6$

c) $93 \div 7$ d) $\frac{213}{31}$

e) $0.012 \div 0.0023$ f) $\frac{(2.43 - 2.39)}{0.047}$

7) What is the cost of 17 litres of petrol at 47.8 p per litre? Give your answer to the nearest penny.

8) What is the cost of 144 kg of tomatoes, if 1 kg costs 74 p? Give your answer to the nearest £.

9) I buy 46 yards of material for £632. What is the cost per yard, given to the nearest penny?

10) To hire a tennis court costs £1.30 per hour. What is the cost per minute, given to the nearest penny?

11) I travel at 55 m.p.h. for $3\frac{1}{2}$ hours. How far have I travelled? Give your answer to the nearest 10 miles.

12) I drove for a distance of 200 miles at a speed of 70 m.p.h. How long did it take me, expressed to the nearest hour?

13) £10,000 is shared out equally between 17 people. How much does each get, expressed to the nearest £10?

3.2 Significant Figures and Decimal Places

The left-most non-zero digit of a number is the *first significant figure*. The next digit, even if it is zero, is the second significant figure.

When a number is rounded to *2 significant figures* the digits to the right of the second significant figure are rounded up or down.

The third digit after the decimal point of a number is in the *3rd decimal place*.

When a number is rounded to *3 decimal places* the digits to the right of the third decimal place are rounded up or down.

❒ **3.2.1 Examples.** *Foundation, levels 4, 5*

1) Express the following to 3 significant figures:

a) 234,576   b) 1,003   c) 0.00000012743.

Solution a) The third significant figure is the 4. The figures to the right of it must be rounded up, giving:

235,000

b) The third significant figure is the 0 in front of the 3. The 3 is rounded down, to give:

1,000

c) The third significant figure is the 7. Round down the remaining figures:

0.000000127

2) Express the following to 2 decimal places:

a) 12.358   b) 5.1029

Solution a) The digit in the second decimal place is the 5. The next figure is 8, so round up:

12.36

b) The digit in the second decimal place is the 0. The next figure is 2, so round down:

5.10

☐ **3.2.2 Exercises.** *Foundation, levels 4, 5*

1) Round the following to 2 significant figures:

a) 2,347   b) 656,265   c) 0.004654   d) 0.0249

e) -3.47   f) 1.03   g) 2.09   h) 497

i) 602   j) 5.096   k) 4.000098

2) Round 4,738 to   a) 3 sig. figs.   b) 2 sig. figs.   c) 1 sig. fig.

3) Round the following to 3 decimal places:

a) 3.5764   b) 5.2349   c) 23.03823   d) 65.4758

e) 0.3457   f) 0.1234   g) 0.00476   h) 0.00012

i) 0.12034   j) 0.26976

4) Round 0.13582 to   a) 4 dec. places   b) 3 dec. places   c) 2 dec. places.

5) Evaluate the following, giving your answer to 2 decimal places.

a) $4.564 + 2.372$ b) $2.4638 - 0.9564$ c) 2.432×3.276

d) $\dfrac{5}{7}$ e) $2.574 \div 0.945$ f) 1.35^2

g) $\dfrac{(1.032 - 1.029)}{7}$

6) 17 cm^3 of a chemical compound weigh 25 grams. What is the density of the compound? Give your answer to 2 decimal places.

7) What is the weight of 288 ball-bearings, each of which weighs 2.3 grams? Give your answer to 2 significant figures.

8) What is the cost of 27 litres of petrol at 48.3 p per litre? Give your answer to the nearest £.

9) The figure shows the display of a calculator after 25 has been divided by 7. Express the result to

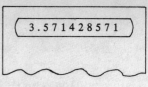

   a) 4 significant figures   b) 2 decimal places.

Fig 3.1

Accuracy of Measurement

If a measurement is given to 1 decimal place, then the true value could be up to 0.05 on either side of the value given.

☐ **3.2.3 Example.** *Intermediate, Level 7*

When rounded to the nearest £10, the amount of money in my bank account is £120. What are the greatest and the least possible amounts that there could be?

  Solution   An amount over £125 would be rounded up to £130. An amount below £115 would be rounded down to £110.

  The greatest possible amount is £125, and the least £115

☐ **3.2.4 Exercises.** *Intermediate, Level 7*

1) A temperature is given as 34°, rounded to the nearest whole number. What is the greatest possible value of the temperature? What is the least?

2) The weight of a car is given as 910 kg, rounded to the nearest 10 kg. What is the greatest possible weight of the car?

3) The population of a town is given as 80,000, to the nearest 10,000. What are the greatest and the least possible values of the population?

4) The weight of a book is given as 0.54 kg, rounded to 2 decimal places. What are the greatest and the least possible values for the weight?

5) The distance between two cities is given as 320 miles, to 2 significant figures. What are the greatest and the least possible values for the distance?

6) A distance is given as 3.5 cm, to 1 decimal place. Between what limits must the true weight lie?

7) A time of a race is given as 9.74 seconds, to 2 decimal places. Between what limits must the true time lie?

Combining Errors

When two quantities are added or subtracted, the errors are added.

When two quantities are multiplied or divided, we can find the bounds within which the result must lie.

The greatest possible value of X ÷ Y is found by taking the greatest possible value of X divided by the least possible value of Y.

☐ **3.2.5 Examples.** *Higher, level 10*

1) The weight of a full bottle is 960 grams, and the weight of its contents is 740. Both figures are given to the nearest 10 grams. Between what limits must the weight of the empty bottle lie?

  Solution   The error in both measurements is up to 5 grams. So the error in their difference is 5 + 5 = 10 grams. The weight of the empty bottle could be 10 grams on either side of 220 grams.

  The weight lies between 210 and 230 grams

2) A rectangle was measured as 13.5 cm long and 26.7 cm wide. What is the area of the rectangle? Give your answer to an appropriate degree of accuracy.

Solution The area is $13.5 \times 26.7 = 360.45$ cm^2. This gives an answer to 2 decimal places, while the original lengths were only given to 1 decimal place.

The length of the rectangle could be as high as 13.55. The width could be as high as 26.75. Its area could then be as high as 362 cm^2.

The answer can only be given to 2 significant figures:

The area is 360 cm^2

☐ **3.2.6 Exercises.** *Higher, Level 10*

1) The times for a relay race were 15.3 seconds, 16.8 seconds, 14.9 seconds, 13.6 seconds, each time being given to 1 decimal place. Between what limits does the total time lie?

2) The weight of a car is 560 kg to the nearest 10 kg, and the weights of the driver and passenger are 73 kg and 64 kg to the nearest kg. What is the least possible weight of the car with the two people inside?

3) The distance to a town is 12 miles. I walk 5 miles towards it. If both distances are given to the nearest mile, what is the least distance I now have to walk to reach the town?

4) There is £840 in my bank account. I spend £230. If both sums are given to the nearest £10, what is the greatest possible amount left in my account?

5) An airplane flies for 8 hours (to the nearest hour) at a speed of 600 m.p.h. (to the nearest 50 m.p.h.) What is the least possible distance it could have covered?

6) I must drive 140 miles (to the nearest 10 miles) at a speed of 45 m.p.h. (to the nearest 5 m.p.h.) What is the greatest possible time the journey could take?

7) Evaluate the following, giving your answer to an appropriate degree of accuracy.

   a) $3.5 + 4.6 + 7.3 + 5.573$ b) $8.3 + 9.4 - 10.345$

   c) 7.32×4.17 d) 34.3×83.4

   e) $2 \div 0.07$ f) $23.6 \div 8.45$

3.3 Checks by Approximation

The result of a calculation can be checked by rounding each number to 1 significant figure.

☐ **3.3.1 Example.** *Foundation, level 6*

Evaluate $4.83 + 3.21 + 5.92$. By rounding each number to 1 significant figure, check that your answer is correct.

Solution Use a calculator or pencil and paper to do the addition.

$$4.83 + \ 3.21 + 5.92 = 13.96$$

Rounding the numbers to 1 significant figure, the sum becomes:

$$5 + 3 + 6 = 14$$

This verifies that the answer above is correct.

❏ **3.3.2 Exercises.** *Foundation, level 6*

1) Evaluate the following. Make checks to verify that your answers are correct.

    a)   3.2 + 5.6   b)   8.3 + 2.8   c)   6.29 + 9.18

    d)   7.35 − 2.28   e)   98.34 − 22.19 f)   1012 − 879

    g)   8.2 × 9.7   h)   6.3 × 2.2   i)   97 × 62

    j)   8.2 ÷ 1.8   k)   8.7 ÷ 2.8   l)   69 ÷ 2.1

2) You buy items costing £8.88 and £7.10. Approximately how much change do you expect from a £20 note?

3) A full bottle weighs 893 grams, and the liquid inside weighs 762 grams. What is the approximate weight of the empty bottle?

4) The four runners of a relay race took 22.54 seconds, 25.83 seconds, 29.41 seconds, 24.71 seconds. What was the approximate total time for the race?

5) A car travels at 59.3 m.p.h. for a journey of 123.4 miles. Approximately how long does the journey take?

6) A single ball bearing weighs 1.132 grams. Approximately how much do 969 of the ball bearings weigh?

❏ **3.3.3 Exercises.** *Intermediate, level 8*

1) Evaluate the following. Then perform a test of accuracy.

    a)   (2.37 × 4.92) ÷ 1.69   b)   (3.12 + 5.39) × (4.19 + 4.39)

    c)   (93.2 + 67.3) ÷ (0.31 + 0.51)   d)   4.97 ÷ (2.17 + 3.20)

2) Sums of £3,856, £2,038 and £5,891 are to be shared equally between 63 people. Approximately how much does each person get?

3) Cans of soup contain 475 grams of liquid, and the weight of the metal is 48 grams. What is the approximate weight of 9,578 cans?

3.4 Appropriate Limits of Accuracy

A measurement should be given to an *appropriate* number of decimal places or an *appropriate* number of significant figures.

❏ **3.4.1 Example.** *Foundation, level 6*

The radius of the earth is 6,400,000 metres, to 2 significant figures. Find the circumference of the earth, giving your answer to an appropriate degree of accuracy.

Solution The formula for the circumference of a circle of radius r is $2\pi r$. Using this formula, the circumference of the earth is:

$$2 \times \pi \times 6,400,000 = 40,212,386 \text{ to the nearest metre.}$$

This answer gives a misleading impression of accuracy. It is given to 8 significant figures. The radius of the earth is only given to 2 significant figures, so the answer can only be given to 2 significant figures.

The circumference of the earth is 40,000,000 metres.

☐ **3.4.2 Exercises.** *Foundation, level 6*

1) 9 inches of rain fell in one week. What was the average rainfall per day? Give your answer to an appropriate degree of accuracy.

2) A ball-bearing weighs 3.2 grams, to 1 decimal place. How much do 111 similar ball-bearings weigh?

3) 6 cakes weigh 400 grams. How much does each cake weigh?

4) The distance of the earth from the sun is 90,000,000 miles. Assuming the orbit of the earth to be a circle, how long is the earth's orbit? (A circle with radius r has length $2\pi r$.)

5) What would be appropriate levels of accuracy for the following measurements:

   a) The distance from London to Paris.

   b) The population of Birmingham.

   c) Your weight.

   d) Your height.

   e) Your mother's height.

   f) The top speed of Concorde.

   g) The price in June 1992 of a gallon of four-star petrol.

Common errors

1) **Rounding**

   a) When you round a number to the nearest 100 do not discard the 0's. That would mean that you have *divided* the number by 100.

   $$3,468 \text{ to the nearest 100 is 3,500, not 35.}$$

   b) When you are asked to give the answer to a calculation in a rounded form, do the rounding right at the end of your calculation. For example:

   $$3.4 + 4.3 = 7.7 = 8 \text{ to the nearest whole number.}$$

   Do not do the rounding before. This would give the wrong answer:

   $$3.4 + 4.3 = 3 + 4 = 7.$$

   c) If a quantity is given as 240 to the nearest 10, then the greatest possible value it could have is 245, not 250. Similarly the least value it could have is 235, not 230.

2) **Significant figures and decimal places**

   a) The first significant figure is the first *non-zero* figure. So the first significant figure of 0.0023 is 2, not 0.

   b) After the first significant figure, zeros count as well as non-zero digits. The first 2 significant figures of 103 are 1 and 0, not 1 and 3.

   c) If you are rounding to 2 decimal places, and the digit in the second place is 0, then leave it in. This shows the degree of accuracy of the calculation or measurement.

   $$\text{To 2 decimal places, } 2.497 = 2.50, \text{ not } 2.5.$$

   d) Save the rounding until the very end of your calculation. A useful rule is to work to more decimal places than you are asked to give. So if your final answer is to be given to 2 decimal places, use at least 3 decimal places until the end of your calculation.

e) If two quantities are subtracted, then the errors are not subtracted, but added.

To find the greatest value of one quantity divided by another, do not take the greatest values of both quantities. Take the greatest value of the top and the least value of the bottom.

3) **Appropriate accuracy**

Do not give your answer to too many decimal places or too many significant figures.

If a cake weighing 1 kg is divided into 3 equal portions, then your calculator will tell you that each portion is 0.33333333 kg. It would be misleading to give this as the answer. Round it to 0.3 or 0.33 kg.

Checks of accuracy are very useful, but don't expect too much of them. They ensure, for example, that your answer is not 10 times to big or 10 times too small.

Chapter 4

Indices

4.1 Squares, Cubes, Square Roots

When a number is multiplied by itself, the result is the *square* of the number.

$$3 \times 3 = 3^2$$

If a number is multiplied by itself 3 times, the result is the cube of the number.

$$4 \times 4 \times 4 = 4^3$$

49 is the square of 7. So 7 is the *square root* of 49.

$$49 = 7^2, \text{ so } 7 = \sqrt{49}.$$

To use a calculator to find a square root, press the number and then the square root button. (Do not press the = button). To find 1024, press:

$$\boxed{1}\ \boxed{0}\ \boxed{2}\ \boxed{4}\ \boxed{\sqrt{}}$$

The answer 32 will appear.

❑ 4.1.1 Examples. *Foundation, level 5*

1) Evaluate the following:

   a) 5^2 b) 3^3 c) $\sqrt{64}$.

   Solution   Use the definitions.

$$\textbf{a)} \quad \mathbf{5^2 = 5 \times 5 = 25}$$

$$\textbf{b)} \quad \mathbf{3^3 = 3 \times 3 \times 3 = 27}$$

   c)   Note that $8^2 = 64$. Hence:

$$\mathbf{\sqrt{64} = 8}$$

2) a)   What is the area of a square whose side is 4 inches?

   b)   A square has area 36 cm^2. What is the side of the square?

   Solution   a)   The area of the square is found by squaring the side. This gives:

$$\textbf{Area} = \mathbf{4^2 = 16 \text{ square inches.}}$$

      b)   When the side is squared, the result is 36 cm^2. This gives:

$$\textbf{The side is } \mathbf{\sqrt{36} = 6 \text{ cm.}}$$

❑ 4.1.2 Exercises. *Foundation, level 5*

1) Evaluate the following:

   a) 3^2 b) 8^2 c) 2^2 d) 9^2

e) 10^2 f) 12^2 g) 100^2

2) Evaluate the following:

a) $\sqrt{4}$ b) $\sqrt{9}$ c) $\sqrt{16}$ d) $\sqrt{64}$

e) $\sqrt{100}$ f) $\sqrt{121}$

3) Evaluate the following:

a) 2^3 b) 4^3 c) 10^3 d) 1.5^2

e) $(\frac{1}{2})^2$ f) $(\frac{1}{3})^2$ g) 0.5^3

4) Work out the following:

a) $\sqrt{10,000}$ b) $\sqrt{81}$ c) $\sqrt{144}$ d) $\sqrt{1}$

e) $\sqrt{\frac{1}{4}}$ f) $\sqrt{\frac{9}{4}}$

5) Complete the following by filling in the boxes:

a) $7^2 + 6^2 = 49 + \Box = \Box$ b) $5^2 - 4^2 = 5^2 - \Box = \Box$

c) $\sqrt{9} + \sqrt{36} = 3 + \sqrt{36} = \Box$ d) $\sqrt{81} - \sqrt{49} = \sqrt{81} - 7 = \Box$

6) A square lawn has side 12 feet. What is the area?

7) A card-table has a square top, of side 40 cm. What is the area of the top?

8) A square of carpet is 4 m^2 in area. What is the side of the carpet?

9) A square of material is 25 cm^2. What is the side of the material?

10) A cube has side 3 cm. What is the volume of the cube?

11) The volume of a cube is 64 cm^3. What is the side of the cube?

4.2 Integer Powers

When a number is multiplied by itself n times, the result is the *n'th power* of the number.

$$3 \times 3 \times 3 \times 3 \times 3 = 3^5 = 243.$$

The *negative* power of a number is 1 over the positive power.

$$3^{-2} = \frac{1}{3^2} = \frac{1}{9}$$

When powers are multiplied, the indices are *added*.

$$3^4 \times 3^5 = 3^9$$

When powers are divided, the indices are *subtracted*.

$$5^6 \div 5^3 = 5^3.$$

The index notation can be used to represent an integer as a power of primes.

$$360 = 5 \times 3 \times 3 \times 2 \times 2 \times 2 = 5 \times 3^2 \times 2^3$$

□ 4.2.1 Examples. *Intermediate, level 7*

1) Evaluate the following:

   a) 2^5 b) 4^{-2}

   Solution   Use the definition:

   $$a) \quad 2^5 = 2 \times 2 \times 2 \times 2 \times 2 = 32$$

   $$b) \quad 4^{-2} = \frac{1}{4^2} = \frac{1}{16}$$

2) Express as a single power of 5:

   $$\frac{5^3 \times 5^5}{5^6}$$

   Solution   The indices are added along the top line of the fraction, and the bottom index is subtracted. This gives:

   $$5^{3+5-6} = 5^2$$

□ 4.2.2 Exercises. *Intermediate, levels 7, 8*

1) Evaluate the following:

   a) 3^4 b) 4^4 c) 1^6 d) 2^{10}

   e) $(\frac{1}{2})^3$ f) 0.3^3 g) 1.5^3 h) $(-2)^3$

2) Evaluate the following:

   a) 2^{-1} b) 3^{-3} c) 5^{-3} d) 4^{-3}

   e) 1^{-7} f) $(\frac{1}{2})^{-1}$ g) $(\frac{1}{3})^{-2}$

3) Simplify the following:

   a) $3^2 \times 3^4$ b) $2^4 \times 2^3$ c) $4^3 \div 4^2$

   d) $3^5 \times 3^{-4}$ e) $7^4 \times 7^{-3}$

4) Simplify the following:

   a) $4^3 \div 4^{-2}$ b) $3^5 \div 3^{-1}$ c) $(\frac{1}{2})^3 \times (\frac{1}{2})^{-3}$ d) $(\frac{3}{2})^5 \div (\frac{2}{3})^4$

5) Express as a single power of 2:

   $$\frac{2^5 \times 2^5}{2^3 \times 2^6}$$

6) Express as a single power of 3:

   $$\frac{3^8}{3^7 \times 3^4}$$

7) Use your calculator to find the following, giving your answer to 3 decimal places:

a) $\sqrt{2}$ b) $\sqrt{3}$ c) $\sqrt{65}$ d) $\sqrt{0.2}$

e) $\sqrt{0.00005}$.

8) Find $\sqrt{1\,000}$, giving your answer to:

a) 1 significant figure   b) 2 significant figures

c) 5 significant figures.

9) Evaluate the following, giving your answer to 2 decimal places:

a) $2 + \sqrt{3}$ b) $35 - \sqrt{7}$

c) $\dfrac{\sqrt{2} + \sqrt{5}}{3}$ d) $\dfrac{5}{\sqrt{6} - \sqrt{3}}$

10) Insert $\sqrt{}$ signs to make the following true:

a) $9 = 3$ b) $4 = 16$ c) $100 = 10,000$

11) Insert indices to make the following true:

a) $4 = 2^*$ b) $8 = 2^*$ c) $27 = 3^*$

d) $3^* = 81$ e) $100,000 = 10^*$

12) Express the following as products of prime numbers, using the index notation.

a) 8   b) 12   c) 75

d) 36   e) 1024   f) 180

13) What is the largest power of 2 which divides 56?

14) Which is the largest odd number which divides 360?

4.3 Fractional Indices and Laws of Indices

Any non-zero number raised to the power 0 is 1.

$$a^0 = 1 \text{ (Provided that } a \neq 0)$$

A fraction index $1/n$ gives the nth root.

$$a^{1/n} = \sqrt[n]{a}$$

Indices obey the following rules:

$$a^n \times a^m = a^{n+m}$$

$$a^n \div a^m = a^{n-m}$$

$$(a^n)^m = a^{nm}$$

$$a^{-n} = 1/a^n$$

$$(ab)^n = a^n b^n.$$

☐ **4.3.1 Examples.** *Higher, level 9*

1) Evaluate the following:

   a) 3^0 b) $4^{1/2}$

   Solution   Use the definitions:

$$a) \quad 3^0 = 1$$

$$b) \quad 4^{1/2} = \sqrt{4} = 2$$

2) Simplify the following:

   a) $5^3 \times 25^2 \div 125^2$ b) $16^{3/4}$

   Solution   a)   The numbers concerned here are all powers of 5. Re-write as:

$$5^3 \times (5^2)^2 \div (5^3)^2$$

$$5^3 \times 5^4 \div 5^6$$

   Add and subtract the indices:

$$5^{3 + 4 - 6} = 5^1 = 5$$

   b)   Re-write $16^{3/4}$ as $(16^{1/4})^3$

   The fourth root of 16 is 2. This gives:

$$2^3 = 8$$

3) Solve the equation $3^x = 9^x + 1$.

   Solution   Write 9 as a power of 3.

$$3^x = (3^2)^{x+1}$$

$$3^x = 3^{2x+2}$$

   Equate the powers of 3, to obtain $x = 2x + 2$. $x = -2$

☐ **4.3.2 Exercises.** *Higher, level 9*

1) Evaluate the following:

   a) $9^{1/2}$ b) $100^{1/2}$ c) $8^{1/3}$

   d) $49^{-1/2}$ e) $64^{-1/3}$ f) $1000^{1/3}$

2) Evaluate the following:

   a) $8^{2/3}$ b) $9^{3/2}$ c) $1000^{2/3}$

   d) $\left(\dfrac{1}{4}\right)^{-2}$ e) $\left(\dfrac{1}{4}\right)^{-2} - \left(\dfrac{1}{2}\right)^{-2}$ f) $0.01^{-1/2}$

3) Simplify the following:

   a) $4^3 \times 2^6 \div 8^3$ b) $27^2 \times 9^2 \times 3^{-7}$ c) $10^{1/2} \times 100^{1/4}$

4) Simplify the following:

a) $x^{\frac{1}{2}} \times x^{\frac{1}{2}}$ b) $y^{\frac{1}{4}} \times y^{\frac{3}{4}}$ c) $3^x \times 9^x \div 27^x$

5) Solve the following equations:

a) $2^x = 4^{x-1}$ b) $25^{2x+1} = 125^{x+2}$ c) $8^x = 32$

4.4 Standard Form

Standard form is used for very large or very small numbers. A number is in standard form when there is only one digit to the left of the decimal point. For example:

$$256,700 = 2.567 \times 10^5$$

Numbers in standard form can be entered in a calculator. To enter 3×10^{12} press the following:

$$\boxed{3} \;\; \boxed{\text{EXP}} \;\; \boxed{1} \;\; \boxed{2}$$

If the power of 10 is negative, press the ± button after the power.

☐ 4.4.1 Examples. *Intermediate, level 8*

1) Write the following in standard form:

a) 34,560,000   b) 0.000000045

Solution   a)   Move the decimal point 7 places to the left. This is balanced by increasing the power of 10 by 7.

$$3.456 \times 10^7$$

b)   Move the decimal point 8 places to the right. This is balanced by decreasing the power of 10 by 8.

$$4.5 \times 10^{-8}$$

2) Evaluate the following, leaving your answers in standard form.

a) $3 \times 10^6 \times 4 \times 10^5$ b) $4 \times 10^9 + 5 \times 10^8$

Solution   a)   Multiply the two numbers together:

$$3 \times 4 \times 10^6 \times 10^5$$

$$12 \times 10^{11}$$

This is numerically correct, but it is not in standard form.

$$1.2 \times 10^{12}$$

b)   Before these numbers can be added, they must have the same power of 10.

$$40 \times 10^8 + 5 \times 10^8$$

$$45 \times 10^8$$

This is numerically correct, but it is not in standard form.

$$4.5 \times 10^9$$

□ **4.4.2 Exercises.** *Intermediate, Level 8*

1) Express the following in standard form:

   a) 23,000   b) 876,000,000   c) 0.000123   d) $\dfrac{1}{1000}$

2) Evaluate the following, leaving your answer in standard form:

   a) 1000×100

   b) $3 \times 10^4 \times 7 \times 10^4$

   c) $\dfrac{900}{10\,000}$

   d) $1.6 \times 10^8 \div 2$

   e) $5.2 \times 10^{-5} \times 2 \times 10^{-8}$

   f) $1.2 \times 10^5 \div 4 \times 10^2$

   g) $5 \times 10^5 \times 4 \times 10^{-3}$

   h) $(5 \times 10^3)^2$

   i) $(2 \times 10^4)^4$

   j) $(4.2 \times 10^{-3})^4$.

   k) $(1.25 \times 10^{-4})^{-2}$.

3) Evaluate the following, leaving your answer in standard form:

   a) $4 \times 10^5 + 8 \times 10^6$

   b) $3.3 \times 10^6 + 2.7 \times 10^5$

   c) $9.5 \times 10^5 + 8 \times 10^4$

   d) $7.2 \times 10^7 - 8 \times 10^6$

   e) $4 \times 10^8 - 3 \times 10^9$

   f) $9 \times 10^{-3} + 8 \times 10^{-4}$

4) Light travels at 1.86×10^5 miles per second. How far does it travel in an hour? How far does it travel in a year?

5) The weight of a carbon atom is 2×10^{-27} grams. How many carbon atoms are there in 10 grams of pure carbon?

6) On the number line below indicate the numbers

   a) 2   b) 1.1×10^1 c) 2×10^{-1} d) -2×10^{-1}

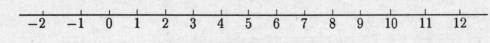

Fig 4.1

Common errors

1) **Squares and square roots.**

   Be careful not to confuse the square of a number with twice that number.

   $$a^2 \neq 2a.$$

   Similarly, $\sqrt{a} \neq \dfrac{1}{2}a$

2) **Arithmetic of indices**

   There are very many common errors when calculating with indices. Usually these occur because *taking powers* has been confused with *multiplying or dividing*. Here are some common mistakes to watch out for:

   $$5^3 \times 5^2 \neq 5^6$$

   $$5^3 + 5^2 \neq 5^5$$

   $$5^3 \times 4^2 \neq 20^5$$

   $$(3^5)^2 \neq 9^{10}$$

35

3) Standard form

a) Be sure that there is only one digit to the left of the decimal point. If there is more or less than one then the number is not in standard form.

b) The errors of indices mentioned above also occur when dealing with standard form. Do not make the following mistakes:

$$2 \times 10^5 \times 3 \times 10^6 \neq 6 \times 10^{30}$$

$$4 \times 10^6 + 5 \times 10^6 \neq 9 \times 10^{12}$$

c) You can only add or subtract numbers in standard form when they have the same power of 10. (Just as you cannot directly add centimetres and kilometres.) The numbers must be changed so that they do have the same power of 10, and then they can be added or subtracted.

Chapter 5

Ratio

Numbers can be compared by *ratios, scales,* and *proportions*.

These are all ways of expressing fractions.

5.1 Ratios and Scales

If two numbers A and B are in the *ratio* 3:1, that means that A is 3 times the size of B.

The ratio does not have to be a whole number. If C and D are in the ratio 5:4, then C is $\frac{5}{4}$ times the size of D.

The *scale* of a map compares the lengths on the map to the lengths on the ground. A scale of 1 cm. to 2 km means that 1 cm. on the map represents 2 km. on the ground.

☐ 5.1.1 Examples. *Foundation, levels 5, 6*

1) A school contains 350 boys and 250 girls. Find the ratio of the number of boys to the number of girls, giving your answer in as simple a form as possible.

Solution   The ratio is 350:250. Ratios obey the same rule as fractions. Both terms can be divided by 50. The ratio is then:

$$7:5$$

2) Two partners Smith and Jones invest in a business in the ratio 4:5. If Smith put in £20,000, how much did Jones put in?

Solution   Jones invests $\frac{5}{4}$ of the amount that Smith invested. Hence Jones invested;

$$\frac{5}{4} \times £20,000 = £25,000$$

3) The scale of a map is 1 cm for 2 km. A distance on the map is measured as 5 cm. How far is it on the ground?

Solution   For this scale, 1 cm represents 2 km. So 5 cm represents 5 times as much. The distance on the ground is:

$$5 \times 2 = 10 \text{ km.}$$

☐ 5.1.2 Exercises. *Foundation, Levels 5 and 6*

1) Joan has £20 and John has £15. What is the ratio between these amounts?

2) A girl is 5 ft high and her father is 6 ft high. What is the ratio of their heights?

3) A man is 32 and his daughter is 8. What is the ratio of their ages?

4) A recipe for pastry requires 8 oz of flour to 4 oz of butter. What is the ratio of flour to butter?

5) There are twice as many boys as girls in a certain school. Express this as a ratio of the number of boys to the number of girls.

6) Three quarters of the population of a nation live in towns. Express this as a ratio of the town-dwellers to the country-dwellers.

7) For a two-stroke motorcycle the fuel consists of petrol and oil in the ratio 20:1. How much petrol should be bought with $\frac{1}{2}$ litre of oil?

8) Joan and Jackie have weights in the ratio 5:4. If Joan weighs 50 kg, how much does Jackie weigh?

9) An alloy consists of copper and tin in the ratio 5:2. How much copper goes with 12 kg of tin?

10) Two friends Jane and Sandra share a flat, and agree to pay the rent in the ratio 8:5. If Sandra contributes £20 per week, how much does Jane pay?

11) The scale of a map is 1 cm per 1 km. A distance on the map is measured as 3 cm. How far is it on the ground?

12) The scale of a map is 1 cm for 2 km. The distance between two towns is 10 km. How far apart are the towns on the map?

13) The model of a aircraft is in the scale 1:80. If the real aircraft is 40 metres long, how long is the model?

14) An architect's model of a block of flats is in the scale 1:50. If the model is 0.5 metres wide, how wide is the block of flats?

15) A model ship is $\frac{1}{2}$ metre long, and the real boat is 50 metres long. What is the scale of the model?

16) A photograph is enlarged in the ratio 2:3. If the photo was 6 cm high before, how high is it after the enlargement?

❏ 5.1.3 Examples. *Intermediate, level 7*

1) A legacy of £8,000 is divided between two people in the ratio 3:5. How much does each get?

Solution The sum is divided up into 3 + 5 = 8 equal shares.

Each share is worth £8,000 ÷ 8 = £1,000.

The first person gets 3 of these shares, and the second person gets 5 shares.

The shares are £3,000 and £5,000

2) Gunpowder contains salt-petre, charcoal and sulphur in the ratio 6:1:1. How much salt-petre does 40 pounds of gunpowder contain?

Solution This example shows that ratios may be used to compare the sizes of three or more things. In this case there are 6+1+1 = 8 shares, each of which is $\frac{40}{8}$ = 5 pounds. The salt-petre has 6 of these shares, and hence it weighs:

$6 \times 5 = 30$ pounds.

❏ 5.1.4 Exercises. *Intermediate, Level 7*

1) In a will money is left to Alfred and Bonnie in the ratio 3:4. If the total sum of money was £21,000, how much did Alfred get?

2) An amalgam mixes mercury and silver in the ratio 3:2. How much silver is needed to make 10 kg of amalgam?

3) A line is divided in the ratio 2:13. If the line is 30 cm long, how long is the larger part?

4) A tennis club has men and women members in the ratio 4:5. If there are 270 members, how many of these are women?

5) Mr Robinson and Mr Gillespie buy a race horse; Mr Robinson contributes £8,000 and Mr Gillespie contributes £4,000.

    a) What is the ratio of their contributions?

    b) The horse wins a race. If they divide the prize money of £6,000 in the same ratio, how much does Mr Robinson get?

6) Building works for a school are to be funded by the local council and by the government in the ratio 4:1. If the building costs £50,000, how much does the government pay?

7) An exam consists of two papers. The first lasts 1 hour and the second $1\frac{1}{2}$ hours. If marks are awarded in the same ratio as the time, how many marks out of 100 are accounted for by the first paper?

8) Money is left to three sons in the ratio 3:4:5. If the total sum of money left is £24,000, how much does each son get?

9) A mixed fruit drink contains orange juice, pineapple juice and lemon juice in the ratio 6:3:1. How much pineapple juice is there in 800 cc of the mixture?

10) A man finds that he divides his day between work, play and sleep in the ratio 5:3:4. For how many hours per day is he asleep?

11) The three movements of a concerto are in the ratio 5:2:3. If the concerto lasts for 20 minutes, how long are the movements?

Ratios of Areas and Volumes

If two lengths are in the ratio $a:b$, then the corresponding areas are in the ratio $a^2:b^2$, and the corresponding volumes are in the ratio $a^3:b^3$.

5.1.5 Examples. *Higher, level 9*

A map is in the scale 1:50,000. A lake on the map covers 5 cm^2. How big is the real lake?

Solution   The lake on the map can be regarded as being 1 cm by 5 cm. The scale of the map means that 1 cm represents $\frac{1}{2}$ km, so the map lake represents $\frac{1}{2}$ km by $2\frac{1}{2}$ km on the ground. Hence the area of the lake is:

$$\frac{1}{2} \times 2\frac{1}{2} = 1\frac{1}{4} \text{ km}^2$$

2) A model of a room is in the scale 1:10. If the model is 1.2 m^3 in volume, what is the volume of the real room?

Solution   The height, the length and the width of the real room are each 10 times as big as in the model. Hence the volume of the real room is $10 \times 10 \times 10$ times as big as the model's volume.

$$\text{Real volume} = 10^3 \times 1.2 = 1,200 \text{ m}^3$$

5.1.6 Exercises. *Higher, level 9*

1) The scale of a map is 1 cm to $\frac{1}{2}$ km. A field on the map is 1 cm^2. How large is the real field?

2) The scale of a map is 1 cm to 1 km. How large on the map is a forest of 20 km^2?

3) The scale of a map is 1:20,000. An area on the map is 3 cm^2. What real area does it correspond to?

4) A map of England is in the scale 1:1,000,000. What area on the map corresponds to a county of 500 km^2?

5) An architect's model of a house is in the scale 1:10. If the living-room is 600 m^3, what is the volume of the model of the living-room?

6) How much does 1 cm correspond to on maps with the following scales:

   a)   1:100,000   b)   1:50,000   c)   2:50,000?

7) On a map a road of length $\frac{1}{2}$ km is represented by a distance of 2 cm. What is the scale of the map?

8) Find the scale of a map, if 7 km is represented by $3\frac{1}{2}$ cm.

9) A model of a ship is in the scale 1:200. If the model has volume 500 cm^3, what is the volume of the ship?

10) A model plane is in the scale 1:60.

   a)   The model is 8 cm long. How long is the plane?

   b)   The area of the wings is 10 m^2. What is the area of the wings of the model?

   c)   The volume of the fuselage of the model is 70 cm^3. What is the volume of the real fuselage?

   d)   The model has 2 wings. How many wings are there on the real plane?

11) A model locomotive is in the scale 1:90.

   a)   The locomotive is 15 metres long. How long is the model?

   b)   The front of the model is 20 cm^2 in area. What is the area of the front of the real locomotive?

   c)   The locomotive weighs 80 tonnes. Assuming that the model is made out of exactly the same materials as the real locomotive, what is the weight of the model?

12) A child has a model of Superman. The model is 10 cm high, while Superman is 2 metres high.

   a)   What is the scale of the model?

   b)   Superman's vest is 0.8 m^2 in area. What is the area of the vest on the model?

   c)   The model has ten fingers. How many fingers does Superman have?

13) A photograph is enlarged in the ratio 2:3. If sky occupies 10 cm^2 of the original photo, how much area does it occupy of the enlargement?

14) A map shows a park of area 3 km^2 as a region of 12 cm^2. What is the scale of the map?

5.2 Proportion

If two quantities increase at the same rate, so that if one is doubled then the other is doubled, the quantities are *proportional*.

If two quantities increase at opposite rates, so that if one is doubled then the other is halved, the quantities are *inversely proportional*.

□ **5.2.1 Examples.** *Foundation, level 6*

1) A pastry recipe requires 8 oz of flour for 4 oz of butter. How much butter should be used to go with 16 oz of flour?

   Solution   Here the weight of flour is proportional to the weight of butter. The amount of flour has been doubled. Hence the amount of butter should also be doubled.

   $$2 \times 4 = \textbf{8 oz of butter}$$

2) A man normally drives to work at 30 m.p.h., and it takes him 10 minutes. On a fine day he decides to cycle instead, and he can cycle at 10 m.p.h. How long does it take him?

   Solution   Here the time taken is inversely proportional to the speed. The speed has been divided by 3, and so the time taken will be multiplied by 3.

   $$\textbf{Time taken} = 3 \times 10 = \textbf{30 minutes.}$$

□ **5.2.2 Exercises.** *Foundation, Level 6*

1) A recipe for mayonnaise tells the cook to use one egg yolk for each 100 ml of oil. How much oil will mix with 3 egg yolks?

2) A journey of 100 miles costs £6. Assuming that the cost per mile is constant, how much will a journey of 200 miles cost?

3) A car travels 34 miles on one gallon of petrol. How far will it get on a tankful of 8 gallons?

4) 40 cm^3 of a metal weigh 120 grams. What is the weight of 30 cm^3?

5) 10 metres of curtain material cost £32. How much do 15 metres cost?

6) The recipe of fig 5.1 for kofte is for 6 people. Write down the ingredients for 9 people.

| 2lb beef 2 Onions |
| 2Tbs ground rice |
| 3Tbs parsley |

Fig 5.1

7) Fig 5.2 shows the ingredients for Welsh Rarebit. Write down the quantities if 12 ounces of cheese are to be used.

| 4oz cheese | 3Tbs milk |
| ¼tsp mustard | 1Tsp flour |

Fig 5.2

8) A certain journey takes 3 hours at 40 m.p.h. How long would it take at 80 m.p.h.?

9) When a sum of money is divided between 8 children, each gets £3. If there were only four children, how much would each get?

10) It takes 3 men 4 days to complete the painting of a house. How long would it take 6 men?

11) After a motorway is completed, the average speed increases from 40 m.p.h. to 60 m.p.h. A certain journey took 3 hours before. How long does it take now?

Laws of Proportionality

If two quantities are proportional (or inversely proportional), then there is an equation connecting them. If y is proportional to x, the equation is of the form $y = kx$, where k is a constant.

Sometimes a quantity is proportional to the *square* of another quantity. (Or inversely proportional to the square.)

The relationship of proportionality is sometimes written using the symbol α. "y is proportional to x" can be written as:

$$y \, \alpha \, x$$

☐ **5.2.3 Examples.** *Higher, Level 9*

1) The extension E of a spring is proportional to the tension T. A tension of 5 N. gives an extension of 4 cm. Find an equation giving E in terms of T, and use it to find the extension for a load of 7 N.

Solution E is proportional to T. So E is a multiple of T. Write:

$$E = kT$$

Here k is a constant to be found from the figures given. $E = 4$ when $T = 5$, so:

$$4 = k \times 5$$

$$k = \frac{4}{5}$$

The equation is $E = \frac{4}{5} \times T$

Put $T = 7$ into this equation:

$$E = \frac{4}{5} \times 7 = 5.6 \text{ cm}$$

2) For a fixed voltage, the current I along a wire is inversely proportional to the resistance R. $I = 4$ amps when $R = 60$ ohms. Find an equation linking I and R, and find the resistance necessary for a current of 6 amps.

Solution   Here the equation is of the form:

$$I = \frac{k}{R}.$$

Put in the values given;

$$4 = \frac{k}{60}$$

$$k = 240.$$

The equation is $I = \frac{240}{R}$.

Put $I = 6$ into this equation:

$$6 = \frac{240}{R}$$

$$R = \frac{240}{6} = 40 \text{ ohms.}$$

3) The weight W of a metal disc is proportional to the square of the radius r. If the weight of a disc with radius 2 cm is 100 grams, find an equation giving W in terms of r. Find the radius of a disc whose weight is 400 grams.

Solution   The formula is:

$$W \alpha r^2.$$

This is equivalent to the equation:

$$W = kr^2.$$

Put $W = 100$ and $r = 2$, to obtain the equation:

$$100 = k \times 4$$

$$k = 25$$

$$W = 25r^2$$

Put $W = 400$ into the equation:

$$400 = 25r^2$$

$$r^2 = 16$$

$$r = 4 \text{ cm.}$$

☐ 5.2.4 Exercises. *Higher, Level 9*

1) y is proportional to x. $y = 5$ when $x = 10$. Find an equation giving y in terms of x, and find y when $x = 5$.

2) R is proportional to S. When $S = 6$, $R = 18$. Find   a) R when $S = 24$, b) S when $R = 90$.

3) For a wire of fixed resistance, the current I is proportional to the voltage V. $I = 7$ amps for $V = 12$ volts.   a)   Find the current if $V = 6$ volts. b) Find the voltage necessary to transmit a current of 70 amps.

4) Distances on a map are proportional to distances on the ground. If 5 cm. corresponds to 20 km, what does 25 cm correspond to?

5) p is inversely proportional to q. For $p = 7$, $q = 10$. Find an equation linking p and q. Use this equation to find q when $p = 4$.

6) F and G are inversely proportional, and F is 6 when $G = 8$.   a)   Find F when $G = 12$.   b)   Find G when $F = 3$.

7) The pressure P of a fixed mass of gas is inversely proportional to the volume V. $P = 50$ kg m^{-2} when $V = 4$ m^3. Find the equation which gives V in terms of P. Find the volume of the gas when the pressure is 40 kg m^{-2}.

8) The wavelength l of certain waves is inversely proportional to the frequency f. When $f = 250,000$ cycles per second the wave length is 1000 metres.

   a)   Find the frequency of waves with length 5,000 m.

   b)   Find the length of waves with frequency 1,000,000 cycles per second.

9) y is proportional to the square of x. $y = 16$ when $x = 2$.

   a)   Find y in terms of x.

   b)   Find y when $x = 3$.

   c)   Find x when $y = 4$.

10) T is proportional to the square of S. When S is 3 then T is 18.

   a)   Find the equation linking T and S.

   b)   Find T when S is 6.

   c)   Find S when T is 162.

11) The energy of a moving body is proportional to the square of the speed. A body moving at 5 m/sec has 50 Nm of energy. Express the energy in terms of the speed. If the body has energy 200 Nm, how fast is it moving?

12) The power generated by an electrical circuit is proportional to the square of the current. When the current is 3 amps the power is 72 Watts. Find an expression for the power P in terms of the current I. Use your expression to find P when $I = 5$ amps.

13) The energy E stored in a spring is proportional to the square of the extension e. Express this fact using the symbol α. If the extension is 3 for an energy of 7, find an equation linking e and E. Find E when $e = 1$.

14) T is inversely proportional to the square of R. $T = 5$ when $R = 2$.

   a) Find the equation giving T in terms of R.

   b) Find T when $R = 4$.

   c) Find R when $T = 80$.

15) B is inversely proportional to the square of C. $B = \frac{1}{2}$ for $C = 2$. Find the expression linking B and C.

   Find B when $C = \frac{1}{2}$, and find C when $B = 8$.

16) The illumination from a light source is inversely proportional to the square of the distance from that light source. If a light gives 4 candle-power at a distance of 4 m, find the illumination at a distance of 6 metres.

17) The resistance R of a fixed length of wire is inversely proportional to the square of the radius r. When the radius is $\frac{1}{2}$ cm, the resistance is 3 ohms. Find R in terms of r. Find the resistance if the radius is $\frac{1}{4}$ cm. What radius should be chosen to give a resistance of $\frac{3}{4}$ ohms?

18) Cylindrical cans are to be made containing a fixed volume. It can be shown that the height is inversely proportional to the square of the radius. A can of height 20 cm has a base radius of 4 cm. What is the base radius of a can of height 45 cm?

19) The square of the time for a planet to revolve around the sun is proportional to the cube of its distance from the sun. The Earth is 93 million miles from the sun. Mars is 142 million miles from the sun. Find the length of the Martian year.

20) y is proportional to a power of x. The following table gives various values of y and x. Find the power, and find y when $x = 6$.

x	2	3	4
y	12	27	48

21) It is thought that T is inversely proportional to a power of S. When S is doubled T decreases by a factor of 8.

   What is the power? If S is divided by 3, what is T multiplied by?

Common errors

1) **Ratios**

    Be careful that you express the ratio in the correct order. If it is said that the ratio of A to B is 6:5, then that means that:

    $$A/B = 6/5.$$

    Be sure not to get this the wrong way round.

2) **Proportional Division**

    When a quantity is divided in the ratio 3:9, then the first part is $\frac{3}{12}$ of the whole, not $\frac{3}{9}$ of the whole.

3) **Area and Volume ratios**

    When one object is a model of another, the scale of the model is the ratio of the lengths. It is not the ratio of the areas or of the volumes. The area ratio is the *square* of the length ratio, and the volume ratio is the *cube* of the length ratio.

4) **Proportion**

    The α symbol is often misused. The statement $y \, \alpha \, x$ is not an equation. It can be converted by putting in a constant k, to give the equation $y = kx$.

    Similarly, it is unnecessary to write $y \, \alpha \, kx$. Either write $y \, \alpha \, x$ or $y = kx$.

5) **Arithmetic**

    Arithmetic errors are very common. Be sure that you do the correct operation when finding the constant k. For example:

    From $4 = k2$ we get $k = 2$. (Not $k = \frac{1}{2}$ or $k = 8$)

    From $4 = k/2$ we get $k = 8$. (Not $k = 2$ or $k = \frac{1}{2}$)

Chapter 6

Percentages

6.1 Percentages of Quantities

A *percentage* is a form of fraction, in which the denominator is always 100. For example, 15 percent (written 15%) is the same as the fraction $\frac{15}{100}$.

☐ **6.1.1 Examples.** *Foundation, levels 4, 5*

1) A car window costs £60 before VAT at $17\frac{1}{2}$%. How much is the VAT?

Solution $17\frac{1}{2}$% represents $\frac{17\frac{1}{2}}{100}$. The VAT is therefore:

$$\frac{17\frac{1}{2}}{100} \times £60 = £10.50$$

2) A woman earning £12,000 is given a 6% pay rise. What is her new salary?

Solution Her rise is $\frac{6}{100} \times £12,000 = £720$. Her new salary is:

$$£12,000 + 720 = £12,720$$

☐ **6.1.2 Exercises.** *Foundation, levels 4, 5*

1) A meal costs £15 plus 10% service charge. How much is the service charge?

2) The cost of some building work is £800. VAT at $17\frac{1}{2}$% is added. How much is the VAT?

3) A house agent asks for 2% of the price of a house. If the house sells for £50,000, how much does the agent get?

4) During a sale a clothing shop takes 20% off the price of its suits. How much is taken off a suit costing £55?

5) An antique dealer reckons that she can make 25% profit on her sales. If she buys a desk for £160, how much profit will she make on it?

6) A brand of whisky contains 40% alcohol. How much alcohol is there in 20 litres of the whisky?

7) An alloy contains 30% of silver. How much silver is there in 40 kg of the alloy?

8) The pass mark for an exam is 45%. If the maximum number of marks is 200, how many must a candidate get to pass?

9) The height of a tree increases by 20%. If it was 10 feet high before, how high is it now?

10) A woman finds that her salary has increased by 15%. If she used to earn £14,000, how much does she earn after the increase?

11) After going on a diet, George finds that his weight has decreased by 25%. If he weighed 16 stone before, how much does he weigh now?

12) A union is asking for wage increases of 7% for all its members. If the total wage bill for a company is now £200,000, how much will it be if it agrees to the increases?

13) A shirt is originally priced at £12, but during a sale all prices are reduced by 20%. How much does the shirt cost now?

14) A pensioner invests her savings of £20,000 at 8% interest. How much income will she receive each year?

15) A man borrows £5,000 at 12% interest. How much interest does he have to pay back each year?

16) The Johnson household takes a daily paper costing 18p for six days of the week, and a Sunday paper costing 32p. There is a delivery charge of 5%. that is the total bill for the week?

17) Kieron wants to pave part of his garden. He buys 32 paving stones costing £2 each and concrete costing £5.60. VAT at $17\frac{1}{2}$% is added to the bill. How much does he have to pay in all?

18) Brenda's telephone bill consists of a rental charge of £16.45 together with 223 units at 5p each. VAT at $17\frac{1}{2}$% is added to the bill. How much in all does she have to pay?

19) A surveyor measures a distance as 125 metres. He knows that his instruments may give an error of up to 2%.

   a) What is the greatest possible error?

   b) What is the greatest possible value of the distance?

   c) What is the least possible value of the distance?

6.2 Conversion of Percentages

☐ 6.2.1 Examples. *Foundation, level 6*

1) a) Convert 25% to a fraction.

   b) Convert $\frac{4}{5}$ to a percentage.

   c) Express 0.55 as a percentage.

   d) Express 15% as a decimal.

   Solution   a)   25% represents 25 over 100. The fraction is:

$$\frac{25}{100} = \frac{1}{4}$$

   b)   Multiply the fraction by 100:

$$\frac{4}{5} \times 100 = \frac{400}{5} = 80\%$$

   c)   Percentages are fractions of 100. To convert decimals to percentages move the decimal point two places to the right.

$$0.55 = 55\%$$

d)   Going the other way, to convert percentages to decimals move the decimal point two places to the left.

$$15\% = 0.15$$

2)   In a school of 800 pupils, 450 are boys. What percentage are girls?

Solution   350 out of the 800 are girls. The fraction of girls is $\frac{350}{800} = \frac{7}{16}$. Convert this to a percentage:

$$\frac{7}{16} \times 100 = 43.75\% \text{ are girls.}$$

2)   A dealer buys a car for £1,200 and sells it for £1,500. What is the percentage profit?

Solution   The profit is £300. Express this as a percentage of the buying price:

$$\frac{300}{1200} \times 100 = 25\%$$

❑ **6.2.2 Exercises.** *Foundation, level 6*

1)   Convert the following percentages to fractions in their simplest form.

   a) 10%   b) 20%   c) 60%   d) 5%

   e) $12\frac{1}{2}\%$ f) 150%   g) 100%

2)   Express the following fractions as percentages:

   a) $\frac{1}{2}$ b) $\frac{1}{4}$ c) $\frac{3}{4}$ d) $\frac{3}{5}$

   e) $\frac{1}{20}$ f) $\frac{17}{20}$ g) $1\frac{3}{4}$

3)   Express the following decimals as percentages:

   a)   0.12   b)   0.74   c)   0.03

   d)   0.6   e)   1.54   f)   2.7

4)   Express the following percentages as decimals:

   a)   23%   b)   54%   c)   5%

   d)   20%   e)   150%   f)   104%

5)   At a concert, 650 out of an audience of 1,000 were women. What percentage were women?

6)   A man works out that of his salary of £12,000 he pays £3,000 in tax. What percentage of his salary goes in tax?

7)   A kilogram of beef contains 150 grams of fat. What percentage is lean meat?

8)   I calculate that I sleep for 8 hours each day. For what percentage of the day am I awake?

9)   Out of 10,000 candidates who take a certain exam, 5,500 will pass. What is the percentage pass rate?

10) An exam is marked out of 250, and the pass mark is 100. What percentage mark is needed to pass?

11) The manager of a bookshop buys books at £1.50 each and sells them at £2.50. What is the percentage profit?

12) After going on a diet, Jane decreased her weight from 10 stone to 9 stone. What was her percentage loss in weight?

13) The price of a car is cut from £3,600 to £2,700. What is the percentage cut?

14) A salary increase raises my wages from £120 to £132. What is the percentage increase?

15) The budget of a council is £130 million, of which £26 million is spent on housing. What is the percentage spent on housing?

16) Tanya puts an apple on her kitchen scales, and finds that it weighs 50g. She knows that the scales may be inaccurate by up to 2 grams. What is the percentage error in her measurement?

6.3 Compound Interest and Reverse Percentages

Suppose a sum of money is invested at 10% compound interest. Then every year the amount invested is multiplied by a factor of $\frac{110}{100} = 1.1$.

Suppose that after an increase of 10%, a sum of money has reached £1331. To find the original sum we *divide* by 1.1, obtaining £1,210.

☐ **Examples 6.3.1.** *Higher, level 9*

1) £400 is left in a building society at 8% compound interest. The interest is allowed to accumulate. How much is there after 10 years?

   Solution When a sum of money is increased by 8%, every £100 is increased to £108. Hence the sum of money is multiplied by a factor of
   $$\frac{108}{100} = 1.08.$$

   After 2 years the sum of money is multiplied by 1.08^2.

   After 3 years the sum of money is multiplied by 1.08^3.

   After 10 years the money will be multiplied by 1.08^{10}. The amount will be:

   $$£400 \times 1.08^{10} = £863.57$$

2) After a salary increase of 8%, a man's salary is £13,500. What was it before the increase?

   Solution In the previous example it was shown that adding 8% was equivalent to multiplying by 1.08.

   So the salary before the rise is £13,500 *divided* by 1.08

   The salary before is £13,500 ÷ 1.08 = £12,500

☐ **6.3.2 Exercises.** *Higher, Level 9*

1) £200 is invested at 10% compound interest. How much is there after 3 years?

2) £1,000 is invested at 5% compound interest for 3 years. How much will it amount to at the end of this period?

3) £400 is put into a government loan, which gives 7% compound interest. How much will there be after 4 years?

4) A car depreciates in value at 15% each year. If it was originally worth £5,000, how much is it worth after 3 years?

5) The population of a country is increasing at 3% each year. If it is 20 million now, how large will it be in 3 years time?

6) £600 is invested at 10% compound interest. How much will there be after

   a) 2 years   b) 10 years?

7) £300 is invested at 12% compound interest. How much will there be after

   a) 4 years   b) 18 years?

8) £1,200 is invested at 9% compound interest. How much will there be after

   a) 2 years   b) 12 years?

9) The population of a country is increasing at 2%. If it is 15 million in size now, how large will it be in 20 years time?

10) The value of a car depreciates at 12% each year. If it cost £10,000 when new, how much will it be worth in 8 years time?

11) A radioactive material decays at the rate of 10% each year. If there is 3 kg now, how much will be left after

   a) 2 years   b) 10 years?

12) I have invested my money at 8% compound interest. If I have £324 now, how much did I have one year ago?

13) A man can invest money at 12%. How much should he invest now, to ensure that he has £1,680 after one year?

14) A dealer sells a car for £2,000, making 25% profit. How much was the car bought for?

15) A population is increasing at 2%; if it is 20,400,000 now, how large was it a year ago?

16) After a 20% wage increase, a man's salary is £13,200. How much did he earn before the increase?

17) An antique dealer buys a chair, but has to sell at a loss of 10%. If she sold it for £450, how much did she pay for it?

18) After a 20% reduction in the sales the price of a pair of trousers is £20. How much did they cost before the reduction?

19) The price of a meal including VAT at $17\frac{1}{2}$% was £18.80. How much was the VAT?

Common errors

1) Conversion

When converting percentages to decimals or fractions, be careful to divide by 100, not by 10. Do not write 5% as 0.5 instead of 0.05.

The same applies when converting fractions or decimals to percentages. 0.3 is 30%, not 3%.

2) Percentage change

The percentage change of a quantity is always the percentage of the *original* value, not the final value.

If an item is bought for £20 and sold for £25, the percentage profit is the percentage of the buying price not the selling price. i.e.:

$$\text{Percentage profit} = \frac{5}{20} \times 100 = 25\%.$$

You must be especially careful when doing reverse percentages. Suppose a dealer sells an item for £50, at a profit of 25%.

The *buying* price is £50 ÷ 1.25 = £40.

If your answer is £37.50 then you have taken 25% off the *selling* price.

Chapter 7

Measures

7.1 Units and Change of Units

There are two systems of units in common use. They are the *metric* system and the *imperial* system.

The most important units are as below. Abbreviations are in brackets.

Length

Metric

The basic unit is the *metre* (m).

1 metre = 100 centimetres (cm) = 1000 millimetres (mm).

1000 m. = 1 kilometre (km).

Imperial

12 inches (in) = 1 foot (ft). 3 feet = 1 yard (yd). 1760 yds = 1 mile.

1 inch is approximately 2.54 cm.

Weight

Metric

The basic unit is the *gram* (g).

1 gram = 1000 milligrams (mg). 1000 grams = 1 kilogram (kg).

Imperial

16 ounces (oz) = 1 pound (lb). 14 lb = 1 stone. 160 stone = 1 ton.

1 pound is approximately 454 grams.

Area

Metric

There are 100^2 = 10,000 cm^2 in 1 m^2.

10,000 m^2 = 1 hectare.

Imperial

There are 4,840 square yards in an acre.

1 acre is approximately 4,047 square metres.

Volume

Metric

The basic unit is the *litre* (l).

1 litre = 100 centilitres (cl) = 1,000 millilitres (ml).

1 millilitre = 1 cubic centimetre. (c.c. or cm^3).

Imperial

The basic units are the *pint* and the *gallon*. 8 pints = 1 gallon.

1 gallon is approximately 4.54 litres.

☐ **7.1.1 Examples.** *Foundation, level 5*

1) Angie measures the length of her bedroom, and finds that it is 15 feet. How many inches is that? Express the length in centimetres and metres.

Solution   There are 12 inches in 1 foot.

The room is 12 × 15 = 180 in.

Using the rate of 2.54 cm per inch:

The room is 180 × 2.54 = 457.2 cm

There are 100 cm in a metre.

The room is 457.2 ÷ 100 = 4.572 m.

2) At a certain time 1 pound (£1) is worth 1.5 dollars ($1.50). The chart of fig 7.1 gives the conversion of £ to $ and from $ to £.

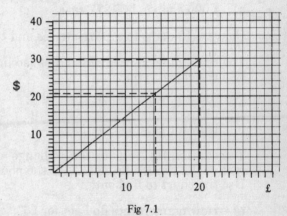
Fig 7.1

a) How much is £14 in dollars?

b) How much is $30 in pounds?

c) After a month the exchange rate changes so that £1 is worth $1.40. Draw a new line on the chart to show the new rate.

Solution   a) Start at 14 on the £ axis, and go up until the line is reached. Now go across to the $ axis, and read off the amount.

£14 is worth $21

b) Start at 30 on the $ axis, go across to the line and down to the £ axis. Read off the result.

$30 is worth £20

c) The new line must start from the bottom left corner as before. £10 is now worth $14, so mark that point in and join up. The result is shown in fig 7.2.

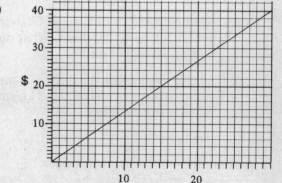
Fig 7.2

3) A farm is 350 hectares. How many acres is that?

Solution   There are 10,000 m² in a hectare.

The farm contains 10,000 × 350 = 3,500,000 m².
There are 4,047 m² in an acre.

The farm contains 3,500,000 ÷ 4,047 = 865 acres.

53

7.1.2 Exercises. *Foundation, level 5*

1) Convert the following to centimetres:

   a) 3 metres   b) 3.6 m.   c) 36 mm.   d) 0.065 km.

   e) 5 inches   f) 3.4 ft.

2) Convert the following to metres:

   a) 40 cm.   b) 213 mm.   c) 1.39 km.   d) 27 in.

   e) 5 yards.

3) Convert the following to grams:

   a) 300 mg.   b) 15.34 kg.   c) 12.3 lb.   d) 0.23 lb.   e) 17 oz.

4) Convert the following to kilograms:

   a) 867 g.   b) 65 g.   c) 3 lb.   d) 4.28 lb   e) 34 oz.

5) Convert the following to litres:

   a) 4,000 cm³   b) 277 cm³   c) 3 gallons   d) 5 pints.

6) Convert 18 acres to square metres and to hectares.

7) The chart of fig 7.3 converts miles to kilometres. Use the chart to find:

   a) What is 14 miles in km?

   b) How far is 26 km. in miles?

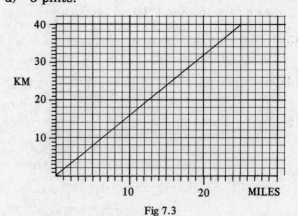

Fig 7.3

8) The chart of fig 7.4 gives the exchange rate of the French Franc (FF) to the pound. Use the chart to find out:

   a) How many francs do I get for £5?

   b) How much is 85 FF worth in pounds?

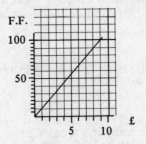

Fig 7.4

9) Before taking her car abroad, Mavis prepares a chart to convert pounds per square inch (lb/in²) to kilograms per square centimetre. (kg/cm²). The chart is shown in fig 7.5.

   a) She keeps her front tyres at 28 lb/in² and her back tyres at 30 lb/in². What do these correspond to in kg/cm²?

   b) After a day's drive she finds at a garage that one tyre has a pressure of 1.8 kg/cm². What is this in lb/in²?

10) Convert to inches:

   a) 5 ft.   b) 16 yards   c) 10 cm.

   d) 24.5 cm.   e) 0.34 metres.

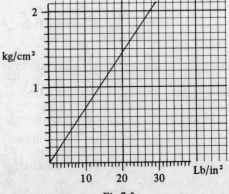

Fig 7.5

11) Convert to pounds:

   a) 3 stone   b) 32 ounces   c) 4,565 g.   d) 23 kg.

12) Convert to acres:

   a) 44,000 m^2 b) 500 hectares.

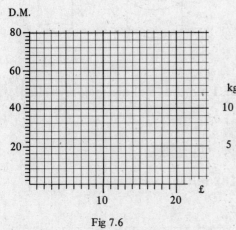

Fig 7.6

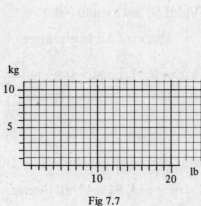

Fig 7.7

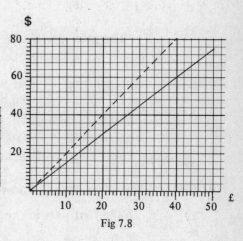

Fig 7.8

13) If there are 2.50 German Marks (DM) to the pound, draw a line on the chart of fig 7.6 to convert from £ to DM. Use it to find the number of DM equal to £16, and the value in pounds of 10 DM.

14) There are approximately 2.2 lb in 1 kg. Draw a line on the chart of fig 7.7 to convert from kg to lb. Use it to find the weight in kg of 20 lb, and the weight in pounds corresponding to 5 kg.

15) Fig 7.8 shows the buying and selling rates for pounds and dollars. To change from £ to $ use the filled-in line, and to change from $ to £ use the dotted line.

   a) If I change £50, how many dollars do I get?

   b) If I now change these back to pounds, how many do I get? How much have I lost?

7.2 Rates and Averages

A *rate* is obtained when one quantity is divided by another. Examples of rates are:

Miles per hour: Price per kilogram: Wages per week.

The *average* speed during a journey is the total speed divided by the total time.

☐ 7.2.1 Examples. *Foundation, level 6*

1) Petrol is sold at £2.15 per gallon. How much will 4 gallons cost? How much petrol can be bought for £10?

Solution Multiply £2.15 by 4.

4 gallons cost 4 × £2.15 = £8.60

Divide £10 by 2.15.

£10 will buy 10 ÷ 2.15 = 4.65 gallons

2) In fig 7.9 two bottles of olive oil are shown. Which brand offers cheaper oil?

Fig 7.9

Solution Puccini brand offers 1,000 c.c. for £2.40. The price per c.c. is:

$$£2.40 \div 1,000 = 0.24 \text{ p.}$$

The price per c.c. for Verdi oil is £1 ÷ 400 = 0.25 p.

Puccini oil is cheaper

3) I drive for 86 miles at 43 m.p.h., then for $1\frac{1}{2}$ hours at 64 m.p.h. Find the total distance travelled and the total time taken. What has been my average speed?

Solution For the second part of the journey I travelled $1\frac{1}{2} \times 64 = 96$ miles.

The total distance was 96 + 86 = 182 miles.

The first part of the journey took 86 ÷ 43 = 2 hours.

The total time was $2 + 1\frac{1}{2} = 3\frac{1}{2}$ hours.

Average speed is total distance divided by total time.

The average speed was $182 \div 3\frac{1}{2} = 52$ m.p.h.

☐ **7.2.2 Exercises.** *Foundation, level 6*

1) Work out how much the following cost:

a) 5 kilograms of flour at 40 pence per kg.

b) 7 metres of rope at 15 p. per metre.

c) 15 litres of petrol at 47.5 p. per litre.

d) 3 m² of cloth at £1.20 per m².

2) I have £32 to spend. How much could I buy of the following:

a) Petrol at £2 per gallon.

b) Beef at £3.20 per pound.

c) Cloth at £4 per m².

d) Flex at 80 p. per metre.

3) a) A 5 lb chicken cost £4.30. What is the cost per pound?

b) 12 gallons of oil cost £72. What is the cost per gallon?

c) To hire a tennis court for an hour costs £1.80. What is the cost per minute?

d) 7 metres of electrical flex cost £1.33. What is the cost per metre?

4) A candle is 20 cm. high. It burns at a rate of 4 cm per hour.

a) How long will it burn?

b) After 3 hours, how much of the candle is left?

c) How long will it be before the candle is half finished?

5) Roger makes a water-clock by punching a hole in the bottom of a 600 c.c. can and filling it with water. The water drips out at a steady rate, and the can is empty in 30 minutes.

a) What is the rate of water loss in c.c. per minute?

b) What is the rate of water loss in c.c. per second?

c) How much water is there left after 20 minutes?

6) A long-playing record rotates at 33 revolutions per minute. If it rotates 726 times, how long does it last?

7) James drives for 100 miles, and it takes him $2\frac{1}{2}$ hours. What has been his average speed?

8) I walk for two hours at 4 km. per hour, then for 1 hour at 5 km. per hour. How far have I gone? What has been my average speed?

9) George buys 10 pencils at 12p each and 5 pencils at 15p each. What is the average price of the pencils?

10) I buy 6 cans of orange juice containing 1 litre each, and 24 cans containing 400 c.c. each. What is the average volume of the cans?

11) A car travels at a steady speed of 45 m.p.h. How far does it go in 3 hours? How long does it take to travel 180 miles?

12) Dolores is driving down the M1 towards London. She sees the sign (i) at 11 a.m. and the sign (ii) at 1 p.m.

(I)   LONDON 135 MILES

a) What is her average speed?

(II)   LONDON 27 MILES

b) If she keeps up this speed, when will she be in London?

Fig 7.10

13) A car does 12 km per litre of petrol. How many litres will it take to cover 156 km?

14) A car travels 160 miles, and uses 5 gallons of petrol. What is the rate of petrol consumption in miles per gallon? At this rate how far will 8 gallons take the car?

15) The charge for a telephone consists of a standing charge of £16 and 5 p. per unit used. If a subscriber uses 55 units, how much is the total bill?

16) The standing charge for an elecricity bill is £9, and each unit of electricity cost 6 p. If a household uses 800 units, how much is the bill?

17) The instructions for cooking a joint of beef are that it should be roasted for 30 minutes plus 20 minutes per pound. How long should a 6 lb joint stay in the oven?

18) The entrance fee for a squash club is £25, and the annual subscription is £36. Naomi joins in May, when there is only 5 months of the squash year left to run. If the annual subscription is reduced proportionally, how much does she pay?

7.3 Money Matters

Wages are usually expressed as a rate per hour or per week. *Salaries* are usually expressed as a rate per month or per year.

Overtime consists of extra hours of work. Overtime is often paid at 'time and a half', which is $1\frac{1}{2}$ times the usual rate.

Income tax is levied by the government on incomes.

Value added tax (VAT) is a tax taking a fixed percentage of the price of goods.

Goods are bought on *hire-purchase* by a down payment and several monthly or weekly payments.

❏ **7.3.1 Example.** *Foundation, level 6*

Jill sees several jobs advertised. One is for a filing clerk at £8,500 per year, one for a shop assistant at £160 per week, one for a packer at £3.50 per hour for a 40 hour week. Which job will give her the most money?

> *Solution*   The second job will pay 52 × £160 = £8,320 per year.
>
> The third job will pay 52 × 40 × £3.50 = £7,280 per year.
>
> **The filing-clerk job will give the most money.**

7.3.2 Exercises. *Foundation, Levels 4, 5, 6*

1) How much per year are the following equivalent to?

   a) £85 per week   b) £185 per week

   c) £900 per month   d) £4.10 per hour for a 40 hour week

   e) £3.80 per hour for a 35 hour week.

2) How much per week are the following equal to?

   a) £5,824 per year   b) £11,960 per year.

3) Rupert earns £4 per hour for proof-reading. Would he be better paid working in an office at £147.60 for a 36 hour week?

4) Stan earns £3.70 per hour for a 36 hour week. Overtime is paid at time and a half. How much is he paid if he works 43 hours?

5) Pearl earns £3.30 per hour for a 30 hour week. Overtime is at time and a half. How much is she paid for 36 hours?
   If she is paid £118.80 for a week's work, how much overtime has she done?

6) Alf is paid at £4.15 per hour for a 38 hour week. Overtime is at time and a half, except at weekends when it is double time. How much is he paid if he works 46 hours, of which 4 hours are at the weekend?

7) Under an income tax scheme, citizens are given a tax-free allowance, and tax is then charged at 30% on the income above that allowance. How much tax is paid by the following:

   a) A single woman with an allowance of £3,000 and an income of £9,500.

   b) A married man with an allowance of £5,500 and an income of £12,000.

8) A man with an allowance of £2,500 pays £2,400 tax at 30%. What is his income before tax?

9) If the rate of VAT is $17\frac{1}{2}$% find the tax on the following:

   a) A radio costing £40.

b) Paint costing £20.

c) Building work costing £420.

10) Find the total cost for each of the following items bought on hire purchase:

a) A television, for £100 down payment and 12 monthly payments of £24.

b) A car for £550 down payment and 36 monthly payments of £60.

11) Mike sees a television for sale. He can buy it for £65 cash, or on hire purchase for £10 down payment and 12 monthly instalments of £5. How much more will he pay if he buys it by hire purchase?

7.4 Charts and Tables

Information about times, distances, prices etc. is often presented in the form of *charts* and *tables*.

Timetables often give the times of flights or trains in the *24 hour clock*. Afternoon times with the 24 hour clock are 12 hours greater than with the 12 hour clock. For example:

17 35 (24 hour clock) = 35 minutes past 5 p.m. (12 hour clock).

☐ **7.4.1 Examples.** *Foundation, level 3*

1) Fig 7.11 shows the dials of the gas meter in a house. Write down the reading.

After two months an extra 3,547 units have been used. Mark on the dials the new position of the hands.

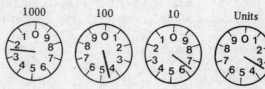

Fig 7.11

Solution   The hand of the 1,000 dial is between 2 and 3. Hence the reading is above 2,000 but below 3,000. The first digit of the reading is therefore 2. The other readings follow similarly.

The reading is 2,463

After two months the reading should be 2,463 + 3,547 = 6,010. The hands of the dials will be in the position shown in fig 7.12.

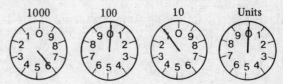

Fig 7.12

2) Fig 7.13 gives part of the timetable for trains from Bedford to Bletchley.

Miles																
0	Bedford Midland . . . . d	0554	0655	0705	0744	. . . .	0824	0846	0957	1057	1205	. . . .	1300	1330	1400	1513	1615
¼	Beford St. Johns . . . . d	0557	0658	0708	0747	. .	0827	0849	1000	1100	1208	. .	1303	1333	1403	1516	1618
3¼	Kempston Hardwick . . . . . d	0600	0707	0717	0756	. . . .	0836	0858	1009	1109	1217	. . . .	1312	1342	1412	1525	1627
5½	Stewartby . . . . d	0610	0711	0721	0800	. .	0840	0902	1013	1113	1221	. .	1316	1346	1416	1529	1631
6¾	Millbrook (Bedfordshire). . . d	0614	0715	0725	0804	. .	0844	0906	1017	1117	1225	. .	1320	1350	1420	1533	1635
8¼	Lidlington . . . . d	0618	0719	0729	0808	. .	0848	0910	1021	1121	1229	. .	1324	1354	1424	1537	1639
10	Ridgmont . . . . d	0622	0723	0733	0812	. . . .	0852	0914	1025	1125	1233	. . . .	1328	1358	1428	1541	1643
11¾	Aspley Guise. . . . . d	0626	0727	0737	0816	. . .	0856	0918	1029	1129	1237	. .	1332	1402	1432	1545	1647
12¾	Woburn Sands. . . . . d	0629	0730	0740	0819	. . .	0859	0921	1032	1132	1240	. . . .	1335	1405	1435	1548	1650
14¾	Bow Brickhill . . . . d	0634	0735	0745	0824	. .	0905	0926	1037	1137	1245	. .	1340	1410	1440	1553	1655
15¾	Fenny Stratford. . . . . . . d	0637	0738	0748	0827	. . . .	0908	0929	1040	1140	1248	. . . .	1343	1413	1443	1556	1658
16¾	Bletchley. . . . . d	0640	0741	0751	0830	. .	0912	0932	1043	1143	1251	. .	1346	1416	1446	1559	1701

Fig 7.13

a)   If I take the first train of the day, how long will the whole journey take?

b)   I want to arrive in Fenny Stratford before 4 p.m. What is the last train I could take from Stewartby?

Solution a) The time taken is the difference between the departure time for Bedford Midland and the arrival time for Bletchley. The train leaves at 6 minutes to 6 and arrives at 40 minutes past 6.

Time taken = 46 minutes.

b) 4 p.m. corresponds to 16 00 in the 24 hour clock. The last train which gets to Fenny Stratford before this arrives at 15 56.

I must catch the 15 29 at Stewartby.

3) Fig 7.14 show the prices of the Hovercraft service between Dover and Calais. The Newman family consists of two adults and two children aged 8 and 3: their car is 4.3 metres long. How much will it cost them to cross the Channel at Tariff C?

Solution The car will cost £53. Each adult costs £12, the elder child costs £6 and the younger goes free. The total cost is:

HOVER*SPEED*

CALAIS ▼

FARES
Standard Single Fares

E	D	C	B	
Tariff £	Tariff £	Tariff £	Tariff £	
				Cars, Minibuses and Campers
21.00	33.00	43.00	54.00	Up to 4.00m (13′ 1″) length/Motorcycle combination
21.00	40.00	53.00	63.00	Up to 4.50m (14′9″) in length
21.00	43.00	58.00	66.00	Up to 5.50m (18′) in length
11.00	11.00	11.00	11.00	Over 5.50m per extra metre or part thereof
				Caravans and Trailers (See Half Price Offer alongside)
15.00	25.00	25.00	25.00	Up to 4.00m (13′ 1″) in length
15.00	25.00	25.00	25.00	Up to 5.50m (18′) in length
11.00	11.00	11.00	11.00	Over 5.50m per extra metre or part thereof
11.00	12.00	13.00	14.00	**Motorcycles, Scooters and Mopeds**
12.00	12.00	12.00	12.00	**Each Adult**
6.00	6.00	6.00	6.00	**Each Child** (4 but under 14 years)

Fig 7.14

£53 + 2 × £12 + £6 = £83

☐ **7.4.2 Exercises.** *Foundation, level 3*

1) Write the following times in terms of the 24 hour clock:

a) 11 a.m.   b) half past 7 in the morning   c) 4 p.m.   d) Noon.

e) Quarter to 8 in the evening   f) 5 minutes past 11 at night.

2) Write the following times in terms of the 12 hour clock:

a) 08 00   b) 10 15   c) 12 15

d) 16 45   e) 20 30.

3) Write down in terms of the 24 hour clock the times shown:

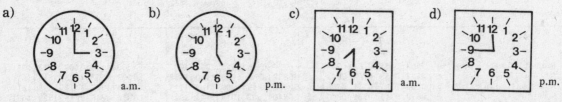

a) ... a.m.   b) ... p.m.   c) ... a.m.   d) ... p.m.

Fig 7.15

4) How much time has passed between the following times:

a) 6.30 a.m. and 11 a.m.

b) 9.45 a.m. and 1 p.m.

c) 04 45 and 16 23.

Calendar   **1986**

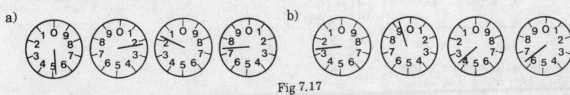

January	February	March
M T W T F S S	M T W T F S S	M T W T F S S
1 2 3 4 5	1 2	1 2
6 7 8 9 10 11 12	3 4 5 6 7 8 9	3 4 5 6 7 8 9
13 14 15 16 17 18 19	10 11 12 13 14 15 16	10 11 12 13 14 15 16
20 21 22 23 24 25 26	17 18 19 20 21 22 23	17 18 19 20 21 22 23
27 28 29 30 31	24 25 26 27 28	24 25 26 27 28 29 30
		31

5) Fig 7.16 shows the calendar for 1986. In that year Rachel started a holiday job on Monday 21 July and ended it on Friday 5 September.

   a) How many weeks did she work?

   b) In each month she was paid on the last Thursday she worked. On which days was she paid?

6) In 1986, how many Sundays were there between the last Friday in January and the first Friday in May?

April	May	June
M T W T F S S	M T W T F S S	M T W T F S S
1 2 3 4 5 6	1 2 3 4	1
7 8 9 10 11 12 13	5 6 7 8 9 10 11	2 3 4 5 6 7 8
14 15 16 17 18 19 20	12 13 14 15 16 17 18	9 10 11 12 13 14 15
21 22 23 24 25 26 27	19 20 21 22 23 24 25	16 17 18 19 20 21 22
28 29 30	26 27 28 29 30 31	23 24 25 26 27 28 29
		30

July	August	September
M T W T F S S	M T W T F S S	M T W T F S S
1 2 3 4 5 6	1 2 3	1 2 3 4 5 6 7
7 8 9 10 11 12 13	4 5 6 7 8 9 10	8 9 10 11 12 13 14
14 15 16 17 18 19 20	11 12 13 14 15 16 17	15 16 17 18 19 20 21
21 22 23 24 25 26 27	18 19 20 21 22 23 24	22 23 24 25 26 27 28
28 29 30 31	25 26 27 28 29 30 31	29 30

October	November	December
M T W T F S S	M T W T F S S	M T W T F S S
1 2 3 4 5	1 2	1 2 3 4 5 6 7
6 7 8 9 10 11 12	3 4 5 6 7 8 9	8 9 10 11 12 13 14
13 14 15 16 17 18 19	10 11 12 13 14 15 16	15 16 17 18 19 20 21
20 21 22 23 24 25 26	17 18 19 20 21 22 23	22 23 24 25 26 27 28
27 28 29 30 31	24 25 26 27 28 29 30	29 30 31

Fig 7.16

7) What is the electricity reading given by the dials below?

   a)   b)

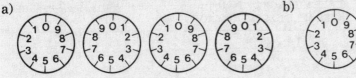

Fig 7.17

8) Fill in the dials below to show readings of   a) 7354   b) 1009.

   a)   b)

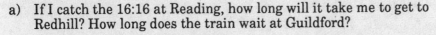

Fig 7.18

9) The table shows the distances between several towns in England. Use the table to find the following:

   a) The distance between York and Leeds

   b) The distance between Bristol and London.

   c) The direct distance from Birmingham to York, and the distance if one goes through Leeds.

10) Fig 7.20 (overleaf) shows part of the timetable for the service from Reading to Redhill.

   a) If I catch the 16:16 at Reading, how long will it take me to get to Redhill? How long does the train wait at Guildford?

   b) If I arrive at Reading station at 5 p.m., when do I arrive at Dorking?

   c) When must I arrive at Sandhurst station to ensure that I reach Betchworth by 10 in the evening?

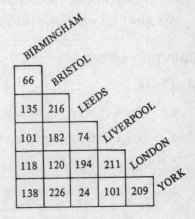

Fig 7.19

Reading	148	d	16 16	16 55	..	..	17 08	17 43	18 15	..	18 36	18 45	19 10	..	..	19 16	20 10	20 14	..	21 16	..	22 16	23 25	...	...
Earley	148	d	16 21	....	..	..	17 13	..	..	..	..	18 50	..	..	..	19 21	..	20 21	..	21 21	..	22 21	23 31	...	...
Winnersh	148	d	16 25	..	..	..	17 17	..	..	..	..	18 54	..	..	..	19 25	..	20 25	..	21 25	..	22 25	..	...	...
Wokingham	148	d	16 31	17 06	..	..	17 27	17 53	18 25	..	..	19 00	..	..	..	19 31	..	20 31	..	21 31	..	22 31	23 38	...	...
Crowthorne		.d	16 38	17 12	..	..	17 29	17 59	18 31	..	..	19 08	..	..	..	19 38	..	20 38	..	21 38	..	22 38	23 44	...	...
Sandhurst (Berks) .		.d	16 42	17 15	..	..	17 31	18 03	18 34	..	..	19 12	..	..	..	19 42	..	20 42	..	21 42	..	22 42	23 47	...	...
Blackwater		.d	16 46	17 19	..	..	17 37	18 07	18 38	..	..	19 16	..	..	..	19 46	..	20 46	..	21 46	..	22 46	23 51	...	...
Farnborough North .		.d	16 51	17 24	..	..	17 42	18 12	18 43	..	..	19 21	..	..	..	19 51	..	20 51	..	21 51	..	22 51	23 56	...	...
North Camp		.d	16 55	17 29	..	..	17 47	18 16	18 48	..	..	19 25	19 31	..	..	19 55	20 31	20 55	..	21 55	..	22 55	23 59	...	...
Ash	148	d	17 00	..	..	..	17 51	18 21	..	..	..	19 30	..	..	..	20 00	..	21 00	..	22 00	..	23 00	00 05	...	...
Wanborough	148	d	17 04	..	..	..	17 56	18 25	..	..	..	19 34	..	..	..	20 04	..	21 04	..	22 04	..	23 04	00 09	...	...
Guildford	148	d	17 12	17 42	..	..	18 03	18 33	19 02	....	19 07	19 42	19 45	19 42	....	20 12	20 43	21 12	..	22 12	..	23 13	00 17	...	...
156 London Waterloo		a	17 57	18 25	..	..	18 57	19 25	19 55	..	..	20 25	..	..	..	20 55	..	21 55	..	22 55	..	23 59	..		
156 London Waterloo		d	16 20	..	..	..	17 32	17 50	..	..	..	18 52	..	..	..	19 20	..	20 20	..	21 20	..	22 20			
Guildford		.d	17 17	17 43	..	..	18 15	18 35	19 11	..	19 09	19 53	19 46	19 53	..	20 18	20 50	21 17	..	22 17	..	23 17	..		
Shalford		.d	17 22	..	..	..	18 21	18 40	..	..	..	19 58	..	..	20 23	..	21 22	..	22 22	..	23 22	..			
Chilworth		.d	17 26	..	..	..	18 25	18 44	..	..	..	20 02	..	..	20 27	..	21 26	..	22 26	..	23 26	..			
Gomshall		.d	17 34	..	..	..	18 33	18 52	..	..	..	20 10	..	..	20 35	..	21 34	..	22 34	..	23 34	..			
Dorking Town		.d	17 41	..	..	..	18 41	18 59	..	..	..	20 17	..	..	20 42	..	21 41	..	22 41	..	23 41	..			
Deepdene		.d	17 44	..	..	..	18 44	19 02	..	..	..	20 20	..	..	20 45	..	21 44	..	22 44	..	23 44	..			
Betchworth		.d	17 49	..	..	..	18 49	19 07	..	..	..	20 25	..	..	20 50	..	21 49	..	22 49	..	23 49	..			
Reigate		.d	17 55	18 07	18 39	..	18 55	19 13	..	19 42	..	20 31	20 42	20 56	..	21 56	..	22 55	..	23 55	..			
Redhill		.d	18 00	18 12	18 43	..	19 00	19 18	19 38	19 47	..	20 14	20 36	20 47	21 02	21 17	22 00	..	23 00	..	23 59	..		

Fig 7.20

11) The Cohen family, consisting of two adults, Daniel (14), Jonathan (10) and Ruth (8) are going on holiday to Greece. How much would it cost them to stay in each of the following hotels for 2 weeks?

a)

HOTEL DIMITRA			
Cost per person	per week £250	per 14 days £320	Reductions per day 2-5 yrs £1 6-11 yrs £0.60 12-16 yrs £0.40

b)

HOTEL IANNIS			
Cost per person	7 days £220	14 days £295	Reductions per week 2-5 yrs £10 6-16 yrs £8

Fig. 7.21

12) The table shows the cost for hiring a car. How much will it cost Kevin to hire a car for one day, in which he covers 180 miles?

13) Fig 7.23 shows some of the programmes on television on a certain evening.

DAILY RATE £10.00
FREE MILEAGE UP TO 100M
5p PER MILE EXCESS MILEAGE

Fig. 7.22

6.45	News
7.30	Comedy Theatre
8.30	Zoo news
8.55	Film: "The Blob"
10.35	News

Fig 7.23

a) How long does the film last?

b) Jason watches television from 7.15 for $2\frac{1}{4}$ hours. What programmes has he seen? How much of the film was there left to run?

Common errors

1) Units

a) When converting from one unit to another, be careful that you go the right way.

To go from inches to centimetres *multiply* by 2.54.

To go from centimetres to inches *divide* by 2.54.

b) Be very careful with area and volume units. There are 100 cm in a metre, but there are 10,000 cm^2 in a m^2. There are 1,000,000 cm^3 in a m^3.

2) Averages

The average cost of several items is the total cost divided by the number of items.

If I buy some items at 40 p. each and some other items at 60 p. each, then the average cost is not necessarily 50 p. To find the average cost divide the total cost by the total number of items.

3) Rates

Mistakes are often made with the quantities distance, speed and time. Make sure that you divide and multiply in the right situations, as follows:

$$\text{Speed} = \text{Distance} \div \text{Time.}$$

$$\text{Distance} = \text{Speed} \times \text{Time.}$$

$$\text{Time} = \text{Distance} \div \text{Speed.}$$

4) Clocks

a) Recall that there are 60 minutes in an hour. So 7.55 is 5 minutes to 8, not 45 minutes to 8.

b) The 24 hour clock adds on 12 hours to afternoon times. A common mistake is to think that 16 00 represents 6 in the afternoon. The correct time is 4 in the afternoon.

Chapter 8

Algebraic expressions

Algebraic expressions contain letters as well as numbers. Some examples are:

$$(F - 32) \times \frac{5}{9}$$

(Which converts temperature in Fahrenheit to Centigrade)

$$\sqrt{a^2 + b^2}$$

(Which is the formula for the hypoteneuse of a right-angled triangle)

8.1 The Use of Letters for Numbers

❏ **8.1.1. Examples.** *Foundation, level 5*

1) We are given the following sequence of equations:

$$5 \times 1 = 5 : 5 \times 2 = 10 : 5 \times 3 = 15 : 5 \times 4 = 20 : 5 \times x = y.$$

Find the values of x and y.

Solution The sequence 1, 2, 3, 4, x is completed by putting:

$$x = 5.$$

The value of y is given by:

$$y = 5 \times 5 = 25$$

2) Anne has 10 more pounds than Bill. Bill has x pounds.

   a) How much does Anne have?

   b) How much do they have in total?

Solution a) Add 10 to Bill's money.

Anne has x + 10 pounds

   b) Add x and x + 10.

Together they have $2x$ + 10 pounds

3) What is the cost of 5 sweets, which are z pence each?

Solution Multiply z by 5.

The cost is $5z$ pence

❏ **8.1.2. Exercises.** *Foundation, level 5*

1) In the sequence 2, 4, 8, x, 32, what is x?

2) A multiplication table gives the following:

$$7 \times 1 = 7 : 7 \times 2 = 14 : 7 \times 3 = 21 : 7 \times p = q.$$

Find p and q.

3) George weighs 7 kg more than Edward. If Edward weighs x kg, what does George weigh?

4) Mr Smith earns £x per hour. Mr Jones earns £2 more than him. How much does Mr Jones earn in an hour?

5) A plank 10 metres long has x metres sawn off. How much is left?

6) Two angles add up to 180°. One is $d°$. What is the other?

7) A machine can produce x items in one hour. How many items can it produce in 8 hours?

8) A room is 10 metres long and b metres broad. What is the area?

9) A record lasts for 20 minutes. There are two tracks on it, and the first track lasts for x minutes. How long is the second track?

10) A machine can produce 100 components per hour. How many can it produce in h hours?

11) Shoes cost £10 per pair. How much do p pairs cost?

12) £P is shared equally between 8 people. How much does each person get?

13) x and y are two numbers which add up to 10. Write down an equation in x and y.

14) N is 5 greater than M. Write down an equation giving N in terms of M.

15) Find the values of x which will make the following equations true:

a) $x + 3 = 7$ b) $5 + x = 19$ c) $x \times 2 = 6$ d) $4 \times x = 24$

☐ **8.1.3 Examples.** *Intermediate, level 7*

1) I buy 10 cakes at x pence each and 15 buns at y pence each. How much do I spend in all?

Solution   The cakes cost $10x$ p. and the buns $15y$ p. The total is:

$$10x + 15y \text{ pence}$$

2) t bricks weigh 50 kg in all. How much do s bricks weigh?

Solution   Each brick weighs $\dfrac{50}{t}$ kg. So s bricks will weigh:

$$s \times \frac{50}{t} \text{ kg.}$$

3) A car travels at x m.p.h. for 1 hour, then at y m.p.h. for 2 hours. How far has it travelled? What has been the average speed?

Solution $x \times 1$ miles is travelled in the first stage, then $y \times 2$ in the second stage. The total distance is:

$$x + 2y \text{ miles}$$

The average speed is the total distance divided by the total time. This gives:

$$\frac{x + 2y}{3} \text{ m.p.h.}$$

☐ **8.1.4 Exercises.** *Intermediate, level 7*

1) A room is L metres long and B metres broad. What is the area and the perimeter?

2) A motorist drives for t hours at 68 m.p.h. If he covers x miles, find x in terms of t.

3) A dealer buys a car for £m and sells it for £n. What is the profit?

4) For an n-sided figure, to obtain the sum of the interior angles we multiply the number of sides by 180° and then subtract 360°. Write the sum of the angles in terms of n.

5) How many minutes are there between 12 o'clock and x minutes past 2 o'clock?

6) A darts game ends after scoring 301. If a player has scored s, how much more must he score?

7) The cost of hiring a television is as follows: £30 down-payment plus £20 for each month. How much does it cost to hire it for m months?

8) A child thinks of a number, doubles it, and then adds 5. If the original number was N, what is the final result?

9) In a rectangle which is s cm by r cm a square of side t cm is cut out. What is the remaining area?

10) At a theatre, a cheap seat costs £4.50 and an expensive one costs £6. How much do x cheap and y expensive seats cost?

11) I buy 10 shirts at £x each and 8 shirts at £y each. What is the total cost of the shirts? What is the average cost?

12) A man is paid £x per hour at standard rates, and double rates for weekends. How much did he earn in a week in which he worked 46 hours, of which 8 were at the weekend?

13) The angles of a triangle are $x°$, $y°$ and $z°$. The sum of the angles in a triangle is 180°. Write x in terms of y and z.

14) In an innings a batsman hit x sixes, y fours and z singles. What was his total score?

8.2 Substitution

An algebraic expression contains letters in place of numbers. When numbers are *substituted* for the letters, the numerical value of the expression is found.

☐ 8.2.1 Examples. *Foundation, level 6*

1) Find the value of $(F - 32) \times \dfrac{5}{9}$ when $F = 50$.

Solution   Put $F = 50$ in the expression. The result is:

$$(50 - 32) \times \frac{5}{9} = 18 \times \frac{5}{9} = 10$$

2) It is given that $p = 3q + 2r$.

Find p when $q = 5$ and $r = 7$.

Solution   Put $q = 5$ and $r = 7$ in the expression.

$$p = 3 \times 5 + 2 \times 7 = 15 + 14 = 29$$

☐ 8.2.2 Exercises. *Foundation, level 6*

1) If $M = 3N + 2$, find M when $N = 5$.

2) Current, voltage and resistance are related by the formula $I = \dfrac{V}{R}$. Find I when $V = 200$ and $R = 40$.

3) In the formula $v = at + b$, find v when $a = 2$, $b = 30$, $t = 6$.

4) If $A = lb$, find A when $l = 3$ and $b = 6$.

5) If $A = \frac{1}{2}bh$, find A when $b = 4$ and $h = 7$.

6) In the formula $R = 2h - 4$, find R when $h = 3$.

7) If $Q = \frac{1}{4}(10 - x)$, find Q when $x = 2$.

8) The area of a circle is given by $A = \pi r^2$. Find A when $r = 7$, taking $\pi = \frac{22}{7}$.

9) From the formula $I = \frac{PRT}{100}$ find I when $P = 100$, $R = 8$ and $T = 4$.

10) If $R = \frac{5s}{t}$ find R when $s = 30$ and $t = 25$.

11) The area of a trapezium is given by the formula:

$$A = H(x + y)/2.$$

Find A when $H = 7$, $x = 3$ and $y = 5$.

12) Find the value of $3X + 4Y$ when $X = 2$ and $Y = 5$.

❐ 8.2.3 Examples. *Intermediate, level 8*

1) In the formula $r = 2s - 7$, find r when $s = -3$.

Solution Put $s = -3$ into the formula:

$$r = 2(-3) - 7 = -13$$

2) It is given that $\frac{1}{R} = \frac{1}{V} - \frac{1}{U}$. Find R when $V = 3$ and $U = 4$.

Solution Put $V = 3$ and $U = 4$.

$$\frac{1}{R} = \frac{1}{3} - \frac{1}{4} = \frac{1}{12}$$

$$R = 12$$

3) From the formula $c = \sqrt{a^2 + b^2}$ find c when $a = 7$ and $b = 24$.

Solution Put $a = 7$ and $b = 24$.

$$c = \sqrt{7^2 + 24^2}$$

$$c = \sqrt{49 + 576} = \sqrt{625} = 25$$

❐ 8.2.4 Exercises. *Intermediate, Level 8*

1) In the formula $s = ut + \frac{1}{2}a\,t^2$, find s when $u = 5$, $t = -2$ and $a = 7$.

2) The quantities P, V, T and k are related by $PVT = k$.

a) Find k when $P = 10$, $V = 120$, $T = 50$.

b) With this value of k, find V when $P = \frac{1}{3}$ and $T = 80$.

3) Illumination is given by the formula $I = \frac{c}{d^2}$. If $c = 72$ and $d = 2$ find I.

4) The sum of the first n mumbers is $\frac{1}{2}n(n + 1)$. Find the sum of the first 15 numbers.

5) The volume of a cone is given by $V = \pi r^2 h/3$. Taking π to be $\frac{22}{7}$, find the volume of a cone for which $r = 3$ and $h = 14$.

6) In the formula $T = a + (n - 1)d$, find T for $a = 3, n = 9, d = -8$.

   If $T = 33, a = 3, d = 5$ then find n.

7) Use the formula $y = \frac{(1 + x)}{(1 - x)}$ to find y when $x = \frac{1}{2}$.

8) C is given in terms of x and a by the formula $C = a(1 + x)^3$. Find C when $a = 5$ and $x = -2$.

9) r, u and v are related by $\frac{1}{r} = \frac{1}{u} - \frac{1}{v}$. Find r when $u = 5$ and $v = 6$.

10) If $H = n(n + 1) + \frac{1}{4}$ find H when $n = \frac{1}{2}$.

11) Use the formula $A = \frac{1}{2}(a + b)h$ to find A when $a = 7, b = 3$ and $h = 7$.

   Find b if $A = 9, h = 6$ and $a = 5$.

12) If $a = \sqrt{25 - h^2}$, find a when $h = 3$.

13) d is given in terms of t and m by the formula: $d = t(m - 1)/m$. Find d when $t = 6$ and $m = 3$.

14) The surface area of a cylinder is given by the formula: $A = 2\pi r(r + h)$. Use $\pi = \frac{22}{7}$ to find A when $r = 3$ and $h = 4$.

8.3 The Use of Brackets

The order in which algebraic operations are done is important. Taking powers (i.e. squaring or cubing etc) is done first, then multiplying or dividing, then adding or subtracting. If the operations are to be done in a different order then they must be put in *brackets*.

The process of eliminating brackets from an expression is called *expansion*.

❐ **8.3.1 Examples.** *Intermediate, level 8*

1) Expand $3(x + y)$

   Solution When the brackets are removed both terms are multiplied by 3.

$$3x + 3y$$

2) Expand and simplify $2(3x + 5) + 4(2x - 7)$.

   Solution Multiply through to obtain:

$$6x + 10 + 8x - 28$$

The x terms can be collected together, and the number terms can be collected:

$$14x - 18$$

3) Expand and simplify $(x - 5)(x + 8)$.

Solution Hold the first pair of brackets, and multiply out:

$$(x - 5)x + (x - 5)8$$

Multiplying out again,

$$x^2 - 5x + 8x - 40$$

$$x^2 + 3x - 40$$

4) Expand out $(a + b)(a - b)$

Solution Hold the second pair of brackets, and multiply through:

$$a(a - b) + b(a - b)$$

Multiply through again:

$$a^2 - ab + ba - b^2$$

The two middle terms cancel, leaving:

$$a^2 - b^2$$

(Note: this formula is known as the *difference of two squares*)

5) Expand and simplify $(2x - 3)(3x - 5)$

Solution Hold the second pair of brackets and multiply out:

$$2x(3x - 5) - 3(3x - 5)$$

$$6x^2 - 10x - 9x + 15$$

$$6x^2 - 19x + 15$$

6) Expand and simplify $(p + q)(r + s)$

Solution Hold the second bracket and multiply out the first.

$$p(r + s) + q(r + s)$$

Now multiply through to obtain:

$$pr + ps + qr + qs$$

□ **8.3.2 Exercises.** *Intermediate, level 8*

Expand the following and simplify as far as possible:

1) $5(x + 7)$ 2) $4(a - b)$

3) $7(2x + y)$ 4) $2(3x - 5y)$

5) $3(x - 4) + 5(x + 6)$ 6) $7(y - 3) - 2(3 + y)$

7) $6(z - 3) - 8(z - 5)$ 8) $5(2w + 3) + 9(3w - 4)$

9) $3(x + y) + 4(x - y)$ 10) $2(a + b) + 3(2a + 3b)$

11) $4(x + y) - 2(x + y)$

12) $3(c - 2d) - 2(c + 4d)$

13) $3(5r - 2s) - 2(2r - 3s)$

14) $4(p + 2q) - 2(p - 3q)$

15) $(x + 1)(x + 5)$

16) $(y + 3)(y + 7)$

17) $(p + 3)(p - 4)$

18) $(z - 4)(z + 6)$

19) $(w - 7)(w - 11)$

20) $(q - 2)(q - 5)$

21) $(x + y)(x - y)$

22) $(2x + y)(2x - y)$

23) $(3p - 2q)(3p + 2q)$

24) $(6r - 4s)(3r + 2s)$

25) $(x + 8)(3x - 5)$

26) $(3y - 4)(2y + 7)$

27) $(3a - 2b)(3a + 2b)$

28) $(2w + 5z)(3w - 2z)$

29) $(x + 3)^2$

30) $(2y - 3)^2$

31) $(x + 2y)^2$

32) $(3z - 2w)^2$

33) $(a + b)(c + d)$

34) $(x - y)(z + w)$

35) $(2t + s)(3r + p)$

36) $(f - 3g)(j - 5k)$

37) $(2y + 3x)(5z - 7w)$

38) $(5s + 1)(3r - 2t)$

8.4 Changing the Subject of a Formula

The *subject* of a formula is the letter which is expressed in terms of the other letters. When the subject is changed the formula is re-arranged so that a different letter is expressed in terms of the others.

❏ **8.4.1 Examples.** *Intermediate, level 8*

1) If $y = mx + c$, put x in terms of y, m and c.

Solution Subtract c from both sides, to obtain $y - c = mx$. Then divide both sides by m, to obtain:

$$\frac{(y - c)}{m} = x.$$

2) If $C = (F - 32) \times \frac{5}{9}$, express F in terms of C.

Solution Multiply both sides by 9, to get $9C = (F - 32) \times 5$.

Divide both sides by 5, to get $\frac{9C}{5} = F - 32$.

Add 32 to get:

$$\frac{9C}{5} + 32 = F.$$

❏ **8.4.2 Exercises.** *Intermediate, level 8*

In each of these questions, change the subject to the letter in brackets.

1) $x = 3y + 2$ (y)

2) $a = b - c + d$ (d)

3) $F = \frac{9C}{5} + 32$ (C)

4) $rt = 3sp$ (s)

5) $y = 3ax^2$ (a)

6) $j = 59 - mn$ (m)

7) $b = \frac{1}{2}(c + a)$ (a)

8) $d + 3 = b/2a$ (b)

9) $3V = 4r - 2$ (r)

10) $\frac{(X + 2Z)}{5} = 4Y$ (X)

11) $2(x + y) = 4z$ (x)

12) $5p(q - r) = 7$ (p)

❏ **8.4.3 Examples.** *Higher, level 10*

1) Make r the subject of the formula $V = \frac{\pi r^2 h}{3}$.

     Solution   Multiply by 3 and divide by πh, to get $3V/\pi h = r^2$. Now square root both sides to get:

$$\sqrt{\frac{3V}{\pi h}} = r$$

2) Make x the subject in the equation: $ax + b = cx - d$.

     Solution   Add d to both sides, and subtract ax from both sides. The equation is now $b + d = cx - ax$. Write this as

$$b + d = (c - a)x.$$

     Now divide both sides by $(c - a)$. The answer is:

$$x = \frac{b + d}{c - a}$$

❏ **8.4.4 Exercises.** *Higher, level 10*

In each of these equations, change the subject to the letter in brackets.

1) $y = 4ax^2$ (x)

2) $V = \frac{4\pi r^3}{3}$ (r)

3) $T + R = S^2 - 3$ (S)

4) $l = \sqrt{h^2 + 4r^2}$ (h)

5) $h + 3 = 3m\sqrt{n}$ (n)

6) $t = 2\pi\sqrt{\frac{l}{g}}$ (l)

7) $a = \frac{c + 1}{b - 3}$ (b)

8) $\frac{1}{f} = \frac{1}{u} + \frac{1}{v}$ (u)

9) $mg - T = ma$ (m)

10) $x - 5a = 3x + 2b$ (x)

11) $ap + bq = cp + dq$ (q)

12) $a(x + 3) = b(7 - x)$ (x)

13) $m = \frac{R}{R + r}$ (R)

14) $y = \frac{x + 1}{x - 3}$ (x)

8.5 Factorization

Factorization is the opposite operation to expansion. When an expression is factorized, it is written as the product of two or more simpler expressions.

❑ **8.5.1 Examples.** *Intermediate, level 8*

1) Factorize $3xt - 6xs$.

   Solution   The letter x can be taken outside a pair of brackets.

$$x(3t - 6s)$$

   Both the terms inside the brackets are divisible by 3. Take 3 outside the brackets:

$$3x(t - 2s)$$

2) Factorize $x^2 + 3x$.

   Solution   Here both terms contain an x. The x can be taken outside a pair of brackets to obtain:

$$x(x + 3)$$

❑ **8.5.2 Exercises.** *Intermediate, level 8*

Factorize the following as far as possible

1) $xy + xz$

2) $pq - 3q$

3) $3s + rs$

4) $qt + rt$

5) $rt - 5rs$

6) $3xy - 4xz$

7) $2x + 4y$

8) $3t - 9r$

9) $2rs + 4rq$

10) $9xy - 3zy$

11) $6fg + 9fh$

12) $7t + 21tq$

13) $a^2 + ab$

14) $z - z^2$

15) $5x^2 - 15x$

16) $21y^2 - 14y$

17) $8z^2 + 4cz$

18) $3d + 9cd^2$

Recall that factorization is the opposite of expansion. In section 8.3 there were the following expansions:

(A) $(a + b)(a - b) = a^2 - b^2$ (Example 4 of 8.3.1)

(B) $(x - 5)(x + 8) = x^2 + 3x - 40$ (Example 3 of 8.3.1)

(C) $(2x - 3)(3x - 5) = 6x^2 - 19x + 15$ (Example 5 of 8.3.1)

(D) $(p + q)(r + s) = pr + ps + qr + qs$ (Example 6 of 8.3.1)

Each of these expansions corresponds to a factorization.

Some expressions factorize as a perfect square. For example:

$$x^2 - 6x + 9 = (x - 3)^2$$

The method of *completing the square* expresses a quadratic expression as a perfect square with a constant added.

$$x^2 - 6x + 11 = (x - 3)^2 + 2$$

❑ **8.5.3 Examples.** *Higher, level 10*

1) Factorize $4t^2 - s^2$.

Solution   This expression consists of one square subtracted from another. It matches the right-hand side of (A) above. Hence it can be factorized to the left-hand side of (A). The *a* term corresponds to 2*t*, and the *b* term corresponds to *s*.

$$(2t + s)(2t - s)$$

2)   Factorize $y^2 + 5y + 6$.

Solution   This expression is similar to the right-hand side of (B) above. To match the left-hand side of (B) the factorization must be:

$$(y + a)(y + b)$$

Here *a* and *b* are numbers whose product *ab* is 6 and whose sum ($a + b$) is 5. Try the factors of 6 until you find the pair whose sum is 5. The factors must be 2 and 3:

$$(y + 2)(y + 3)$$

3)   Factorize $6x^2 - x - 2$.

Solution   This is similar to the right-hand side of (C). When it is factorized it must be:

$$(ax + b)(cx - d)$$

Here *a*, *b*, *c* and *d* are numbers such that $ac = 6$ and $bd = 2$. The first pair could be 6×1 or 3×2, and the second pair must be 2×1. Try the possible arrangements until the correct values are found:

$$(2x + 1)(3x - 2)$$

4)   Factorize $xa - yb + ya - xb$.

Solution   This is similar to the right-hand side of (D). Re-arrange the terms so that the first two have a common factor:

$$xa + ya - yb - xb$$

$$(x + y)a - (y + x)b$$

Now there is a common factor of ($x + y$). Take it out:

$$(x + y)(a - b)$$

5)   Complete the square of the expression $x^2 + 4x - 5$. Hence find the least value of this expression.

Solution   The expansion of $(x + 2)^2$ is $x^2 + 4x + 4$. This is 9 greater than the expression.

$$x^2 + 4x - 5 = (x + 2)^2 - 9$$

A square cannot be negative. So the least value of the expression is reached when $(x + 2)^2$ is zero.

The least value of $x^2 + 4x - 5$ is –9

☐ **8.5.4 Exercises.** *Higher, level 10*

Factorize the following as far as possible:

1) $p^2 - q^2$ 2) $t^2 - 1$

3) $4 - n^2$ 4) $q^2 - 9$

5) $9 - s^2$ 6) $7 - 7s^2$

7) $2 - 8x^2$

8) $9 - 81a^2$

9) $49t^2 - 9s^2$

10) $16m^2 - 25n^2$

11) $y^2 - 1/4$

12) $1/9 - z^2$

13) $x^2 + 3x + 2$

14) $x^2 + 4x + 4$

15) $x^2 + 7x + 12$

16) $x^2 - 11x + 24$

17) $x^2 + x - 6$

18) $y^2 - y - 12$

19) $z^2 + 5z - 6$

20) $w^2 - 4w - 12$

21) $x^2 - 5x - 24$

22) $p^2 + 7p - 30$

23) $2x^2 + 5x - 3$

24) $3x^2 - 10x - 8$

25) $5y^2 + 9y + 4$

26) $6x^2 + 35x - 6$

27) $6a^2 + 11ab + 5b^2$

28) $6p^2 - 5pq - 6q^2$

29) $st - sq + rt - rq$

30) $ax + by - ay - bx$

31) $6st - 4s + 3t - 2$

32) $2xz - 3y - 6x + yz$

33) Complete the square of the following expressions. Hence find their least values.

a) $x^2 - 2x + 5$ b) $x^2 + 4x + 1$ c) $x^2 - 10x + 11$

d) $x^2 + 6x + 13$ e) $x^2 + 3x + 1$ f) $x^2 - 5x - 3$

8.6 Algebraic Fractions

The rules for numerical fractions apply to algebraic fractions. When fractions are multiplied, their numerators are multiplied and their denominators are multiplied. When fractions are added, they must first be put over a common denominator.

☐ 8.6.1 Examples. *Higher, level 10*

1) Express as a single fraction: $\dfrac{2a}{5} \times \dfrac{3}{x}$

Solution Multiply together the tops and bottoms.

$$\frac{6a}{5x}$$

2) Simplify $\dfrac{2a^2}{3} \div \dfrac{4a}{9}$

Solution To divide by $\dfrac{4a}{9}$ is the same thing as to multiply by $\dfrac{9}{4a}$. The expression can be written:

$$\frac{2a^2}{3} \times \frac{9}{4a} = \frac{18a^2}{12a}$$

$6a$ divides both top and bottom. Cancel to obtain:

$$\frac{3a}{2}$$

3) Simplify $\dfrac{2p}{3} + \dfrac{5p}{2}$.

Solution Here both terms must be put over the common denominator of 6.

$$\frac{4p}{6} + \frac{15p}{6}$$

$$\frac{4p + 15p}{6} = \frac{19p}{6}$$

4) Express as a single fraction: $\dfrac{1}{x-1} + \dfrac{2}{x+3}$

Solution The common denominator of this fraction is $(x-1)(x+3)$. Put both terms over this:

$$\frac{(x+3) + 2(x-1)}{(x-1)(x+3)}$$

$$\frac{3x+1}{(x-1)(x+3)}$$

☐ **8.6.2 Exercises.** *Higher, level 10*

Express the following as single fractions:

1) $\dfrac{a}{b} \times \dfrac{c}{d}$

2) $\dfrac{3x}{5} \times \dfrac{4x}{6}$

3) $\dfrac{2x}{3y} \div \dfrac{y}{2x}$

4) $\dfrac{x}{y} \div \dfrac{z}{w}$

5) $\dfrac{7ab}{4c} \times \dfrac{2c}{ab}$

6) $\dfrac{5pq}{r} \div \dfrac{10p}{qr}$

7) $\dfrac{x}{3} + \dfrac{2x}{5}$

8) $\dfrac{3a}{2} - \dfrac{a}{4}$

9) $\dfrac{r}{4} + \dfrac{(r-1)}{3}$

10) $\dfrac{3(x-2)}{4} - \dfrac{5x}{3}$

11) $\dfrac{(7b-1)}{2} + \dfrac{(3b+1)}{3}$

12) $\dfrac{4(c-1)}{3} - \dfrac{3(c+1)}{4}$

13) $\dfrac{2}{x} + \dfrac{3}{xy}$

14) $\dfrac{3}{a} + \dfrac{4}{bc}$

15) $\dfrac{1}{3x} \times \dfrac{1}{(x-1)}$

16) $\dfrac{3z}{(z-1)} \times \dfrac{2z}{(z+1)}$

17) $\dfrac{(x+2)}{3} \div \dfrac{(x-1)}{2}$

18) $\dfrac{2}{(1-y)} \div \dfrac{3}{(2-y)}$

19) $\dfrac{3}{x} + \dfrac{4}{(x+1)}$

20) $\dfrac{5}{z} - \dfrac{2}{(1-z)}$

21) $\dfrac{z}{3} + \dfrac{2}{(z+1)}$

22) $\dfrac{5}{(3y+2)} + \dfrac{4}{y}$

23) $\dfrac{2}{(y-1)} + \dfrac{1}{(y+3)}$

24) $\dfrac{1}{(3w+1)} - \dfrac{1}{(w+2)}$

25) $\dfrac{3}{(2p-1)} + \dfrac{5}{(2p+3)}$

26) $\dfrac{7}{(3-2q)} - \dfrac{2}{(1-q)}$

Common errors

1) **Collecting**.

   It is incorrect to add together terms unless they are of the same sort, with the same letters occurring.

   $5a^2b$ and $7a^2b$ can be added to give $12a^2b$.

   But $5a^2b + 7ab^2$ cannot be simplified in this way.

2) **Brackets**.

   a) Do not ignore brackets:

   $$a(b + c) \neq ab + c$$

   b) Do not forget to put brackets in when they are necessary.

   From $\dfrac{y}{4} = x + 3$ go to $y = 4(x + 3)$.

   Do not go to $y = 4x + 3$. This gives incorrect values.

3) **Multiplying negative terms.**

   Remember the basic rules of:

   $$minus\ times\ minus\ is\ plus - \times - = +$$
   $$plus\ times\ plus\ is\ plus + \times + = +$$
   $$minus\ times\ plus\ is\ minus - \times + = -$$

   It is quite common to get these wrong, when expanding out something like $(x - y)(p - q)$. The yq term is positive.

4) **Expansion**.

   When multiplying out brackets, *all* the terms in the first pair must multiply *all* the terms in the second pair. Do not just multiply together the first two and the last two.

   $$(a + b)(x + y) \neq ax + by.$$

   The correct expansion is $ax + ay + bx + by$

   In particular, be careful when expanding $(a + b)^2$ or $(a - b)^2$

   $$(a + b)^2 = a^2 + 2ab + b^2 \neq a^2 + b^2$$
   $$(a - b)^2 = a^2 - 2ab + b^2 \neq a^2 - b^2$$

5) **Factorizing**.

   a) The following are common mistakes when factorizing:

   $$49x^2 - 1 \neq (49x - 1)(49x + 1)$$
   $$x^2 + 5x + 6 \neq (x + 2x)(x + 3x)$$

   b) $x^2 + a^2$ is not the difference of two squares. It cannot be factorized at all.

   c) Make sure you factorize fully. It is true that:

$$6st + 3sr = s(6t + 3r)$$

But a further factor of 3 can be taken out.

d) When completing the square of an expression like $x^2 + 8x + 2$, consider the expansion of $(x + 4)^2$, not of $(x + 8)^2$.

6) **Changing the subject of a formula**.

a) Frequently the wrong operation is done when making x the subject of a formula.

If a term has been *added* to x, we must *subtract* it away.

If a term is *multiplying x*, then we must *divide* by it.

In each case we do the opposite operation to get x on its own. We must also be sure to do them in the correct order.

b) When x is made the subject of a formula, it must be expressed as a function of the other terms. It cannot occur on both sides of the equation. If your answer is of the form $x = 5c + 7xy$, then you have not made x the subject.

7) **Algebraic Fractions**.

a) When multiplying a fraction by another term, it is incorrect to multiply both numerator and denominator by that term.

$$5 \times \frac{a}{b} \neq \frac{5a}{5b}$$

It may be helpful to think of 5 as $\frac{5}{1}$, which gives the correct expression:

$$\frac{5}{1} \times \frac{a}{b} = \frac{(5 \times a)}{(1 \times b)} = \frac{5a}{b}$$

b) Wrong addition of fractions. When adding fractions, we cannot simply add the numerators and the denominators.

$$\frac{a}{b} + \frac{c}{d} \neq \frac{(a + c)}{(b + d)}$$

c) Sums of fractions cannot be simply inverted.

From $\frac{1}{f} = \frac{1}{h} + \frac{1}{k}$ it does not follow that $f = h + k$.

Chapter 9

Equations

An *equation* states that one mathematical expression is equal to another. Examples of equations are:

$$x + 3 = 56; \quad y^2 + 3y - 2 = 0; \quad 2x + 3y = 6$$

Usually an equation contains an *unknown*. To solve the equation is to find the value or values of this unknown.

There is one basic rule to obey when solving equations.

Do to the left what you do to the right.

So if 5 is added to the left hand side of an equation, 5 must be added to the right also. If the left is divided by 4.7, the right also must be divided by 4.7.

9.1 Equations with one Unknown

❏ 9.1.1 Examples. *Foundation, level 6*

1) Solve the equation $x - 2 = 4$.

   Solution   Add 2 to both sides.

$$x = 6$$

2) Solve the equation $3x + 13 = 40$.

   Solution   Subtract 12 from both sides:

$$3x = 27$$

   Divide both sides by 3:

$$x = 9$$

3) Solve the equation $3y - 4 = 2y + 1$, checking that your answer is correct.

   Solution   Subtract $2y$ from both sides.

$$3y - 2y - 4 = 1$$
$$y - 4 = 1$$

   Add 4 to both sides,

$$y = 1 + 4 = 5$$

   Check the answer by putting $y = 5$ in the original equation.

   Left hand side $= 3 \times 5 - 4 = 11$
   Right hand side $= 2 \times 5 + 1 = 11$

❏ 9.1.2 Exercises. *Foundation, level 6*

Solve the following equations:

1) $x + 3 = 5$ 2) $y - 4 = 13$

78

3) $12 - x = 5$ 4) $7 - x = 9$

5) $13 = 7 + x$ 6) $9 = 13 - z$

7) $2x + 5 = 11$ 8) $5z - 3 = 7$

9) $5 - 3a = 2$ 10) $7 + 3b = 10$

11) $9 = 3 + 2x$ 12) $5 = 17 - 3z$

13) $5(x - 2) = 15$ 14) $4(b + 3) = 16$

15) $\dfrac{z}{7} = 21$ 16) $\dfrac{12}{y} = 4$

17) $2x - 10 = 18 + x$ 18) $3y + 1 = 8 + 2y$

19) $5m + 1 = 19 - 4m$ 20) $5n + 2 = 3n + 12$

21) $3z + 5 = z - 7$ 22) $6x - 2 = 19 - x$

9.2 Solution by Trial and Improvement

Sometimes more complicated equations can be solved approximately by trying different values until we get close to the solution.

❑ **9.2.1 Example.** *Foundation, levels 6, 7*

Solve the equation $x^3 = 31$, giving an answer to 1 decimal place.

Solution First of all try whole number values.

$$1^3 = 1. \text{ Too small.}$$

$$2^3 = 8. \text{ Too small.}$$

$$3^3 = 27. \text{ Too small.}$$

$$4^3 = 64. \text{ Too big.}$$

So the solution lies between 3 and 4.

$$3.1^3 = 29.791. \text{ Too small}$$

$$3.2^3 = 32.768. \text{ Too big.}$$

So the solution lies between 3.1 and 3.2. To find which it is closer to, find the value of x^3 at 3.15.

$$3.15^3 = 31.255875. \text{ Too big.}$$

So the solution is between 3.1 and 3.15.

To 1 decimal place, the solution is 3.1

❑ **9.2.2 Exercises.** *Foundation, levels 6, 7*

1) Solve the following by trial and improvement, giving your answers to 1 decimal place.

     a) $x^2 = 29$ b) $x^2 = 17$ c) $x^2 = 23$

     d) $x^3 = 68$ e) $x^3 = 0.7$ f) $x^3 = 60$

g) $x^2 + x = 13$ h) $x^3 + x^2 = 32$ i) $x^3 - x = 23$

2) Solve the following, giving your answers to 2 decimal places.

a) $x^2 + x = 0.5$ b) $x^3 + 2x = 1.2$ c) $x^2 - x = 0.5$

9.3 Simultaneous Equations

Simultaneous equations consist of two equations with two unknowns.

❏ 9.3.1 Examples. *Intermediate, level 7*

1) Solve the equations: [1] $x + 4y = 8$

                               [2] $x + 5y = 12$

Check your answer.

Solution   The way to solve this is to get rid of the x term.

Subtract [1] from [2]:

$$[2] - [1] = [3] \qquad y = 12 - 8 = 4$$

The value of y can be put into either of the original equations to find x.

$$x + 4 \times 4 = 8$$

$$x = 8 - 16 = -8$$

Equation [1] was used to find the value of x. The working can be checked by putting the values in [2]

$$-8 + 5 \times 4 = -8 + 20 = 12. \text{ Correct.}$$

2) Solve the equations: [1] $3x - 5y = 1$

                               [2] $2x + 7y = 11$

Solution   Here both equations must be multiplied by a number in order to match either the x terms or the y terms. If the y terms are matched:

$$[3] = 7 \times [1] \quad 21x - 35y = 7$$

$$[4] = 5 \times [2] \quad 10x + 35y = 55$$

Eliminate y by adding.

$$[5] = [3] + [4] \; 31x = 62$$

$$x = 2$$

Substitute in either [1] or [2] to find that

$$y = 1.$$

❏ 9.3.2 Exercises. *Intermediate, Level 7*

Solve the following pairs of simultaneous equations.

1) $x + y = 3$ 2) $2z + w = 3$
    $x - y = 7$ $3z + w = 9$

3) $3x + 2y = 4$
    $5x - 2y = 12$

4) $3x + y = 7$
    $3x - 4y = -13$

5) $5q - 3p = 11$
    $q + p = 7$

6) $7x + 5y = 4$
    $3x + 2y = 7$

7) $3z - 5w = 8$
    $z + 3w = 12$

8) $4q - 3p = 5$
    $7q - 6p = 5$

9) $3x + 2y = 14$
    $x - 3y = 1$

10) $2x - 4y = 14$
    $3x + y = 7$

11) $3x + 2y = 20$
    $-5x + 3y = 11$

12) $4s + 5r = 1$
    $2s - 3r = 17$

13) $2x + 3y = 13$
    $7x - 5y = -1$

14) $6p - 7q = -1$
    $5p - 4q = 12$

15) $\dfrac{x}{6} + \dfrac{y}{3} = 8$
    $\dfrac{x}{4} - \dfrac{y}{9} = 1$

16) $\dfrac{m}{6} + \dfrac{2n}{3} = 6$
    $\dfrac{m}{10} - \dfrac{2n}{5} = -2$

9.4 Quadratic Equations

If an equation contains the square of the unknown then it is a *quadratic equation*. Sometimes the equation can be solved by factorizing. Sometimes the *formula* must be used.

$$\text{For} \quad ax^2 + bx + c = 0,$$

$$x = \frac{-b \pm \sqrt{b^2 - 4ac}}{2a}$$

☐ 9.4.1 Examples. *Higher, level 10*

1) Solve the equation $x^2 - 3x + 2 = 0$

    Solution: This quadratic factorizes, to become:

    $$(x - 2)(x - 1) = 0.$$

    Either $x - 2 = 0$ or $x - 1 = 0$. There are the two solutions:

    $x = 2$ or $x = 1$

2) Solve the equation $y^2 + 17y = 60$

    Solution   The left hand side should not be factorized until there is a 0 on the right hand side. So subtract 60 from both sides:

    $$y^2 + 17y - 60 = 0.$$

    This factorizes, to become:

    $$(y + 20)(y - 3) = 0$$

    $y = -20$ or $y = 3$

3) Solve the equation $z^2 + 13z + 2 = 0$, giving your answer to 3 decimal places.

    Solution: This equation does not factorize. So the formula must be used. Here $a = 1$, $b = 13$, $c = 2$. Substitute these values in to obtain:

$$z = \frac{-13 \pm \sqrt{13^2 - 4 \times 1 \times 2}}{2 \times 1}$$

$$z = \frac{-13 \pm \sqrt{161}}{2}$$

$$z = -0.156 \text{ or } z = -12.844$$

❏ 9.4.2 Exercises. *Higher, level 10*

Solve the following equations. Where relevant give your answers to three decimal places.

1) $x^2 + 5x - 6 = 0$

2) $y^2 + 7y + 12 = 0$

3) $x^2 - 10x + 9 = 0$

4) $x^2 + 19x + 60 = 0$

5) $x^2 - 28x - 60 = 0$

6) $4x^2 - 4x - 3 = 0$

7) $z^2 - 3z = 10$

8) $p^2 = 8p - 15$

9) $(x - 2)(x + 1) = 18$

10) $y(y - 3) = 5y - 12$

11) $\frac{6}{y} = y + 5$

12) $z + \frac{12}{z} = 7$

13) $4x^2 + 10x + 3 = 0$

14) $y^2 - 5y + 2 = 0$

15) $z^2 + 3z - 5 = 0$

16) $q^2 - q - 3 = 0$

17) $x + \frac{1}{x} = 3$

18) $(y + 1)(y + 2) = 28$

9.5 Problems which Lead to Equations

Often problems can be translated into equations.

❏ 9.5.1 Example. *Foundation, Level 6*

The rental for a telephone is £12 per quarter, and the call charges are 2 p per unit. A bill was £25. Let x be the number of units used. Form an equation in x and solve it.

Solution   The total cost is $2x + 1200$ pence. The bill is £25, so this gives the equation:

$$2x + 1200 = 2500$$

This can be solved to give:

$$x = 650.$$

❏ 9.5.2 Exercises. *Foundation, Level 6*

1) The hire of a car is £20 per day plus a charge of 10 p per mile. Letting x be the number of miles covered, find an expression giving the charge in terms of x.

   If the total cost of hiring was £37, how many miles were covered?

2) I think of a number x, double it and then add 5. If the result is 37, what was the original number?

3) Two partners in a business agree that the first partner shall receive £1500 more of the profits than the second partner.

Letting x be the amount received by the second partner, find the total profit in terms of x.

If the total profit was £27,700, how much did each partner get?

4) The three angles of a triangle are $x°$, $3x° - 30°$, $60° - x°$. Find x, given that the sum of the angles in a triangle is 180°.

5) Jane tells John: "Think of a number, multiply it by 3, subtract 7 and then take away the number you first thought of."

If John though of x, what number has he ended with?

If John ends with 11, what was his original number?

6) A car is driven at x mph for $1\frac{1}{2}$ hours, then at $x + 20$ mph for $\frac{1}{2}$ hour. The total distance covered is 90 miles. Find x.

❒ **9.5.3 Examples.** *Intermediate, level 7*

1) A man is three times as old as his son. In 12 years time he will be twice as old. Let the son be aged y years. Find an equation in y and solve it.

Solution The father is $3y$ years. In 12 years time the son will be $y + 12$, and the father will be $3y + 12$.

The father will then be twice as old. This gives the equation:

$$3y + 12 = 2(y + 12)$$

Expand out and solve to obtain

y = 12 years.

2) 3 kg of plums and 2 kg of apples cost £3.40, while 1 kg of plums and 4 kg of apples cost £2.80. Find the cost of 1 kg of apples.

Solution Here there are two unknown. Let xp be the price of 1 kg of plums, and yp be the price of 1 kg of apples. There are two bits of information, which lead to the simultaneous equations;

$$3x + 2y = 340$$
$$x + 4y = 280$$

These can be solved to give $y = 50$ (and $x = 80$.)

1 kg of apples costs 50p

❒ **9.5.4 Exercises.** *Intermediate, level 7*

1) A woman is six times as old as her daughter. 3 years ago she was 11 times as old. Let the daughter's age be y years. Form an equation in y.

Solve the equation to find how old the daughter is.

2) The sum of three consecutive numbers is 441. Letting n be the smallest of the three, write down the other numbers in terms of n. Find an equation in terms of n, and solve it.

3) A girl buys x magazines at 50p each and $x - 3$ magazines at 40p each. She spends £6 in all. Find x.

4) Money is left in a will, so that Anne gets twice as much as Brian, and Brian gets three times as much as Christine. If Christine gets £x, express the amounts received by Brian and Anne in terms of x. If the total amount left was £1800, find how much each received.

5) Adrian takes a holiday job at £x per week. After 4 weeks his pay increases by £10 per week. He then works for 5 more weeks. Write in terms of x the total amount he has earned.

   If the total is £617, find x.

6) Jane has £x, and Bill has £y. Jane is £372 richer than Bill. Together they have £628. How much does Jane have?

7) At a concert 500 tickets were sold: the cheaper ones cost £2.50 and the more expensive ones £4.50. The total receipts were £1,610. Let x be the number of cheap tickets, and y the number of expensive tickets. Form two equations in x and y, and hence find how many cheap tickets were sold.

8) In a pub, brandy is 10 p. more expensive than whisky. A round of 3 brandies and 5 whiskies cost £7.66. Let x be the cost of a whisky, and y the cost of a brandy. Find the cost of a whisky.

9) 3 pounds of butter at x p per pound and 4 pints of milk at y p per pint cost £3.84. 5 pounds of butter and 7 pints of milk cost £6.48. Form two equations in x and y and solve them.

10) Chuck saves money by putting every 50 p and every 20 p coin he receives in a box. After a while he find that he has 54 coins, amounting to £17.10. How many 50 p coins does he have?

□ **9.5.5 Example.** *Higher, level 10*
A rectangle is 3 cm longer than it is wide. The area is 108 cm². Find the length.

Solution   Let the length be x cm. The width must be $x - 3$ cm. The area is 108, which gives the equation:

$$x(x - 3) = 108$$

Expand and re-arrange to get:

$$x^2 - 3x - 108 = 0$$

This has solutions $x = 12$ or $x = -9$. The negative answer is not meaningful here. The answer is;

$$x = 12 \text{ cm}$$

□ **9.5.6 Exercises.** *Higher, level 10*

1) The manager of a shoe shop spends £S on shoes at £10 a pair, and £$(S + 100)$ on boots at £15 a pair. If she obtains 50 pairs in all, find S.

2) A train runs for $x + 10$ miles at 40 mph, then for $x - 60$ miles at 60 mph. The total time for the run was 3 hours. Find x.

3) From Leeds to London is 180 miles. A car travels from London to Leeds at 60 m.p.h., and starting at the same time a lorry travels from Leeds to London at 30 m.p.h. How far from London do they pass each other?

4) A positive number x is 156 less than its square. Find the number.

5) The area of a circle is 3π greater than the perimeter. Find the radius.

6) The sides of a right-angled triangle are $2x + 1$ cm, $2x - 1$ cm, and x cm. Find x.

7) The area of a rectangle is 72 cm², and its perimeter is 34 cm. Find the lengths of the sides.

8) A piece of wire of length 80 cm is cut in two, and the segments are each bent into a square. The total area is 208 cm². Find the sides of the squares.

9) A boy bats in 6 matches. In the 7th match he makes 8 runs and his average decreases by 1. How many runs has he scored in all?

10) A man can row in still water at 4 m.p.h. He rows upstream for 7 miles, then downstream for the same distance. The total time taken was 8 hours. Find the speed of the current.

Common errors

1) **Arithmetic**

All the common errors which arise in basic algebra are very easy to make when solving equations. Be very careful when multiplying out brackets, or when multiplying two negative numbers together. See Chapter 8 for a list of some common algebraic errors.

2) **Equations in one unknown**

The unknown cannot appear on both the sides of the answer. If your solution is of the form $x = 3 - 2x$ then you have not solved the equation at all, you have merely re-written it in another form. The solution of the equation must be a statement which gives us the *numerical value* of the unknown.

3) **Solving equations**

a) It is common to do the wrong operations when solving equations. The following examples show which operations should be used:

For $x + 3 = 5$, *subtract* 3 from both sides.

For $y - 4 = 7$, *add* 4 to both sides.

For $3z = 9$, *divide* by 3.

For $\frac{w}{8} = 10$, *multiply* by 8.

b) Suppose you are solving an equation by trial and improvement, and have found that the solution lies between 5.7 and 5.8. You won't find the solution to 1 decimal place until you have tested at 5.75.

4) **Simultaneous equations**

a) Do not try to get by with one equation for two unknowns. Nearly always you must have two equations for two unknowns.

b) Incorrect algebra. One common error is to disobey the basic rule of equations. In other words, to go from:

$$2x + 3y = 4$$

$$\text{to } 4x + 6y = 4$$

Here the right hand side was not multiplied by the same thing as the left hand side. the correct step would be to get:

$$4x + 6y = 8$$

c) Another common error occurs when adding or subtracting equations. Remember the basic rule that *two minuses make a plus*. So subtracting a negative number is the same as adding a positive one. For example, in the situation:

$$[1] \quad 3x + 4y = 7$$

$$[2] \quad 3x - 5y = 2$$

When we subtract [2] from [1] we obtain:

$$4y - (-5y) = 5$$

$$4y + 5y = 5$$

5) **Quadratic equations**

a) Incorrect assignment of a, b, c. For the equation:

$x^2 - 2x + 2 = 0$, then $a = 1$ (not 0) and $b = -2$ (not 2).

b) Incorrect algebra. Mistakes are often made when b or c are negative. Remember that "minus times minus is plus". For the equation:

$$x^2 - 3x - 8 = 0, \text{ both } b \text{ and } c \text{ are negative.}$$

The formula gives:

$$x = \frac{+3 \pm \sqrt{9 - 4 \times 1 \times (-8)}}{2}$$

$$x = \frac{3 \pm \sqrt{41}}{2}$$

$$x = 4.70 \text{ or } x = -1.70.$$

c) Errors in calculation. When using the formula for a quadratic, remember that the $2a$ term must divide the whole of the top line. So press the $=$ button on your calculator before you divide by $2a$.

d) In the formula, the $\pm$ means that we add or subtract $\sqrt{b^2 - 4ac}$. Do not multiply.

6) **Problem solving**

a) It is important to state clearly what your letters stand for. It is not enough to say: *Let x be the plums*. We must make it clear whether x is the price, or the weight, or the number of the plums. In Example 2 of 9.5.3 we said: *Let xp be the price of 1 kg of plums.*

b) It is very common to make mistakes with the quantities *Distance*, *Time* and *Speed*.

Travel for t hours at s m.p.h., the distance covered is ts miles.

Travel for d miles at s m.p.h., the time taken is d/s hours.

If in doubt, think of an actual journey. If we travel 100 miles at 40 m.p.h., it will take us $2\frac{1}{2}$ hours.

Chapter 10

Inequalities

10.1 Order

☐ 10.1.1 Examples. *Foundation, level 6*

1) Find the largest of 0.12, $\frac{1}{8}$, $\frac{2}{15}$.

   Solution   Write all three as decimals.

$$0.12: \qquad \frac{1}{8} = 0.125: \qquad \frac{2}{15} = 0.13333....$$

The largest is $\frac{2}{15}$.

2) Which of the sizes of soap, shown in fig 10.1, offers better value?

   Solution   The cost of soap in the economy size packet is:

   $$\frac{40}{150} = 0.267 \text{ p. per gram.}$$

   The cost of soap in the family size packet is:

   $$\frac{60}{250} = 0.24 \text{ p. per gram.}$$

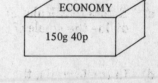

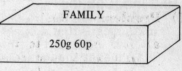

Fig 10.1

The Family size offers better value.

☐ 10.1.2 Exercises. *Foundation, level 6*

1) Write down the larger of $6\frac{3}{4}$ and 6.7.

2) Write down the smaller of 4.34 and $4\frac{1}{3}$.

3) Arrange in increasing order:

$$1.3, \ 1\frac{1}{4}, \ 1.42, \ \frac{9}{7}.$$

4) Arrange in decreasing order:

$$0.0011, \ 0, \ \frac{1}{100}, \ \text{a millionth, a thousandth.}$$

5) Which of the jam-jars of fig 10.2 offers better value?

6) Jason has a roll of 28 holiday snaps he wishes to print. Which is the

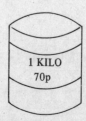

Fig 10.2

cheaper way of paying for them?
(See fig 10.3).

7) Which is the cheaper way for Linda to buy her motor-cycle? How much will it cost her? (See fig 10.4)

£3.20 per roll or 12p per print

Fig 10.3

| 1. £50 deposit, and 24 payments of £15 |
| 2. £100 deposit, and 12 payments of £25 |

Fig 10.4

8) The first half of a journey took $\frac{1}{2}$ hour and was 20 miles in distance: the second half took 20 minutes and was 17 miles long. During which half was the average speed greater?

9) Five days a week Kevin takes the tube to and from work. His fares are 60p. each way. Is it worth his while to buy a weekly ticket at £5.40?

10) Samantha sees two jobs advertised. One pays £123 per week, the other £6,400 per year. Allowing 52 weeks in the year, which job is better paid?

11) George is paid £125 per week. He is offered either an extra £17 per week or a 12% pay rise. Which should he take?

12) While on holiday in Holland, Mary sees perfume advertised at 50 Guilders for 20 c.c. She knows that in England the same perfume will cost £16 for 25 c.c. If there are 3.9 guilders to the pound, is the perfume cheaper in Holland or in England?

13) A tool set contains spanners whose sizes go up in steps of an eighth of an inch up to 1 inch. Arrange the sizes in order.

$$1, \frac{7}{8}, \frac{1}{4}, \frac{1}{2}, \frac{3}{8}, \frac{5}{8}, \frac{3}{4}, \frac{1}{8}.$$

10.2 Inequalities

$a < b$ means that a is *less than b*.

$a \leq b$ means that a is *less than or equal to b*.

$a > b$ means that a is *greater* than b.

$a \geq b$ means that a is *greater or equal to b*.

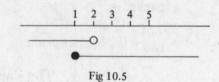

Fig 10.5

Inequalities can be illustrated on a *number line*. A bar above or below the line gives the range of values of x. If the endpoint is included, the bar ends with a filled-in dot. It the endpoint is not included, it ends with a hollow dot. In fig 10.5 the inequalities $x < 2$ and $x \geq 1$ are shown.

◻ **10.2.1 Exercises.** *Intermediate, level 7*

1) Which of the following are true?

   a) $3 < 5$ b) $5 < 3.6$ c) $\frac{1}{3} \leq 0.3$

   d) $10^2 > 90$ e) $\frac{1}{8} \leq 0.125$.

2) Arrange the following in increasing order, using the < sign.

$$4.6, 4.45, \frac{19}{4}, \frac{14}{3}.$$

3) Arrange the following in decreasing order, using the > sign.

$$\frac{1}{8}, 0, 0.2, \frac{-1}{2}, \frac{-1}{8}, \frac{-3}{17}, \frac{-1}{4}.$$

4) Find a whole number x for which $x > 3$ and $2x < 12$.

5) Find an integer y for which $y \leq -3$ and $y > -5$.

6) List the positive integers for which:

   a) $0 < x < 6$ b) $2 \leq x \leq 7$ c) $13 < x \leq 17$ d) $7 \leq x < 10$

7) List the whole numbers for which:

   a) $-2 < x < 2$ b) $-7 \leq x \leq -3$ c) $-1 \geq x > -5$ d) $4 > x \geq -2$

8) Identify the inequalities shown on the number lines below.

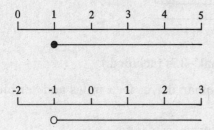
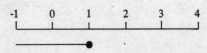
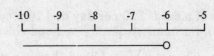

Fig. 10.6

9) Illustrate on number lines the following inequalities:

   a) $x > 2$ b) $x \leq 1$ c) $x \geq 0$ d) $x < -1$

10) illustrate on number lines the following inequalities:

   a) $0 < x < 2$ b) $-1 \leq x \leq 1$ c) $-1 < x \leq 2$ d) $-2 \leq x < 2$

Inequalities in one variable

An *inequality* looks like an equation, with the = sign replaced by one of the order signs. ($<, >, \leq, \geq$).

When solving inequalities, the same basic rule applies as to equations.

Do to the left what you do to the right

There is one additional rule for the solving of inequalities. When multiplying or dividing by a *negative* number, the inequality sign changes round.

$$\text{If } a < b, \text{ then } -2a > -2b$$

$$\text{If } c \leq d, \text{ then } -\frac{1}{2}c \geq -\frac{1}{2}d.$$

❏ **10.2.2 Examples.** *Intermediate, level 8*

1) Find the solution set of the inequality $3x - 2 < 4$, giving your answer in the form $\{x : x \dots \}$.

   Solution   Add 2 to both sides.

$$3x < 6$$

   Divide both sides by 3. As 3 is a positive number, the inequality sign stays the same way round.

$$x < 2$$

   The solution set is:

$$\{x : x < 2\}$$

2) Solve the inequality $2 - 3x \leq 2x + 17$, illustrating your answer on a number line.

Solution   Subtract $2x$ from both sides, and subtract 2 from both sides.

$$-5x \leq 15$$

Divide by –5, being sure to reverse the inequality sign.

$$\mathbf{x \geq -3}$$

This is shown on the number line as follows:

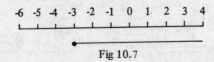

Fig 10.7

(The dot at the left of the line is filled in, to indicate that –3 is included.)

3) The hire of a car is £20 per day and 10 p per mile. A woman drives for x miles and her bill is less than £35. Form an inequality in x and solve it.

Solution   The total bill, of fixed charge plus mileage charge, comes to:

$$20 + \frac{x}{10}.$$

This must be less than £35.

$$20 + \frac{x}{10} < 35$$

$$\frac{x}{10} < 15$$

$$\mathbf{x < 150}$$

☐ **10.2.3 Exercises.** *Intermediate, level 8*

1) Find the solution sets for the following inequalities, giving your answer in the form $\{x : x \ldots\}$:

a) $x - 3 < 4$ b) $2x + 5 > 7$ c) $3x + 1 \geq 10$

d) $2x + 11 < 3$ e) $12 - x < 5$ f) $13 - 3x \geq 4$

2) Solve the following inequalities, illustrating your answers on number lines:

a) $x + 2 < 2x - 10$ b) $4x + 5 < x + 14$

c) $x + 1 < \frac{1}{2}x + 4$ d) $\frac{1}{2}(x-3) \leq 5$

e) $\frac{1}{3}(x + 2) \geq \frac{1}{2}(1-x)$ f) $\frac{1}{2}(3x + 5) < 2(x - 7)$

3) Find the solution sets for the following inequalities, both in the form $\{x : \ldots\}$ and on the number line:

a) $3 < x + 1 < 7$ b) $7 < 2x - 1 < 11$ c) $-4 \leq 3x + 2 \leq 20$

d) $1 \geq 3 - x \geq -8$ e) $7 \leq 3 - 2x < 13$ f) $8 > 2 - 3x \geq -10$

4) A boy buys x apples at 10 p. each. If he has at most £1 to spend, form an inequality in x and solve it.

5) Cakes cost 25 p. each. Jane buys y of them, and spends less than £1.50. Form an inequality in y and solve it.

6) A car does 30 miles per gallon, and there are z gallons of petrol in the tank. If the car goes for 100 miles without re-fuelling, find an inequality in z and solve it.

7) I have hired a car for 5 hours, and the maximum speed is 50 m.p.h. If I drive for m miles, form an inequality in m and solve it.

8) A man buys x stamps at 17 p. and $2x$ stamps at 12 p. He spends less than £5. Form an inequality in x and solve it.

9) The rental charge of a phone is £15 per quarter, and it costs 5p. per unit. A man uses x units, and his bill is less than £50. Form an inequality in x and solve it.

10) A room is 2 metres longer than it is wide. Its width is x metres, and the perimeter is less than 60 m. Form an inequality in x and solve it.

10.3 Inequalities in 2 Variables

The solution set for simultaneous inequalities in 2 variables consists of a region of a plane. Hence the solution set can be illustrated on graph paper.

If the inequality involves $\leq$ or $\geq$, then the boundary is included and is shown by a filled-in line. If the inequality involves $<$ or $>$, then the boundary is not included and is shown by a broken line.

☐ 10.3.1 Examples. *Intermediate, level 8*

1) Illustrate on graph paper the solution set of the inequality $2x + 3y \leq 6$.

Solution First draw the line $2x + 3y = 6$.

The line goes through the points $(0,2)$ and $(3,0)$. Fig 10.8.

Because the inequality requires $2x + 3y$ to be *less* than 6, the solution set is the region *below* the line. Fig 10.9. Note that the boundaries are shown by filled-in lines.

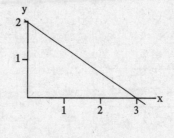

Fig. 10.8

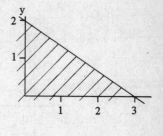

Fig. 10.9

2) Illustrate on graph paper the solution set of the simultaneous inequalities

$$x > \frac{1}{2}, y > \frac{1}{2}, 4x + 3y < 12, 3x + 4y < 12.$$

Find the pairs of whole numbers which satisfy these inequalities.

Solution First draw the lines corresponding to the 4 equalities. Fig 10.10.

The region required is above the line $y = \frac{1}{2}$, to the right of the line $x = \frac{1}{2}$, and below the two slanting lines. Note that the boundaries are shown by broken lines. The pairs of points within the required region are:

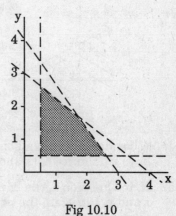

Fig 10.10

(1,1), (1,2), (2,1)

3) A man has £10 to spend on cigars. Superbas cost 50 p each, and Grandiosos cost 75 p. If he buys x Superbas and y Grandiosos, obtain an inequality in x and y. Illustrate this inequality on a graph.

Solution The total amount he spends on cigars is given by:

$50x + 75y$ p.

This must be less than 1,000p.

$50x + 75y \le 1000$

$2x + 3y \le 40$

This is illustrated in fig 10.11.

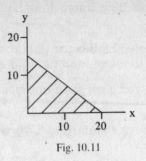

Fig. 10.11

☐ **10.3.2 Exercises.** *Intermediate, level 8*

1) Illustrate on graph paper the regions corresponding to the solution sets of the following:

   a) $x + y \le 1$ b) $2x + 4y \le 8$ c) $3x + 5y \le 15$

   d) $4x + 3y < 12$ e) $x + 2y > 4$ f) $3x + 2y > 12$

   g) $x - y \ge 0$ h) $2x - 3y < 6$ i) $y - 3x \le 6$ j) $5y - 4x > 20$

2) Illustrate the following inequalities on graph paper. Each group should be illustrated simultaneously.

   a) $x \ge 0, y \ge 0, x + y \le 6$ b) $x + y \ge 3, x < 7, y < 8$

   c) $x \ge 0, y \ge 0, 2x + 3y < 1$ d) $x \ge 0, y \ge 0, x + 3y \le 9, 2x + y \le 10$

   e) $y \ge 0, y \le x, 3x + 4y > 12, 3x + 4y < 18$ f) $x - y \ge 1, x - y \le 3, x + y \le 8, x + y \ge 4$

3) Describe the shaded regions of fig 10.12 by inequalities in x and y.

   a)

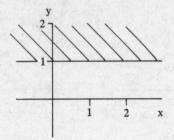

   b)

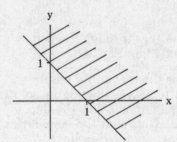

   c)

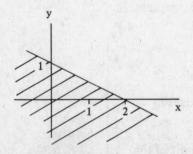

   d)

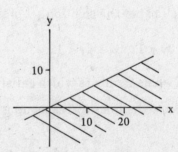

Fig. 10.12

4) A boy has £1 to spend on sweets. Liquorice bars cost 10 p each, and toffee bars cost 15 p. Say he buys x liquorice bars and y toffee bars. Find an inequality in x and y, and illustrate it on a graph.

5) If, in Question 4, the boy's mother tells him not to buy more than 8 bars, find another inequality in x and y. Illustrate the new inequality on the same graph.

6) A patient can be prescribed either pills or tablets. Tablets contain 3 units of vitamin A, and pills contain 4 units. The daily intake must be at least 15 units. If the patient takes x tablets and y pills, obtain an inequality in x and y and illustrate it on a graph.

10.4 Linear Programming

Suppose we have a region defined by inequalities, as in section 10.3. The process of finding the maximum value of a linear function in the region is *linear programming*.

The maximum value of a linear function must occur at one of the corners of the region. So we can evaluate the function at all the corners and pick the greatest value.

☐ 10.4.1 Example. *Higher, level 10*

Fire regulations require that a cinema should admit at most 400 people. Each cheap seat occupies 1 m^2 of space, and each expensive seat occupies 2 m^2. There is 600 m^2 of space available. Cheap seats are to cost £3, and expensive seats £5. Letting x and y be the numbers of cheap and expensive seats respectively, form inequalities in x and y and illustrate them on a graph. What allocation of seats will bring in the most money?

Solution The total number of seats must be at most 400. This gives the inequality $x + y \le 400$.

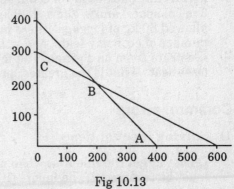

The total space used by the seats must be at most 600 m^2. This gives the inequality $x + 2y \le 600$.

x and y cannot be negative. Hence $x \ge 0$ and $y \ge 0$.

Illustrate the four inequalities on a graph as shown.

The money received by the cinema is £($3x + 5y$).

Fig 10.13

The corners of the region are O(0,0), A(400,0), B(200,200), C(0,300). Evaluate $3x + 4y$ at these points, to obtain 0, 1200, 1600, 1500 respectively. The maximum value is 1600.

The best allocation is 200 cheap and 200 expensive seats

☐ 10.4.2 Exercises. *Higher, level 10*

1) A factory manager is to buy two sorts of machines. Type 1 occupies 3 m^2 and costs £200, type 2 occupies 2 m^2 and costs £400. There is 40 m^2 of space available, and £4000 to spend. Let x be the number of type 1 machines, and y the number of type 2 machines.

Show that the restrictions can be expressed by the inequalities:

$$3x + 2y \le 40 \text{ and } x + 2y \le 20.$$

Illustrate these inequalities on a graph. If type 1 produces 100 items per hour, and type 2 150 items, find how many of each should be bought to maximize production.

2) In a car park there is 6000 m^2 of land available. A car space occupies 15 m^2, and a lorry space 30 m^2. There must be space for at least 50 lorries, and there must be at least twice as many car spaces as lorry spaces. Let c be the number of car spaces, and L the number of lorry spaces.

Show that the restrictions can be expressed by the inequalities:

$$c + 2L \le 400, L \ge 50 \text{ and } c \ge 2L.$$

Illustrate these inequalities on a graph. The parking fees are £2 for a car and £3 for a lorry. How should the spaces be arranged to bring in the most profit?

3) A shop has space for 5 refrigerators. The shop buys them from the manufacturers, at a price of £200 for type A, and £300 for type B, and there is £1200 available. The profits on types A and B are £45 and £70 respectively. If x of type A and y of type B are bought, express the restrictions in the form of inequalities. Illustrate the in equalities on a graph, and find the maximum profit.

4) Food A contains 4 units of protein and 5 units of starch per kg, and food B contains 6 units of protein and 3 units of starch per kg. The minimum daily intake of protein is 16 units, and the minimum daily intake of starch is 11 units. Let X be the number of kg of A to be eaten, and Y the number of kg of B. Obtain inequalities in X and Y, and illustrate them on a graph. What are the values of X and Y which ensure that the least weight of food is eaten?

5) A mail order firm must deliver 900 parcels using a lorry which takes 150 at a time or a van which takes 80 at a time. Each journey costs £5 for the lorry and £4 for the van. The total cost cannot exceed £44 and the lorry must not make more journeys than the van. Let x be the number of lorry journeys, and y the number of van journeys. Find inequalities in x and y and illustrate them on a graph. What should x and y be in order to make the total number of journeys as small as possible?

6) An aircraft has 600 m² of cabin space, and can carry 5000 kg of luggage. An economy passenger gets 3 m² of space, and is allowed 20 kg of luggage. A first class passenger gets 4 m² of space, and is allowed 50 kg of luggage. There must be space for at least 50 economy passengers. Let x be the number of economy seats, and y the number of first class seats. Obtain inequalities in x and y, and illustrate them on a graph. The profit from a first class seat is £100, and is £40 from an economy class seat. What is the greatest profit?

Common errors

1) Solving inequalities

Do not forget the rule that when multiplying or dividing by a negative number the inequality sign must be reversed. If you ignore this rule you will get exactly the wrong answer, i.e. you will get:

$$x < 3 \text{ instead of } x > 3.$$

2) Two-dimensional inequalities

a) Be sure to draw the correct line. The line $2x + 3y = 6$ goes through $(3,0)$ and $(0,2)$.

But do not draw the vertical and horizontal lines through these points. Fig 10.14 shows the correct and incorrect diagrams.

b) Be sure that you take the correct side of the line. If in doubt, take a test point. If your inequality is $2x + 3y \leq 6$, then test by putting $x = 1$ and $y = 1$.

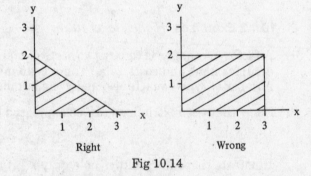

Right   Wrong

Fig 10.14

$2 \times 1 + 3 \times 1 < 6$, so the point $(1,1)$ must lie on the shaded part of your diagram.

c) When dealing with a problem, make sure that you obtain the full inequalities.

If a boy buys x items at 20 p, and y items at 30 p, out of a budget of 60 p, then the inequality is $20x + 30y \leq 60$.

Do not just put $x \leq 3$ and $y \leq 2$. These do not express the full situation.

3) Linear Programming

a) Make sure that your variables stand for the unknown quantities, not for the known restrictions. In the example of 10.4.1, do not let x stand for the number of people and y the amount of floor space. These are known to be 400 and 600 respectively.

b) If you have two inequalities, then the maximum point is often at the intersection of the two boundaries. But not assume this until you have tested – the maximum point may be at where one of the inequality lines crosses an axis.

Chapter 11

Coordinates and graphs

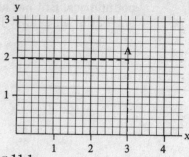

Fig 11.1

11.1 Axes and Coordinates

Axes are two lines, one horizontal, the other vertical. The horizontal line is the *x-axis* and the vertical line is the *y-axis*.

The position of a point along the *x*-axis is its *x co-ordinate*, and its position along the *y*-axis is its *y co-ordinate*.

The *x* co-ordinate of a point is always given first. So at the point A(3,2), *x* = 3 and *y* = 2.

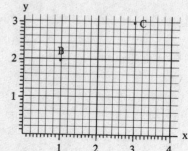

Fig 11.2

The point where *x* = 0 and *y* = 0 is the *origin*.

❑ 11.1.1 Examples. *Foundation, levels 4, 5*

1) a) Give the co-ordinates of the point B as shown in fig 11.2.

   b) Mark on the grid the point C(3,3).

   Solution a) B is 1 along the *x*-axis, and 2 up the *y*-axis.

   B is at (1,2)

   b) Go 3 along the *x*-axis, and 3 up the *y*-axis. Mark in C as shown in fig 11.2.

2) a) Give the coordinates of the point X on Fig 11.3.

   b) Mark on Fig 11.3 the point Y with coordinates (–2,–1).

   Solution a) Point X is 1 to the right of the origin, but 2 below it. Hence its *x*-coordinate is +1 and its *y*-coordinate –2.

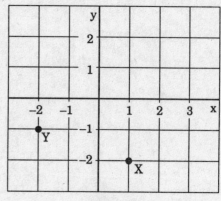

Fig 11.3

   b) To plot Y, go 2 to the left and 1 down. Y is shown plotted on the graph.

3) On the graph of Fig 11.4 mark the point Z with coordinates (1.6,0.2)

   Solution Note that on the paper there are 5 divisions between whole numbers. Hence each division corresponds to $\frac{1}{5}$ = 0.2. So the *x*-coordinate of Z is 3 divisions to the right of 1, and its *y*-coordinate is 1 division above 0.

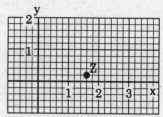

Fig 11.4

❑ 11.1.2 Exercises. *Foundation, levels 4, 5*

1) Fig 11.5 shows a map of a village. The church is in the square C5. Write down:

   a) The square containing the station.

   b) The squares through which Church Road passes.

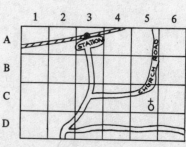

Fig 11.5

96

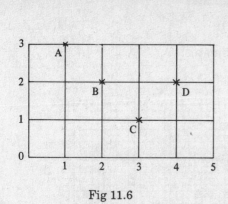

Fig 11.6

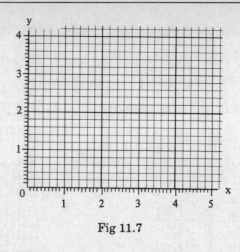

Fig 11.7

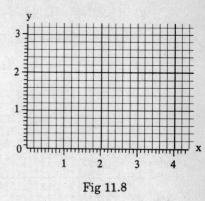

Fig 11.8

2) Write down the co-ordinates of the points A, B, C, D of fig 11.6.

3) Make a copy of the grid of fig 11.6. Mark on it the points X(1,1), Y(2,3), Z(4,1).

4) On the graph paper of fig 11.7 mark in the points P(1,3), Q(5,3), R(5,1), S(1,1).

   Join up the sides to make a rectangle PQRS.

   What are the co-ordinates of the join of the diagonals PR and QS?

5) On the graph of fig 11.8 mark the points A(1,3), B(1,2), C(4,2). Join the sides AB and BC.

   Find the point D on the graph so that ABCD is a rectangle. Write down its co-ordinates. Mark on the graph the midpoint of AC. What are its co-ordinates?

6) Fig 11.9 represents a region of sea, in which the wind is coming from the North East. A sailing dinghy starts from (0,0), and has to tack against the wind. It sails 2 squares North, then 2 squares East, then 2 squares North, and so on.

   Draw its route on the diagram, up to the point (6,6).

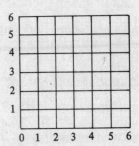

Fig 11.9

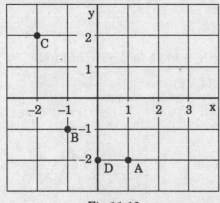

Fig 11.10

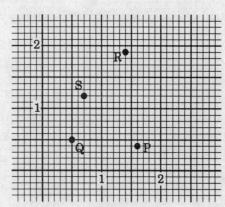

Fig 11.11

7) Write down the coordinates of the points A, B, C, D on Fig 11.10.

8) Write down the coordinates of the points P, Q, R, S in Fig 11.11.

9) Draw axes on graph paper, taking the origin near the centre of the page. Mark the points with the following coordinates:

   a) (-2,1)   b) (3,-2)   c) (-2,-2)

10) On the graph of Fig 11.12 plot the points with the following coordinates:

    a) (1.2,2.4)   b) (−1.8,0.5)   c) (2,-1.3)   d) (0.7,-0.9)

11) On the graph of Fig 11.13 plot the points with the following coordinates:

    a) (6,12)   b) (17,8)   c) (20,5)   d) (16,13)

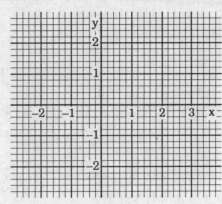
Fig 11.12

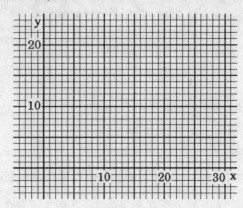
Fig 11.13

Three-Dimensional Coordinates

Two-dimensional coordinates are used to describe points in a plane. If we want to describe points in space, then we need three coordinates.

The x-axis and y-axis are horizontal, the z-axis is vertical. The position of a point is given by coordinates (x,y,z) relative to these axes.

Fig 11.14

❑ **11.1.3 Examples.** *Intermediate, level 7*

1) An aircraft is 10 km North, 5 km East and 2 km high. Regarding these as x, y, z coordinates respectively, give its position in terms of coordinates. If it is flying North at 8 km per minute, give its coordinates 3 minutes later.

Solution Write down its position in (x,y,z) form:

The position is (10,5,2)

3 minutes later it will have travelled 24 km North. Add 24 to the x coordinate:

3 minutes later it is at (34,5,2)

2) A regular pyramid has a square base with corners at $(0,0,0)$, $(2,0,0)$, $(2,2,0)$, $(0,2,0)$. If the pyramid has height 3 units, give possible coordinates for the vertex.

Solution The centre of the square base is at $(1,1,0)$, halfway between $(0,0,0)$ and $(2,2,0)$.

Because the pyramid is regular, the vertex must be either above or below the centre of the square base. The vertex is 3 units away along the z-axis.

The vertex is at (1,1,3) or (1,1,–3)

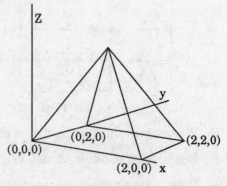
Fig 11.15

❑ **11.1.4 Exercises.** *Intermediate, level 7*

1) A box is 20 cm high, 30 cm wide, 40 cm long. Taking the origin at one of the bottom corners, find the coordinates of the top corners.

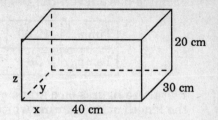
Fig 11.16

2) A room is 5 m long, 4 m wide, 3 m high. Taking the origin at a corner of the floor, give the coordinates of the opposite corner on the ceiling. Give the coordinates of the centre of the ceiling.

3) One face of a cube has corners at (0,0,0), (1,0,0), (1,1,0), (0,1,0). Find the possible coordinates of the corners of the opposite face.

4) The vertices of a solid are at (0,0,0), (2,0,0), (2,3,0), (0,3,0), (0,0,1), (2,0,1), (2,3,1), (0,3,1). What is this solid? What is its volume?

5) The vertices of a solid are at (0,0,0), (3,0,0), (0,4,0), (0,0,5), (3,0,5),(0,4,5). What is the name for this solid? What is its volume?

6) A pyramid has a square base of side 2 and height 3. It is placed with the centre of the base at the origin. Find possible coordinates for the vertices.

7) A prism has triangular cross-section. One end triangle has vertices at (0,0,2), (1,0,2), (0,1,2). If the prism is 3 long find possible coordinates of the other end triangle.

11.2 Interpretation of Graphs

A connection between two quantities can be shown by a graph. One quantity is measured along the *x*-axis, and the other up the *y*-axis.

For example, a graph with time along the *x*-axis and distance up the *y*-axis is a *distance time* graph.

❑ **11.2.1 Examples.** *Foundation, levels 4, 5, 6*

1) Allen cycles to see his grandmother. On the way his bike gets a puncture. He repairs it and cycles on. Fig 11.17 is a distance time graph showing his journey.

a) How long did he take in all?

b) How far away is his grandmother?

c) How long did he take to mend the puncture?

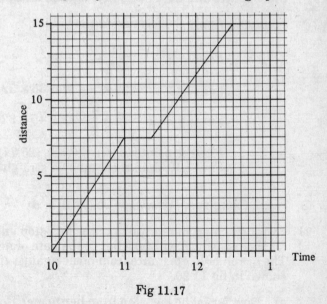
Fig 11.17

Solution a) The total time along the *x*-axis is $2\frac{1}{2}$ hours.

The journey took $2\frac{1}{2}$ hours.

b) The distance along the *y*-axis is 15 miles.

His grandmother lives 15 miles away.

c) He was delayed between 11 o'clock and 11.20.

He mended the puncture in 20 minutes

2) Jasmine has a fever. Her temperature, taken every 4 hours, was recorded as follows.

Time	12.00	16.00	20.00	24.00	04.00	08.00	12.00
Temp.	98	102	104	105	105	102	99

Plot these figures on a graph of temperature against time. Estimate her highest temperature, and the length of time during which her temperature was over 100°.

Solution The graph of temperature against time is shown in fig 11.18. Read off the highest value:

Her highest temperature was 105.1°

Her temperature was over 100° between 14.00 and 11.00.

Her temperature was over 100° for 21 hours.

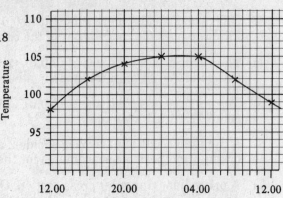

Fig 11.18

3) A temperature in Centigrade is converted to Fahrenheit by the formula

$$F = C \times \frac{9}{5} + 32.$$

Find the F values when C is 0°, 20°, 25°.

Plot your results on a graph, using the Centigrade figures for the x values and the Fahrenheit figures for the y values.

Solution Apply the formula to the values.

For $C = 0$, $F = 0 \times \frac{9}{5} + 32 = 32.$

For $C = 20$, $F = 20 \times \frac{9}{5} + 32 = 36 + 32 = 68$

For $C = 25$, $F = 25 \times \frac{9}{5} + 32 = 45 + 32 = 77$

Plot the points (0,32), (20,68), (25,77) on the graph of Fig 11.19. Join them up by a straight line.

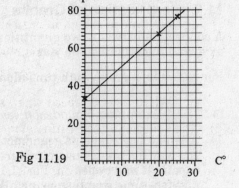

Fig 11.19

□ **11.2.2 Exercises.** *Foundation, levels 4, 5, 6*

1) Every morning Karen walks to the bus-stop and waits for the bus which will take her to the gate of her school. The graph of her distance from home against time is shown in fig 11.20.

a) How far is the bus-stop from her home? How long does it take to walk there?

b) How far is it from the bus-stop to school? How long does the bus take?

c) How long does she have to wait at the bus-stop?

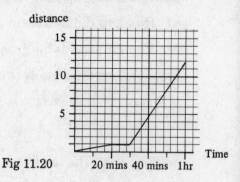

Fig 11.20

2) George sets off for school one morning, but on his way he realises that he has forgotten his books.

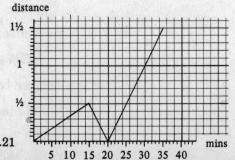

Fig 11.21

He returns home to collect them and sets off for school again. The graph of his distance from home against time is shown in Fig 11.21.

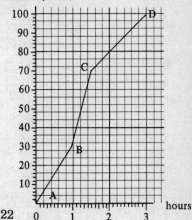

distance (miles)

a) How far had he gone when he turned back?

b) How long did the total journey take?

c) Why are the last parts of the journey steeper than the first part?

3) The Jameson family sets off for their holiday by car. They drive from their home A to the motorway at B. Then they travel along the motorway to C, where they turn off and drive to the hotel at D. The graph of Fig 11.22 shows the distance travelled against time.

a) Find the times taken for the three stages of the journey, AB, BC, CD.

b) Find the distances of the three stages.

c) Find the speeds of the three stages.

d) Find the average speed for the whole journey.

Fig 11.22

4) Dr Forster drives from her surgery to see a patient. On her way back she stops to visit another patient. The graph of her distance from the surgery against time is shown in Fig 11.23.

Fig 11.23

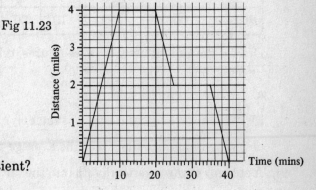

a) How long did she spend with each patient?

b) How far was the first patient from the surgery?

c) What was her speed when she drove to the first patient?

5) The speed of a tube train between stations is given in Fig 11.24.

Fig 11.24

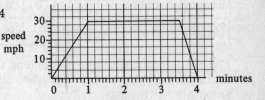

a) How long did it take the train to travel between the two stations?

b) What was the maximum speed?

c) For how long did the train stay at a steady speed?

6) Fig 11.25 (overleaf) shows the graph of the inflation rate in a country over a period of 6 years.

a) For how long was inflation over 10%?

b) When did inflation start to come down?

c) If inflation continued down at a steady rate, when did it reach zero?

7) Simon records the temperature every hour for 24 hours. His result are shown in the graph of Fig 11.26 (overleaf).

a) What was the greatest and the least temperatures?

b) When was the temperature falling most rapidly?

c) For how long was the temperature below 5°?

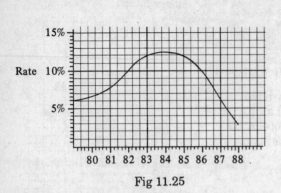

Fig 11.25

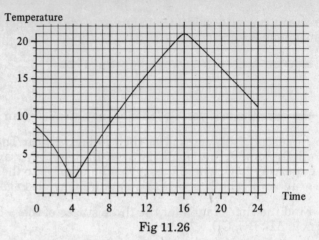

Fig 11.26

8) A house agent keeps records of the average price of houses in his town. The results over 8 years are given by the following table:

Year (31 December)	1978	1979	1980	1981	1982	1983	1984	1985
Price in £1,000's	17	19	20	18	21	26	30	34

Draw a graph to illustrate these figures, taking 1 cm along the x-axis to represent 1 year, and 1 cm up the y-axis to represent £5,000.

a) When did prices fall?

b) When were prices rising most steeply?

c) When approximately was the average price equal to £25,000?

9) A stone is thrown vertically up into the air. Its height afterwards is given by the following table:

Time in Secs.	1	2	3	4	5	6	7	8	9
Height in m.	42	74	96	108	110	102	84	56	18

Plot these figures, taking 1 cm along the x-axis to represent 1 sec, and 1 cm up the y-axis to represent 10 m. Join the dots up as smoothly as you can.

a) What was the greatest height reached? When was it reached?

b) At what times was the stone 100 m above the ground?

c) By continuing the graph until it reaches the x-axis, find how long the stone was in the air.

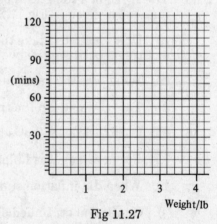

Fig 11.27

10) The cooking instructions for pork tell the cook to roast it for 30 minutes plus 20 minutes per pound. Calculate the roasting time for joints weighing 2 pounds, 3 pounds, 4 pounds.

Plot a graph showing this information, with the weight along the x-axis and the cooking time up the y-axis.

11) An electricity supply company offers its customers two schedules of payment for the power they use:

Schedule A: £20 per quarter plus 5 p. per unit.

Schedule B: £5 per quarter plus 8 p. per unit.

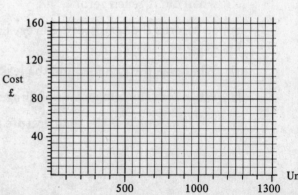

Fig 11.28

Complete the following table for the cost of electricity:

Units used	100	400	800	1200	1500
Cost by A	£25	£	£60	£	£
Cost by B	£13	£37	£	£	£

Plot these figures on the graph of Fig 11.28. Join up the points for Schedule A and for Schedule B.

If you expect to use 200 units, which Schedule should you pay by?

When does it become cheaper to pay by Schedule A rather than by Schedule B?

Gradients of Graphs

The gradient of a straight line is the ratio of the *y*-change to the *x*-change.

The gradient of a distance-time graph represents the *speed*.

The gradient of a curve is the gradient of its *tangent*.

◻ **11.2.3 Example.** *Higher, level 9*

Maurice enters a marathon race. He starts off at a steady rate, then as he gets tired he slows down. The graph of distance against time is shown in Fig 11.29. Find a) his starting speed b) his speed after 2 hours.

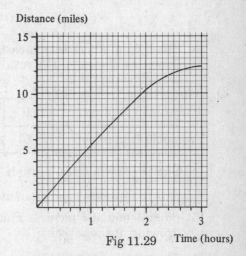
Fig 11.29   Time (hours)

Solution a) For the first hour the graph has a constant gradient. The *y* change is 5.5 miles, and the *x* change is 1 hour.

   He starts off at 5.5 m.p.h.

b) Draw a tangent to the curve at *x* = 2, and complete the triangle as shown in Fig 11.30.

   The *y* change of the tangent is 7 miles, and the *x* change is 1.85 hours.

   After 2 hours he is running at $\dfrac{7}{1.85}$ = 4 m.p.h.

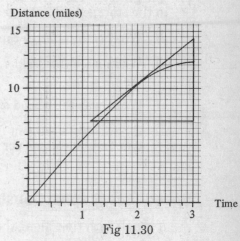
Fig 11.30

◻ **11.2.4 Exercises.** *Higher, level 9*

1) Michael sets off for school, then realises that he might be late and starts to run. The graph of his distance against time is shown in Fig 11.31.

   a) What was his walking speed?

   b) What was his running speed?

   c) If he had continued to walk, how long would he have taken to get to school?

2) In Question 1 of 11.2.2, what speed does Karen walk at? What speed is the bus?

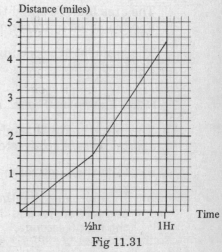
Fig 11.31

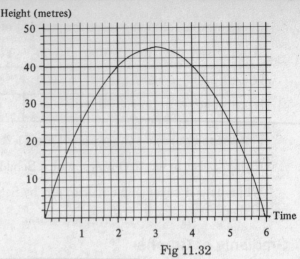

Fig 11.32

3) In Question 2 of 11.2.2, how fast did George walk in the three stages of his journey?

4) Fig 11.32 shows the graph of the height of a stone against time. By drawing suitable tangents find the speeds at 1 sec and at 3 secs.

5) The height of a child is measured over the first 6 months of its life. The results are shown in the following table:

Age in months	0	1	2	3	4	5	6
Height in in.	14	15	15.8	16.5	17.1	17.6	18.0

Plot these values on a graph, taking 1 cm along the x-axis to represent 1 month and 1 cm up the y-axis to represent $\frac{1}{2}$ in. Join up the points as smoothly as you can. Draw tangents to the curve at 2 months and at 5 months. Find the gradients of these tangents, and hence find how fast the baby is growing at these times.

6) The graph of Fig 11.33 shows the amount of petrol in the storage tanks of a garage over a 24 hour period.

a) What is the storage capacity of the tanks?

b) When were the tanks refilled? How long did it take?

c) When was the period of greatest demand?

d) By drawing a tangent find out what was the demand, in gallons per hour, at 09.00.

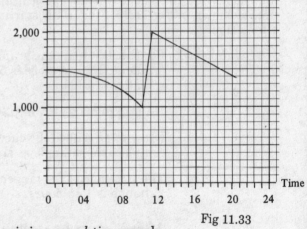

Fig 11.33

Speed-time graphs, Construction of Graphs

A graph with time along the *x*-axis and speed up the *y*-axis is a *speed-time* graph.

The area under a speed-time curve corresponds to the *distance*. This area can often be approximated by *trapezia*.

The gradient of a speed-time curve corresponds to the *acceleration*.

❑ **11.2.5 Examples.** *Higher, level 9*

1) The graph of Fig 11.34 gives the speed of a sports-car as it accelerates from rest.

a) Find the acceleration at 3 secs.

b) By approximating the area by two trapezia and a triangle, find the distance gone in the first 6 seconds.

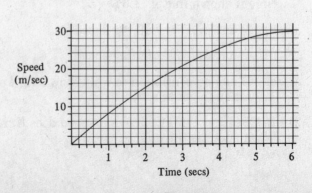

Fig 11.34

Solution a) Draw a tangent to the curve at $x = 3$, as in Fig 11.35. Measure the gradient of this tangent.

$$\text{Acceleration} = \frac{20}{4} = 5 \text{ m/sec}^2$$

b) Draw vertical lines at $x = 2$ and $x = 4$, as in Fig 11.36. The area under the curve is approximately the area in the two trapezia and the triangle.

$$\text{Distance} - \frac{1}{2} \times 15 \times 2 + \frac{1}{2}(15 + 25) \times 2 + \frac{1}{2}(25 + 30) \times 2 = 110 \text{ m.}$$

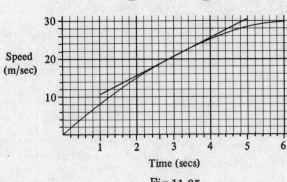

Fig 11.35

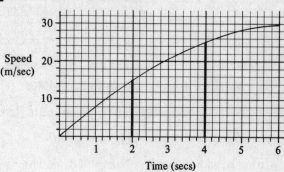

Fig 11.36

2) Fred sets off for school, and walks the half mile to the bus-stop at a steady speed of 3 m.p.h. He waits 20 minutes for a bus, which then travels the 5 miles to school at 30 m.p.h. Draw a distance-time graph for his journey.

Solution The first stage of the journey lasts for 10 minutes, and reaches $\frac{1}{2}$ mile.

While he is waiting at the stop the graph is flat, for a period of 20 minutes.

The final stage lasts for 10 mins and reaches $5\frac{1}{2}$ miles. The result is shown in Fig 11.37.

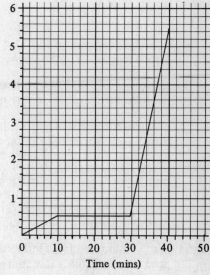

Fig 11.37

☐ **11.2.6 Exercises.** *Higher, level 9*

1) The speed of a tube train between successive stations is given by Fig 11.38.

a) Find the acceleration when it starts and the deceleration when it stops.

b) Find the distance between the stations.

2) The speed of a runner during a sprint race is shown in Fig 11.39.

a) Find the acceleration of the runner after 1 second.

b) By splitting the region into 3 trapezia and a triangle estimate the total distance of the race.

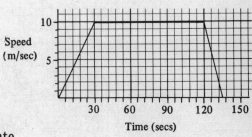

Fig 11.38

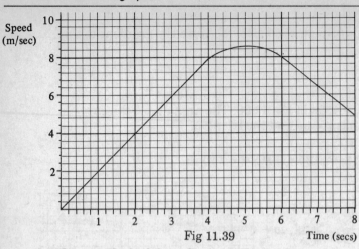

Fig 11.39

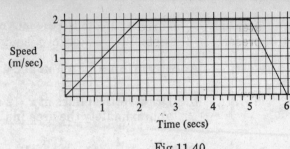

Fig 11.40

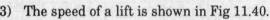

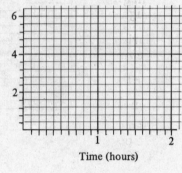

Fig 11.41

3) The speed of a lift is shown in Fig 11.40.

  a) What is the acceleration when it starts to rise?

  b) What is the total height it rises?

4) Jules lives 6 miles due North of Jim. They set out to visit each other at the same time; Jules walks South at $4\frac{1}{2}$ m.p.h. and Jim walks North at $3\frac{1}{2}$ m.p.h.

  On the diagram of Fig 11.41 draw graphs showing their distances from Jim's house. When do they meet? How far North of Jim's house do they meet?

5) Laurie and Daphne are both driving to Bristol along the M4. Laurie's car cruises at 50 m.p.h., and he sets off from London at noon. Daphne's car cruises at 60 m.p.h., and she sets off 20 minutes later.

  On the diagram of Fig 11.42 draw graphs showing their distances from London. When does she pass him? How far from London are they when she passes him?

6) Fred has two inaccurate clocks. At midday the first clock is 15 minutes fast, but it loses 5 minutes per hour. The second clock is 15 minutes slow, but it gains 15 minutes per hour.

  What times do the clocks show at 1 o'clock? On the graph of Fig 11.43 draw lines representing the times shown by the two clocks. When do they show the same time?

7) Val's bathtime is as follows. She fills the bath, then lies in it. After a while she adds more hot water, and lies in it for a few more minutes. She then pulls the plug.

  Draw a graph, not to scale, showing the level of water in the bath against time.

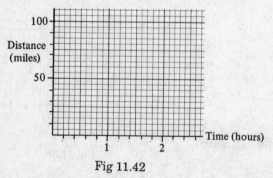

Fig 11.42

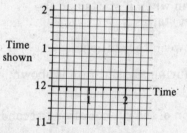

Fig 11.43

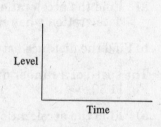

Fig 11.44

8) A tap can deliver 20 litres of water per minute. It is being used to fill a cylindrical tank which holds 80 litres. Half way up the tank is a hole, through which water pours out at 10 litres per minute.

  Draw an accurate graph of the level of water against time.

9) Water is being poured at a steady rate into a conical container. Which of the following pictures represents the graph of the water level against time?

a)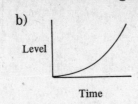

b)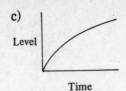

c)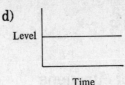

d)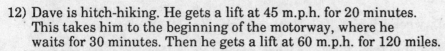

Fig 11.45

10) You read in a newspaper that the population of your town is still growing, but at a progressively slower rate. Draw a rough graph of population against time.

11) During a rainstorm an open cylinder was left out to catch the rain.

For the first 2 hours 6 cm fell at a steady rate. Then the rain slackened, and for the next hour $1\frac{1}{2}$ cm fell. The rain then ceased, and 2 hours after that a plug was removed at the bottom of the cylinder, causing the water to drain away in 10 minutes.

Illustrate the information above, by means of an accurate graph of the water level in the cylinder against time.

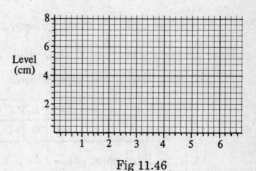

Fig 11.46

12) Dave is hitch-hiking. He gets a lift at 45 m.p.h. for 20 minutes. This takes him to the beginning of the motorway, where he waits for 30 minutes. Then he gets a lift at 60 m.p.h. for 120 miles.

Illustrate the above information, by means of a graph of distance against time.

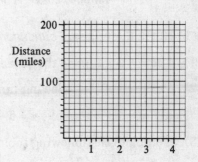

Fig 11.47

Common Errors

1) **Co-ordinates**

   a) The x co-ordinate is always given first. The point at (3,2) is at $x = 3$ and $y = 2$. Be sure not to get these the wrong way round.

   b) Make sure that the scale of your graph is uniform. It is misleading to write down a scale in which the x-values or y-values are not evenly spaced.

   c) Make sure that you get the gradient the correct way up. It must be the y-change over the x-change, not the other way round.

   d) The gradient is the *y-change* over the *x-change*. It is not the *y-value* over the *x-value*.

   e) A tangent should just *touch* a curve. It does not necessarily go through the origin (0,0).

   f) The gradient of a curve at a point is the gradient of the tangent at that point. It is not the gradient of the line from the origin to that point.

2) **Interpretation**

   a) Read the question very carefully, to make sure that you do not draw a distance-time graph instead of a speed-time graph, or the other way round.

   b) To find the *average* speed, divide the total distance by the total time. To find the *instantaneous* speed, find the gradient of the tangent to a distance-time curve. Do not confuse these.

Chapter 12

Functions

12.1 Graphs of Functions

Let y be a function of x. Then for each x there is a corresponding y. If the pairs (x,y) are plotted, the result is the *graph* of y.

☐ **12.1.1 Example.** *Foundation, level 6*

Find the values of x^2 for $x = -2, -1, 0, 1, 2$. Hence plot the graph of $y = x^2$.

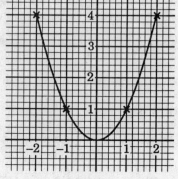

Fig 12.1

> *Solution* Fill in the table below.
>
> | x | -2 | -1 | 0 | 1 | 2 |
> |-----|----|----|---|---|---|
> | x^2 | 4 | 1 | 0 | 1 | 4 |
>
> Plot points at $(-2,4)$, $(-1,1)$ and so on. Join them up by a smooth curve. The result is shown in Fig 12.1.

☐ **12.1.2 Exercises.** *Foundation, level 6*

1) Let $y = 2 + x$. Find the values of y when x is 0, 1, 2, 3. Hence plot the graph of the function.

2) For each of the following functions, find the values of y when x is 0, 1, 2, 3. Hence plot the functions.

    a) $y = 1 + x$ b) $y = 3 - x$ c) $y = 2x + 1$

3) Complete the following table for the function $y = x^2 + 1$. Hence plot the graph of the function.

x	-2	-1	0	1	2
x^2		1		4	
$x^2 + 1$					5

4) Find the values of $2x^2$ for x equal to $-2, -1, 0, 1, 2, 3$. Hence plot the graph of $y = 2x^2$.

☐ **12.1.3 Examples.** *Intermediate, level 8*

1) Complete the table for the function $y = (x + 1)^2$. Draw a graph of the function using a scale of 1 cm per unit.

x	-3	-2	$-1\frac{1}{2}$	-1	$-\frac{1}{2}$	0	1
y	4		$\frac{1}{4}$				

> *Solution* For $x = -2$, $y = (-2 + 1)^2 = (-1)^2 = 1$. Complete the rest of the table similarly. The final result is:
>
> | x | -3 | -2 | $-1\frac{1}{2}$ | -1 | $-\frac{1}{2}$ | 0 | 1 |
> |-----|----|----|----|----|----|---|---|
> | y | 4 | 1 | $\frac{1}{4}$ | 0 | $\frac{1}{4}$ | 1 | 4 |

The graph is shown in Fig 12.2.

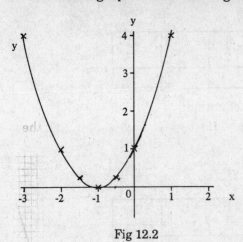

Fig 12.2

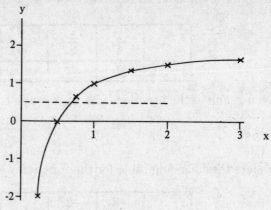

Fig 12.3

2) y is given in terms of x by the formula $y = 2 - \frac{1}{x}$. Complete the following table.

x	$\frac{1}{4}$	$\frac{1}{2}$	$\frac{3}{4}$	1	$1\frac{1}{2}$	2	3
y	-2					$1\frac{1}{2}$	

Draw a graph of the function, using a scale of 1 cm per unit. Use your graph to solve the equation $2 - \frac{1}{x} = \frac{1}{2}$.

Solution For $x = \frac{1}{2}$, $y = 2 - \frac{1}{\frac{1}{2}} = 2 - 2 = 0$. The other values are found similarly. The completed table is:

x	$\frac{1}{4}$	$\frac{1}{2}$	$\frac{3}{4}$	1	$1\frac{1}{2}$	2	3
y	-2	0	$\frac{2}{3}$	1	$\frac{4}{3}$	$1\frac{1}{2}$	$\frac{5}{3}$

The graph is shown in Fig 12.3.

Draw a line for $y = \frac{1}{2}$, and read off where it crosses the graph.

The solution of $2 - \frac{1}{x} = \frac{1}{2}$ is $x = 0.7$

□ **12.1.4 Exercises.** *Intermediate, level 8*

1) Complete the following table for the function $y = (x - 1)^2$.

x	-1	0	$\frac{1}{2}$	1	$1\frac{1}{2}$	2	3
y	4				$\frac{1}{4}$		

Draw a graph of the function, using a scale of 1 cm per unit.

2) Complete the table for the function $y = 1 + \dfrac{2}{x}$.

x	$\dfrac{1}{4}$	$\dfrac{1}{2}$	$\dfrac{3}{4}$	1	$1\dfrac{1}{2}$	2	3
y	9			3			

Draw a graph of the function. Use your graph to solve the equation:

$$1 + \frac{2}{x} = 4.$$

3) Complete the following table for the function $y = x^2 + 2x - 1$.

x	-3	-2	$-1\dfrac{1}{2}$	-1	$-\dfrac{1}{2}$	0	1
y							

Draw the graph of your function. Use your graph to find approximate solutions to the equation:

$$x^2 + 2x - 1 = 0$$

4) Complete the following table for the function $y = x + \dfrac{1}{x}$.

x	$\dfrac{1}{8}$	$\dfrac{1}{4}$	$\dfrac{3}{8}$	$\dfrac{1}{2}$	$\dfrac{3}{4}$	1	$1\dfrac{1}{2}$	2	3
y									

Draw a graph of the function. Use your graph to solve the equation:

$$x + \frac{1}{x} = 3$$

5) Set up a table of values for the function $y = 3x - 2$, using the x-values 1, 2, 3, 4. Draw a graph for the function, using a scale of 1 cm per unit.

6) Set up a table of values for the function $y = x^2 - 3x + 1$, using the x-values 0, 1, $1\dfrac{1}{2}$, 2, 3. Draw a graph of the function using a scale of 1 cm per unit.

Use your graph to solve the equation $x^2 - 3x + 1 = 0$.

7) Draw a graph of the function $y = 3 - x^2$, using the x-values -2, -1, $-\dfrac{1}{2}$, 0, $\dfrac{1}{2}$, 1, 2.

Use your graph to solve the equation $3 - x^2 = 1$.

8) Draw a graph for the function $y = 3 - \dfrac{1}{x}$, using the x-values $\dfrac{1}{4}$, $\dfrac{1}{2}$, 1, 2, 3, 4, 5.

Use your graph to solve the equation $3 - \dfrac{1}{x} = 1\dfrac{1}{2}$.

Gradients and Tangents

The *gradient* of a straight line graph is the ratio of the y-change to the x-change.

The gradient of a curved graph at a point is the gradient of the tangent at that point.

The function $y = 3x + 2$ is sometimes written as:

$$f(x) = 3x + 2$$

$$\text{or as } f\!:\!x \to 3x + 2.$$

☐ **12.1.5 Examples.** *Higher, level 9*

1) A function f is given by $f(x) = x^3 - 2x + 1$. Complete the following table:

x	-2	-1	$-\frac{1}{2}$	0	$\frac{1}{2}$	1	$1\frac{1}{2}$	2
$f(x)$								

Draw a graph of the function, using a scale of 1 cm per unit. By drawing a suitable line find the solutions of the equation:

$$x^3 - 2x + 1 = x + 1$$

Find the gradient at $x = 1$

Solution After it has been completed the table is as follows:

x	-2	-1	$-\frac{1}{2}$	0	$\frac{1}{2}$	1	$1\frac{1}{2}$	2
$f(x)$	-3	2	$1\frac{7}{8}$	1	$\frac{1}{8}$	0	$1\frac{3}{8}$	5

The graph is shown in Fig 12.4.

Draw the line $y = x + 1$. Notice that it crosses the graph at 3 places.

The solutions are $x = -1.7, 0, 1.7$.

Draw a tangent at $(1,0)$. Its gradient is $\frac{2}{2} = 1$

At $x = 1$ the gradient is 1

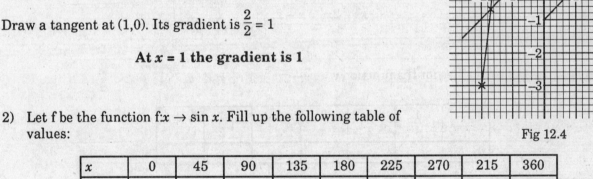

Fig 12.4

2) Let f be the function $f\!:\!x \to \sin x$. Fill up the following table of values:

x	0	45	90	135	180	225	270	215	360
$f(x)$									

Sketch the graph of this function, taking 1 cm to be 45° on the x-axis, and 2 cm to be 1 unit along the y-axis.

Solution Use your calculator to find the values of f. Complete the table:

x	0	45	90	135	180	225	270	315	360
$f(x)$	0	0.71	1	0.71	0	-0.71	-1	-0.71	0

The graph is shown in Fig 12.5.

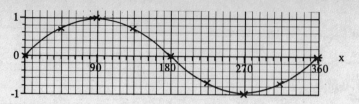

Fig 12.5

□ **12.1.6 Exercises.** *Higher, level 9*

1) Complete the table of values for the function $f(x) = x^3 - x$:

x	-2	-1	$\frac{-1}{2}$	0	$\frac{1}{2}$	1	$1\frac{1}{2}$	2
$f(x)$								

Draw the graph of this function. By drawing an appropriate line solve the equation:

$$x^3 - x = \frac{1}{2}x - 1.$$

Find the gradient at $x = 1$.

2) Complete the table of values for the function $f(x) = x^3 - 2x^2$:

x	-1	$-\frac{1}{2}$	0	$\frac{1}{2}$	1	$1\frac{1}{2}$	2
$f(x)$							

Draw a graph of this function. By drawing an appropriate line solve the equation:

$$x^3 - 2x^2 = x - 2.$$

Find the gradient at $x = 1$.

3) Complete the table of values for the function $y = x^2 - \frac{1}{x}$

x	$\frac{1}{4}$	$\frac{1}{2}$	$\frac{3}{4}$	1	$1\frac{1}{2}$	2
y						

Draw the graph of this function. On the same paper draw the graph of the line $y = x - \frac{1}{2}$.

Show that where the graphs cross x obeys the equation

$$x^3 - x^2 + \frac{1}{2}x - 1 = 0.$$

Write down the solutions to this equation.

Find the gradient at $x = \frac{1}{2}$.

4) Complete the table of values for the function $f(x) = x + \dfrac{1}{x^2}$:

x	$\frac{1}{4}$	$\frac{1}{2}$	$\frac{3}{4}$	1	$1\frac{1}{2}$	2	3
$f(x)$							

Draw the graph of this function, taking 1 cm to be 1 unit along the x-axis and 1 cm to be 4 units up the y-axis.

On the same paper draw the line $g(x) = \dfrac{1}{2}x + 3$. Show that where the lines cross x obeys the equation:

$$x^3 - 6x^2 + 2 = 0.$$

Write down the solutions to this equation.

Find the gradient at $x = 2$.

5) Draw the graphs of the following quadratics, indicating the axes of symmetry of the curves.

a) $y = x^2 - 1$ b) $y = x^2 - 2x + 3$

c) $y = x^2 + 3x - 2$ d) $y = 1 + x - x^2$

6) Draw the graph of $y = x - \dfrac{1}{2}x^3$, taking x-values of $-2, -1, -\dfrac{1}{2}, 0, \dfrac{1}{2}, 1, 2$.

Draw the tangent to the curve at $x = 1$ and find its gradient.

7) Complete the table for the function $c(x) = \cos x$:

x	0	30	60	90	120	150	180	210	240	270	300	330	360
$c(x)$													

Draw the graph of $c(x)$, taking 1 cm per 60° along the x-axis and 2 cm per unit up the y-axis.

Use your graph to solve the following equations:

a) $\cos x = \dfrac{1}{2}$ b) $\cos x = -\dfrac{1}{2}$ c) $\cos x = 0.6$ d) $\cos x = -0.8$

12.2 Straight Line Graphs

Suppose two graphs are plotted on the same paper. Where the graphs cross both the corresponding equations are true. This can be used to solve simultaneous equations.

☐ **12.2.1 Example.** *Intermediate, level 7*

Find three pairs (x, y) such that $y = 2x - 4$. Plot these points on a graph, and join them up with a straight line.

Repeat this process with the equation $3x + 2y = 6$, plotting the points on the same graph.

Use your graph to solve the simultaneous equations:

$$y - 2x = -4$$

$$3x + 2y = 6$$

Solution   3 points which satisfy $y = 2x - 4$ are $(0,-4)$, $(1,-2)$, $(2,0)$. They are plotted on the graph of fig 12.6.

3 points which obey $3x + 2y = 6$ are $(0,3)$, $(2,0)$, $(1,1\frac{1}{2})$. They are also plotted on fig 12.6.

The two equations are both true at the crossing point of the two lines. Read off the co-ordinates of this point:

$$x = 2 : y = 0$$

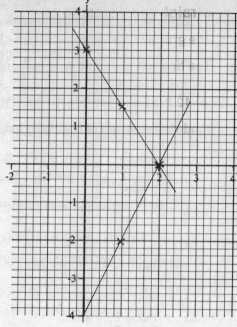

Fig 12.6

☐ **12.2.2 Exercises.** *Intermediate, level 7*

1) Find the gradient of the line segments in fig 12.7.

2) Find three points of the form (x,y) which satisfy the equation:

$$y = 3x + 1.$$

Plot these points on the graph of fig 12.8. Find the gradient of the line.

3) Repeat Question 2, using the same graph paper, for the equation:

$$2y = x + 7.$$

Hence solve the simultaneous equations:

$$y = 3x + 1$$
$$2y = x + 7.$$

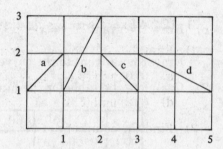

Fig 12.7

4) Use the method of Question 2 to find the gradients of the lines given by the equations:

   a) $y = 4x - 3$ b) $y = x + 1$

   c) $y = \frac{1}{2}x + 3$ d) $4y + 3x = 2$

5) Solve the following simultaneous equations, by means of drawing lines on graph paper and finding the point of intersection:

   a) $y = x + 1$ b) $y = 2x - 4$
        $y = 3x - 5$ $y = 11 - 3x$

   c) $2y + 3x = 12$ d) $y = \frac{1}{2}x - 1$
        $x + 3y = 11$ $4y = 3x - 8$

6) Draw, on the same piece of graph paper, the lines corresponding to the equations $y = x$, $y = x + 3$, $y = x - 1$. What can you say about the lines?

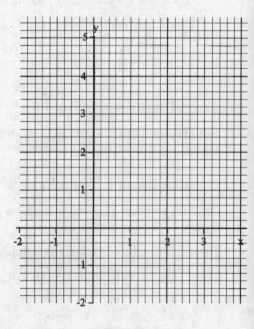

Fig 12.8

Straight line equations

The graph of a straight line must be of the form $y = mx + c$, where m and c are constants.

The line has gradient m, and it crosses the y-axis at $(0,c)$.

☐ **12.2.3 Example.** *Higher, level 9*

Find the equation of the straight line which passes through (1,5) and (3,9)

> *Solution* The line goes through (1,5). Put these values into the equation $y = mx + c$:
>
> $$[1] \quad 5 = m + c.$$
>
> The line also goes through (3,9):
>
> $$[2] \quad 9 = 3m + c.$$
>
> Subtract [1] from [2].
>
> $$4 = 2m.$$
>
> Hence $m = 2$. Substitute to find that $c = 3$.
>
> **The equation is $y = 2x + 3$.**

☐ **12.2.4 Exercises.** *Higher, level 9*

1) Find the equations of the straight lines which go through the following pairs of points:

   a) (0,0) and (2,6)   b) (1,2) and (4,5)   c) (2,1) and (3,−2)

   d) (3,2) and (5,3)   e) (−1,−2) and (2,4)   f) (−2,3) and (2,−5)

2) Find the equations of the straight lines shown in fig 12.9.

  a) 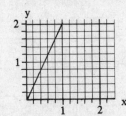 b) 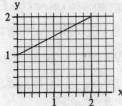 c)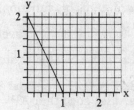

Fig 12.9

12.3 Graph Problems

Problems can often be solved with the aid of a graph.

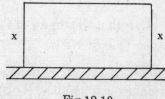

Fig 12.10

☐ **12.3.1 Example.** *Higher, level 9*

A farmer has 100 m. of fencing with which to enclose a rectangular field, one side of which is a stone wall.

Let the two sides perpendicular to the wall be of length x. Show that the area enclosed is $100x - 2x^2$.

Draw the graph of this function, and find the greatest area that the farmer can enclose.

> *Solution* After x metres have been used for the two perpendicular sides, the amount remaining for the parallel side is $100 - 2x$.
>
> The area of the rectangle is the product of the length and the breadth:
>
> $$x(100 - 2x) = 100x - 2x^2 \text{ m}^2$$

The graph of $y = 100x - 2x^2$ is in fig 12.11. The greatest area is the greatest height of the curve.

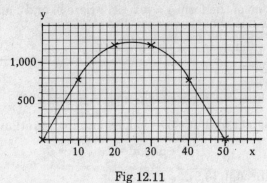

Fig 12.11

The greatest area is 1,250 m^2

☐ **12.3.2 Exercises.** *Higher, level 9*

1) A stone is thrown up in the air, so that x seconds later its distance y above the ground is $2 + 20x - 5x^2$.

   Plot the graph of this function, for x between 0 and 5. Use your graph to answer the following:

   a) What is its greatest height?

   b) When is it 10 m. high?

   c) For how long is it in the air?

2) The *Highway Code* gives values for the distance D (in feet) it takes for a car to stop when it travels at speed s m.p.h. These values fit the formula $D = s + \dfrac{s^2}{20}$.

   Plot the graph of this function. Use your graph to solve the following:

   a) How long does a car travelling at 55 m.p.h. take to stop?

   b) If a car must be able to stop within 200 feet, what is the greatest speed it can do?

3) In Fig 12.12 ABCD is a square with side 10 cm. P, Q, R and S are points along the sides such that AP = BQ = CR = DS = x cm.

   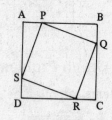

   Show that the area of △APS is $\dfrac{1}{2}x(10 - x)$.

   Show that the area y of PQRS is given by:

   $$y = 100 - 2x(10 - x)$$

   Fig 12.12

   Plot the graph of y against x, taking values of x between 0 and 10. Hence find the value of x which makes y as small as possible.

4) An open cardboard box is to be made by cutting off the corners from a 20 cm square sheet of cardboard and folding the sides over. Fig 12.13.

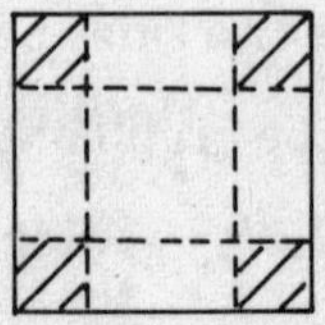

   Let a square of side x cm be cut off from each corner. Show that the base of the box is a square of side $20 - 2x$.

   Show that the volume V of the box is $x(20 - 2x)^2$.

   Fig 12.13

   Draw a graph of V in terms of x, taking values of x from 0 to 10. Hence find the value of x which gives the largest volume.

5) An open rectangular box is to be made with a square base. The volume must be 4 in³.

Let the height be h in. and the base have side x in. Show that $hx^2 = 4$.

Show that the surface area of the box is $A = x^2 + 4xh$.

Fig 12.14

Show that this can be re-written as $A = x^2 + \dfrac{16}{x}$.

Draw a graph of A against x, taking values of x from $\frac{1}{2}$ to 3. Hence find the least possible surface area of the box.

12.4 Shifting Graphs

Suppose we have the graph of $y = f(x)$, and that k is a constant. Related graphs are obtained as follows.

$y = f(x) + k$. Shift the graph up by k.

$y = f(x+k)$. Shift the graph to the left by k.

$y = f(kx)$. Squash the graph horizontally by a factor of k.

$y = kf(x)$. Stretch the graph vertically by a factor of k.

❏ **12.4.1 Example.** *Higher, level 10*

Below is the graph of $y = f(x)$. (Fig 12.15.) Use it to sketch the graphs of   a) $y = f(x-1)$ b) $y = f(2x)$.

  Solution a) Shift the graph to the right by 1. Notice that it now crosses the x-axis at (1,0) and (3,0). (Fig 12.16.)

        b) Squash the graph horizontally by a factor of 2. Notice that it now crosses the x-axis at (0,0) and (1,0). (Fig 12.17.)

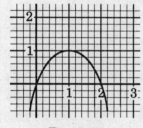 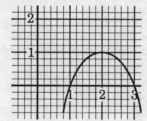 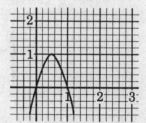

      Fig 12.15   Fig 12.16   Fig 12.17

❏ **12.4.2 Exercises.** *Higher, level 10*

1) On the right is sketched the graph of $y = f(x)$. Sketch the graphs of:

  a) $y = f(x) + 1$

  b) $y = f(2x)$

  c) $y = f(x - 2)$

  d) $y = 2f(x)$

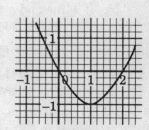

Fig 12.18

2) Fig 12.5 shows the graph of $y = \sin x$. Use it to sketch the graphs of:

  a) $y = \sin 2x$ b) $y = \sin(x + 90°)$ c) $y = 3 \sin x$

3) Fig 12.1 shows the graph of $y = x^2$. Use it to sketch the graphs of:

a) $y = (x - 1)^2$ b) $y = 2x^2$ c) $y = (x - 1)^2 + 1$

4) Below are the graphs of   a) $y = x^2$ b) $y = (x + p)^2$ c) $y = x^2 + q$ d) $y = (x + r)^2 + s$.
   Find p, q, r, s.

a)

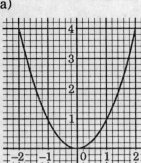

b)

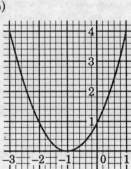

c)

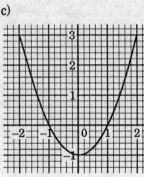

d)

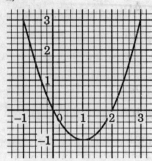

Fig 12.19

Common Errors

1) **Graphs of functions**

   a) Algebraic mistakes are very frequent when working out tables of values. Be very careful when negative numbers are involved. For example, it is easy to get the wrong sign when working out $1 - 2x^2$ for $x = -1$.

   Be very careful also when dividing. When $x = \frac{1}{2}$, $\frac{1}{x}$ is equal to 2, not $\frac{1}{2}$.

   Do not divide by 0. $1 + \frac{1}{x}$ has no meaning when $x = 0$.

   b) Once you have plotted the points of a graph, do not join them up by straight lines. Make as smooth a curve as you can to connect them.

   c) If your scale is not simple, then be very careful when plotting points. It is easy to make mistakes when the scale is, for example, 1 cm per 2 metres.

   d) There are various common errors to do with tangents and scatter diagrams, which are listed in the previous chapter.

2) **Solving equations by graphs**

   a) If you are solving an equation in x by the graph method, then you need only give the *x-value* of the crossing point. The *y*-value is unnecessary.

   b) If a line and a curve cross at more than one point, then give the *x*-values of *all* the points where they cross. They are all solutions of the equation.

3) **Straight line graphs**

   If you know that the line $y = mx + c$ goes through (2,3) say, then that means that x and y are equal to 2 and 3 respectively. So the equation you obtain is:

   $$3 = 2m + c$$

   Do not put $m = 2$ and $c = 3$, to obtain $y = 2x + 3$

4) **Shifting graphs**

   The graph of $y = f(x+k)$ if obtained from the graph of $y = f(x)$ by shifting it to the *left*, not to the right.

Chapter 13

Plane figures

13.1 Angles and Lines

The angles round a point add up to 360°. Note that in Fig 13.1:

130° + 140° + 90° = 360°

The angles along a line add up to 180°. Note that in Fig 13.2:

75° + 105° = 180°

An angle of 90° is a *right-angle*. Lines which meet at a right-angle are *perpendicular*.

An angle between 0° and 90° is *acute*.

An angle between 90° and 180° is *obtuse*.

An angle greater than 180° is *reflex*.

Lines which are in the same direction are *parallel*.

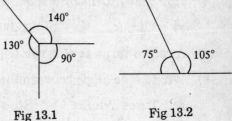

Fig 13.1   Fig 13.2

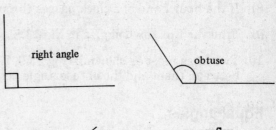

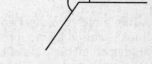

Fig 13.3

☐ **13.1.1 Examples.** *Foundation, levels 4 and 5*

1) Find the angle labelled $x°$ in Fig 13.4.

    Solution   The angles must add up to 360°. Subtract the other angles from 360°:

$$x° = 360° - 140° - 100° = 120°$$

2) The lines of Fig 13.5 are perpendicular. Find the angle labelled x.

    Solution   The angles along the line must add up to 180°. The two angles labelled x must add up to 90°.

$$x = \frac{90°}{2} = 45°$$

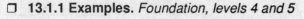
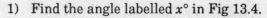

Fig 13.4

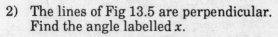

Fig 13.5

☐ **13.1.2 Exercises.** *Foundation, levels 4 and 5*

1) Find the unknown angles of the diagrams in Fig 13.6.

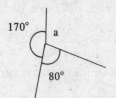
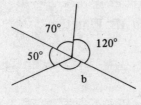
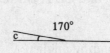
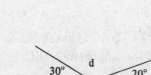

Fig 13.6

2) Find the values of the unknowns in the diagrams in Fig 13.7.

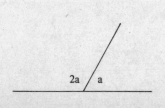

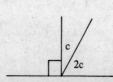

119

Fig 13.7

3) One turn corresponds to 360°. How many degrees correspond to:

   a)   Half a turn   b)   A quarter turn c)   3 turns.

4) How many turns correspond to:

   a)   720°   b)   3,600°   c)   540°   d)   120°?

5) A record turns at 45 revolutions per minute. How many degrees does it turn through in 1 second?

6) What is the angle between the hands of a clock at the following times:

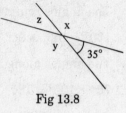

   a)   Three o'clock   b) Six o'clock   c)   Five o'clock ?

7) In ten minutes, through what angle does the minute hand of a clock pass?

8) If the hour hand of a clock passes through 120°, how much time has passed?

9) Find the unknown angles in Fig 13.8.

Fig 13.8

10) Two lines meet as shown in Fig 13.9. Write down the acute angle between them and the obtuse angle between them.

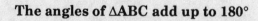

Fig 13.9

Equal angles

When two lines cross, the *opposite* angles are equal.

A line which crosses two parallel lines is called a *transversal*. The angles on either side of the transversal are equal, and are called *alternate* angles.

The angles on the same side of the transversal are equal and are called *corresponding* angles.

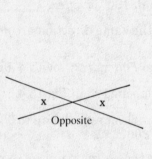

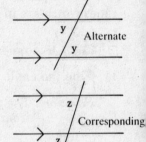

Fig 13.10

☐ **13.1.3 Examples.** *Foundation, level 5*

1) Find the angles labelled *x*, *y* and *z* in Fig 13.11.

   Solution   The angle labelled *x* is opposite to the 50° angle.

$$x = 50°$$

   The angle labelled *y* is alternate to the angle of 50°.

$$y = 50°$$

   The angle labelled *z* is corresponding to the angle of 50°.

$$z = 50°$$

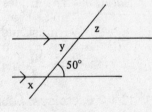

Fig 13.11

2) In Fig 13.12 AC is extended to D, and CE is parallel to AB. Find the sum of the angles in △ABC.

   Solution   BÂC = EĈD (Corresponding angles)

                AB̂C = BĈE (Alternate angles)

So the sum of the angles of △ABC is the sum of the angles BCA, BCE and ECD. These are the angles along a straight line.

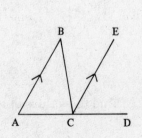

          The angles of △ABC add up to 180°

Fig 13.12

☐ **13.1.4 Exercises.** *Foundation, level 5*

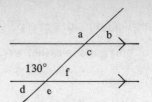

1) Find the angles labelled *a, b, c, d, e, f, g* in Fig 13.13.

Fig 13.13

2) Find the unknown angles in Fig 13.14.

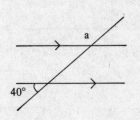 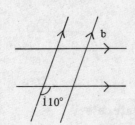 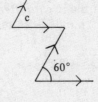 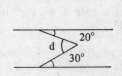

Fig 13.14

3) AB and CD are parallel lines, and BD is a transversal. The bisectors of the angles AB̂D and CD̂B meet at E. If BD̂C = 80°, find BED. Fig 13.15.

4) As far as possible label the angles of Fig 13.16. What is the relationship between *a* and *b*?

5) Find the angle *x* of Fig 13.17.

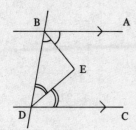 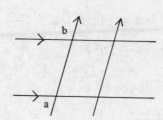 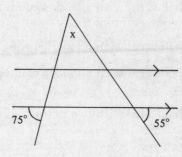

Fig 13.15   Fig 13.16   Fig 13.17

6) In Fig 13.18, which of the diagrams contain a pair of parallel lines?

(a)   (b)   (c)   (d)

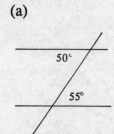 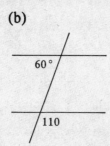 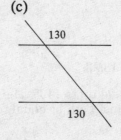 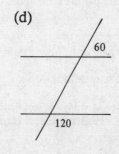

Fig 13.18

13.2 Polygons

A plane figure with straight sides is called a *polygon*.

The names of polygons are as follows:

    3 sides: a *triangle*   4 sides: a *quadrilateral*

    5 sides: a *pentagon*   6 sides: a *hexagon*

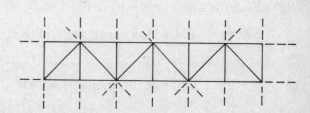

Fig 13.19

If all the angles and sides of a polygon are equal, then it is *regular*.

A pattern of polygons is called a *tessellation* or *tiling*. Fig 13.19 above.

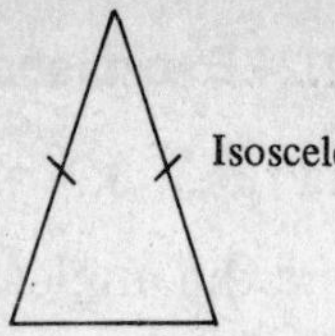

Triangles

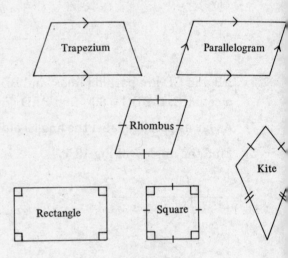

The angles of a triangle add up to 180°.

A triangle with two equal sides is *isosceles*. The base angles of an isosceles triangle are equal.

A triangle with all its sides equal is *equilateral*. Each angle of an equilateral triangle is equal to 60°.

A triangle, one of whose angles is 90°, is a *right-angled* triangle.

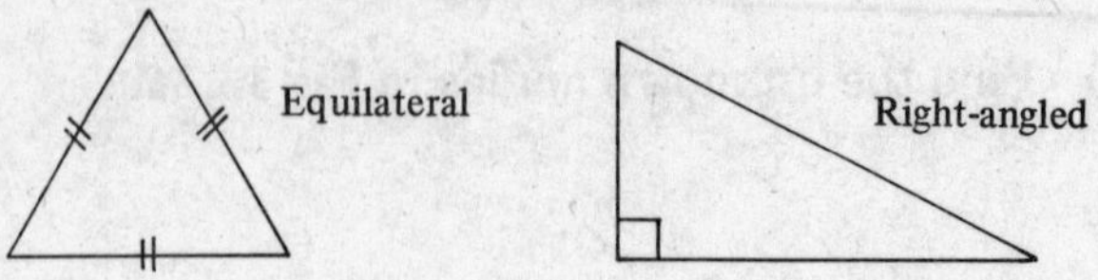

Fig 13.20

Quadrilaterals

The angles of a quadrilateral add up to 360°.

A quadrilateral with one pair of parallel sides is a *trapezium*.

A quadrilateral with two pairs of parallel sides is a *parallelogram*.

A quadrilateral with all sides equal is a *rhombus*.

A quadrilateral with all angles equal to 90° is a *rectangle*.

A quadrilateral with all sides equal and all angles equal is a *square*.

A quadrilateral with two pairs of adjacent sides equal is a *kite*.

Fig 13.21

Symmetry

If a figure is identical on both sides of a line, then it is *symmetrical* about that line.

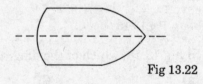

Fig 13.22

□ **13.2.1 Examples.** *Foundation, level 6*

1) Find the angle labelled *x* in Fig 13.23.

   Solution   The two base angles add up to 140°. The triangle is isosceles, so they must both be equal to *x*.

$$x = \frac{140°}{2} = 70°$$

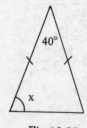

Fig 13.23

2) Find the angle labelled *y* in Fig 13.24.

   Solution   To find the fourth angle of the quadrilateral, subtract the other three angles from 360°:

   $$360° - 110° - 70° - 80° = 100°.$$

   Subtract this angle from 180°.

   $$y = 180° - 100° = 80°$$

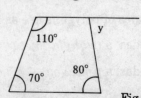

Fig 13.24

❏ **13.2.2 Exercises.** *Foundation, level 6*

1) Find the unknown angles in the triangles of Fig 13.25.

  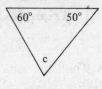 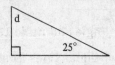

Fig 13.25

2) Find the unknown angles of the quadrilaterals of Fig 13.26.

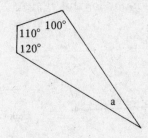 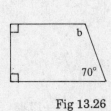 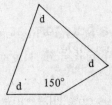

Fig 13.26

3) Find the exterior angles in the diagrams of Fig 13.27.

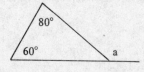 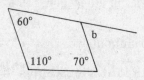

Fig 13.27

4) Find the unknown lengths in the triangles of Fig 13.28.

 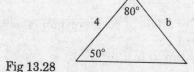 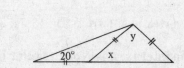

Fig 13.28

5) Find the angles *x* and *y* in Fig 13.29.

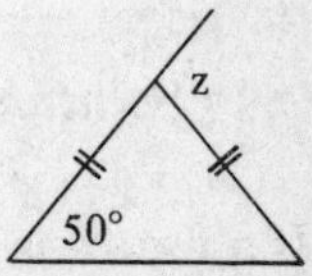

Fig 13.29   Fig 13.30

6) Find the angle *z* in Fig 13.30.

7) ABC is a triangle in which AB = AC and $\hat{BCA} = 75°$. Find $\hat{BAC}$.

8) DEF is a triangle in which $\hat{DFE} = \hat{DEF}$ and $\hat{FDE} = 580$. Find the unknown angles.

9) In the triangle PQR, PQ = PR and $\hat{PQR} = 60°$. What sort of triangle is △PQR?

10) Draw dotted lines to show the lines of symmetry of the shapes in Fig 13.31.

  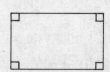 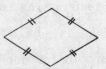

Fig 13.31

11) Complete the tessellations of Fig 13.32.

(a) 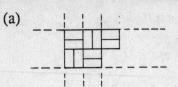 (b)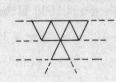

Fig 13.32

12) Draw a quadrilateral which is both a rectangle and a rhombus. What is the name for this figure?

13) Draw a quadrilateral which is both a parallelogram and a kite. What is the name for this figure?

Angles of a polygon

If a polygon has n sides, the sum of its *interior* angles is $180n - 360°$.

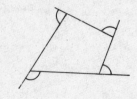

The sum of its *exterior* angles is 360°.

Fig 13.33

☐ **13.2.3 Examples.** *Foundation, level 6*

1) Find the sum of the interior angles of a decagon. (10 sides). If the figure is regular, how big is each exterior angle?

Solution   Put $n = 10$ into the formula above.

The sum of the interior angles is $180 \times 10 - 360 = 1{,}440°$

If the figure is regular, then the exterior angles are equal.

Each exterior angle is $\dfrac{360}{10} = 36°$

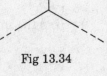

Fig 13.34

2) A tesselation is made out of regular n-sided polygons. Three tiles meet at each vertex. Find n.

Solution   A vertex is shown in Fig 13.34. The interior angle of each polygon is $\dfrac{360}{3} = 120°$.

The internal angle of a regular hexagon is equal to 120°.

The number of sides is 6

☐ **13.2.4 Exercises.** *Foundation, level 6*

1) Find the sum of the interior angles of:

   a) A pentagon   b) a hexagon

   c) an octagon (8 sides)   d) a 20 sided figure.

2) Find the interior angle of a regular figure with:

   a) 5 sides   b) 10 sides   c) 12 sides   d) 30 sides.

3) Find the exterior angle of a regular figure with:

   a) 8 sides   b) 12 sides   c) 18 sides   d) 30 sides.

4) The sum of the interior angles of a polygon is 900°. How many sides does it have?

5) The sum of the interior angles of a polygon is 1260°. How many sides does it have?

6) Each exterior angle of a regular polygon is 72°. How many sides does it have?

7) Each interior angle of a regular polygon is 150°. How many sides does it have?

8) The angles of a pentagon are x, $x + 10°$, $x + 20°$, $x + 30°$, $x + 40°$. Find x.

9) The angles of a hexagon are y, $2y$, $4y$, y, y, $3y$. Find y.

10) Three of the angles of a quadrilateral are equal, and the fourth is 30°. Find the other angles.

11) Which of the following angles could be the interior angles of a regular polygon? In each possible case give the number of sides of the polygon.

    a) 170°   b) 160°   c) 145°   d) 150°   e) 130°

12) The interior angle of a regular polygon is 100° greater than the exterior angle. Find the number of sides of the polygon.

13) The interior angle of a regular polygon is 4 times as great as the exterior angle. Find the number of sides of the polygon.

14) ABCDE is a regular pentagon. Find the angles AB̂C, AD̂E, AĈD. Fig 13.35.

15) ABCDEF is a regular hexagon. Find the angles AB̂C, AĈD, AD̂E, AD̂F. What can you say about the quadrilateral ACDF? Fig 13.36.

16) ABCD is a parallelogram. AD̂B = AB̂D = 40°. Find BĈD. What can you say about ABCD? Fig 13.37.

17) Find the interior angle of a regular pentagon. Show that it is not possible to tile a floor with regular pentagons.

18) Suppose we wish to tile a floor with regular n-sided polygons. What values of n are possible?

19) A floor is tiled with equilateral triangles and with certain regular polygons. Part of the tiling is shown. Find the internal angle of the polygon, and hence find the number of its sides. Fig 13.38.

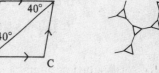

Fig 13.35   Fig 13.36

Fig 13.37   Fig 13.38

20) Extend the following tessellations:

    a)   b)

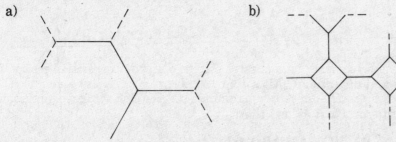

Fig 13.39

13.3 Similarity

Two figures are *similar* if they have the same shape. (But not necessarily the same size).

If two triangles ABC and DEF are similar then there is a fixed ratio between the sides of ΔABC and the sides of ΔDEF.

$$\frac{AB}{DE} = \frac{BC}{EF} = \frac{AC}{DF}$$

The ratio of sides is the same for both triangles.

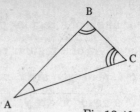
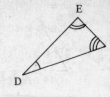

$$\frac{AB}{BC} = \frac{DE}{EF} : \frac{AB}{AC} = \frac{DE}{DF} : \frac{AC}{BC} = \frac{DF}{EF}.$$

Fig 13.40

Two triangles are similar if any of the following sets of conditions are true:

I The angles of one triangle are equal to the angles of the other triangle.

II One pair of angles are equal and the ratios of the enclosing sides are equal

III The ratios of corresponding sides are equal.

Note that if △ABC is similar to △DEF, the order of the letters is important. $\hat{A} = \hat{D}$, $\hat{B} = \hat{E}$, $\hat{C} = \hat{F}$.

Suppose triangles ABC and DEF are similar, with the ratio of their sides $a{:}b$. Then the ratio of their *areas* is $a^2{:}b^2$.

❏ **13.3.1 Examples.** *Intermediate and Higher, levels 8, 9*

1) L and M are the midpoints of the sides AB and AC of the triangle ABC. Find a pair of similar triangles. If BC = 8 find LM.

Solution The triangles ABC and ALM have angle $\hat{A}$ in common.

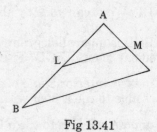

The ratios AL:AB and AM:AC are both equal to $\frac{1}{2}$.

Hence △ABC is similar to △ALM.

It follows that LM:BC is also equal to $\frac{1}{2}$.

Fig 13.41

$$\mathbf{LM = 8 \times \frac{1}{2} = 4}$$

2) L and M are on the sides AB and AC of the triangle ABC. AM = 3, MC = 9, AL = 4, LB = 5.

Find the ratio LM:BC. If △ABC has area 54, what is the area of the quadrilateral BLMC?

Solution AB = 9 and AC = 12.

The triangles ABC and AML have $\hat{A}$ in common.

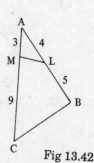

The ratio AL:AM = 4:3 and the ratio AC:AB = 4:3.

Hence △ABC is similar to △AML. It follows that:

$$\mathbf{LM{:}BC = AM{:}AB = 1{:}3}$$

The sides of the triangles are in the ratio 1:3. Hence the areas of the triangles are in the ratio $1^2{:}3^2 = 1{:}9$.

Fig 13.42

It follows that the area of △ALM is 54÷9 = 6.

The area of BLMC is 54 – 6 = 48

❏ **13.3.2 Exercises.** *Intermediate and Higher, levels 8, 9*

1) Which of the following triangles are similar to each other?

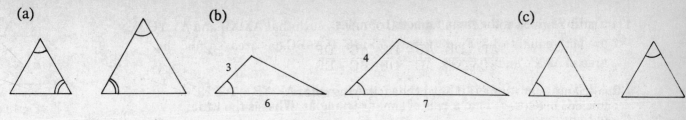

(a)   (b)   (c)

Fig 13.43

2) L and M lie on the sides AB, AC of the triangle ABC, so that LM is parallel to BC. Write down a pair of similar triangles. Fig 13.44

3) L, M and N lie on the sides AB, BC, CA of △ABC. LM, MN, NL are parallel to CA, AB, BC respectively. Find as many pairs of similar triangles as you can. Fig 13.45

4) Fig 13.46 shows a 3 barred gate in which AB, EF, DC are horizontal and AD, BC are vertical. AGC is a straight line. Write down as many similar triangles as you can.

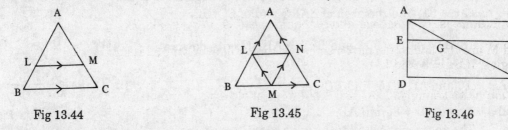

Fig 13.44   Fig 13.45   Fig 13.46

5) The triangles ABC and DEF of Fig 13.47 are similar. They are not drawn to scale.

a) If AB = 4, DE = 2 and AC = 6, find DF.

b) If AB = 3, AC = 2 and DE = 6, find DF.

c) If BC = 9, EF = 12 and DE = 8, find AB.

d) If AB = $\frac{3}{4}$, DE = $\frac{1}{2}$ and BC = $1\frac{1}{2}$, find EF.

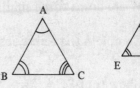

Fig 13.47

6) In the camp-stool of Fig 13.48 the lengths of wood above and below the crossing are 20 cm and 30 cm respectively. If the feet are 45 cm apart, how wide is the seat?

7) In Fig 13.46 above, AE = 40 cm, ED = 80 cm, AB = 240 cm. Find the lengths of EG and GF.

8) In Fig 13.49 a ladder AB is laid against two walls as shown. The higher wall is 3 m high, and the lower wall is 1 m. high. If the foot of the ladder is 0.5 m. from the lower wall, how far is it from the higher wall? How far apart are the walls?

9) You are given an isosceles triangle ABC, which is not equilateral. By drawing at least 4 more triangles show that it is possible to tessellate the plane with triangles congruent to △ABC.

10) You are given a right-angled triangle ABC, which is not isosceles. By drawing at least 4 more triangles show that it is possible to tessellate the plane with triangles congruent to △ABC.

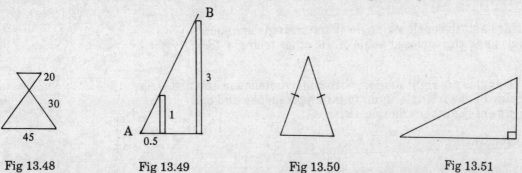

Fig 13.48   Fig 13.49   Fig 13.50   Fig 13.51

127

11) X and Y are on the sides AB and AC of △ABC, such that AX:XB and AY:YC are both equal to 1:3. Find the ratio XY:BC. If △ABC has area 32, find the area of △AXY and of XYCB.

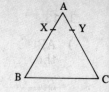

Fig 13.52

12) ABCD is a parallelogram, and E lies on AB so that AE:EB = 1:2. ED and AC meet at F. Find a pair of similar triangles. What is the ratio of their areas?

13) X and Y lie on the sides AB and AC of △ABC. AX = 4, XB = 8, AY = 5, YC = 10.

Write down a pair of similar triangles. Find the ratio XY:BC. If △AXY has area 8, find the area of XYCB.

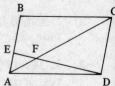

Fig 13.53

14) P and Q lie on the sides AB and AC of △ABC. AP = 12, PB = 3, AQ = 10, QC = 8.

Write down a pair of similar triangles. Which pairs of angles are equal? If △ABC has area 81, find the area of △APQ and of PQCB.

15) In Fig 13.54 BÂD = AB̂C = 90°, and BĈA = AB̂D. Write down a pair of similar triangles.

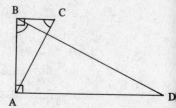

a)   If AB = 10 and AD = 15, find BC.

b)   If BC = 2 and AD = 8, find AB.

Fig 13.54

16) △ABC is right-angled at A. The perpendicular from A meets BC at D.

Write down three similar triangles.

Complete the equations: $\dfrac{BD}{AB} = \dfrac{AB}{--} : \dfrac{CD}{AC} = \dfrac{--}{BC}$.

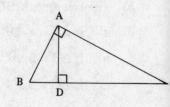

Show that $BC = \dfrac{AB^2}{BC} + \dfrac{AC^2}{BC}$. Multiply this equation by BC. What theorem have you now proved?

Fig 13.55

13.4 Congruence

Two figures are *congruent* if they have the same shape and size.

Two triangles are congruent when they obey any of the following sets of conditions.

I   The sides of one triangle are equal to the sides of the other triangle. (SSS).

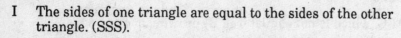

II   One side and two angles of one triangle are equal to one side and two angles of the other triangle. (ASA).

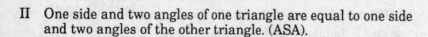

III Two sides and the enclosed angle of one triangle are equal to two sides and the enclosed angle of the other triangle. (SAS).

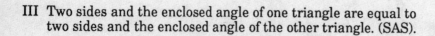

IV Both triangles are right-angled, with the hypoteneuse and one other side of one triangle equal to the hypoteneuse and one other side of the other triangle. (RHS).

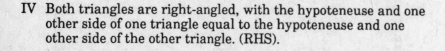

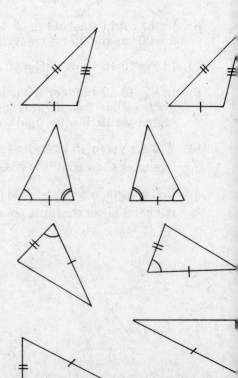

Fig 13.56

□ **13.4.1 Examples.** *Higher, level 9*

1) ABCD is a parallelogram. Join A to C; Find a pair of congruent triangles. What can be said about the sides of the parallelogram?

   Solution   Consider the two triangles ABC and ADC.

   They have the line AC in common.

   Angles BĈA and DÂC are alternate

   Angles BÂC and DĈA are alternate.

   Hence by condition ASA the triangles ABC and CDA are congruent.

   From congruence it follows that AB = DC and AD = BC.

   Fig 13.57

2) ABC is an equilateral triangle. L, M, and N are points on AB, BC, CA respectively such that AL = BM = CN. Show that ΔLMN is also equilateral.

   Solution   Consider the three triangles ALN, BML, CNM.

   Fig 13.58

   As ΔABC is equilateral, it follows that the angles Â, B̂, Ĉ are equal.

   It is given that AL, BM and CN are equal.

   The three sides of ΔABC are equal, hence it follows that AN, CM, and BL are equal.

   The three triangles are congruent, by SAS. It follows that:

   LN = NM = ML. ΔLMN is equilateral.

□ **13.4.2 Exercises.** *Higher, level 9*

1) Which of the following pairs of triangles are congruent?

   (a)   (b)   (c)

   Fig 13.59

2) ABD and CBE are straight lines meeting at B. CB = BD and AB = BE. Find two congruent triangles. Write down two pairs of equal angles. Fig 13.60

3) ABD and CBE are straight lines meeting at B. CB = BE and AB = BD. Find two congruent triangles. Write down two pairs of equal angles. What can you say about AC and DE? Fig 13.61

4) PQ and RS are equal and parallel lines. PR and SQ cross at X, as in Fig 13.62. Find two congruent triangles. Write down two pairs of equal sides.

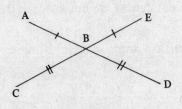

Fig 13.60

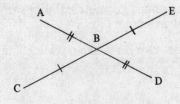

Fig 13.61

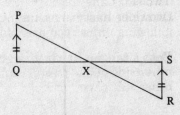

Fig 13.62

5) ABCD is a quadrilateral in which AB = CD and AD = BC. Draw the diagonal AC.

   Find a pair of congruent triangles. Write down which pairs of angles are equal. What can you say about AB and CD, and about AD and BC? What sort of figure is ABCD?

6) ABCD is a parallelogram. The diagonals AC and BD meet at X. Write down as many pairs of congruent triangles as you can.

7) ABCD is a kite in which AB = AD and CB = CD. The diagonals AC and BD meet at X.

   What is the line of symmetry of the figure? Write down 3 pairs of congruent triangles. Which angles are equal to each other?

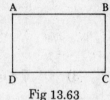

Fig 13.63

8) ABCD is a rectangle. Show that the triangles ABC and BAD are congruent. Deduce that the diagonals AC and BD are equal.

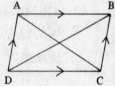

9) ABCD is a parallelogram, in which the diagonals AC and BD are equal. Show that $\triangle ABC$ is congruent to $\triangle BAD$. Deduce that ABCD is a rectangle.

Fig 13.64

10) ABCD is a rhombus. Let the diagonals AC and BD meet at X. Show that $\triangle ABX$ is congruent to $\triangle ADX$. Deduce that AC and BD are perpendicular.

11) ABCD is a parallelogram in which the diagonals AC and BD are perpendicular. Show that ABCD is a rhombus.

12) ABCDE is a regular pentagon. X and Y are on BC and CD respectively, and CX = CY. Show that $\triangle ABX$ is congruent to $\triangle EDY$. Deduce that AX = EY.

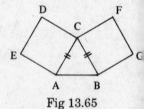

Fig 13.65

13) In Fig 13.65 ACDE and BCFG are squares on the equal sides of the isosceles triangle ABC. Show that AF = BD.

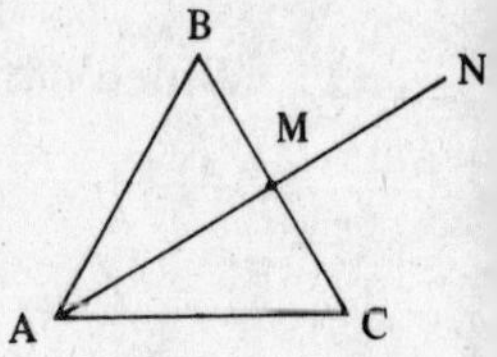

14) M is the midpoint of the side BC of $\triangle ABC$. AM is extended to N, where AM = MN. Show that ABNC is a parallelogram.

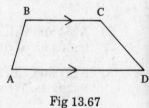

Fig 13.66

15) ABCDE... are adjacent points of a regular 17 sided figure. Show that $\triangle ABD$ is congruent $\triangle BCE$.

16) You are given a trapezium ABCD as shown. By drawing at least 4 more trapezia show that it is possible to tessellate a plane with trapezia congruent to ABCD.

Fig 13.67

Common errors

1) Angles and lines

Do not be hasty in assuming facts about geometrical diagrams. In particular, do not assume the following unless you are told they are true:

That a certain line is straight. That two lines are parallel. That a certain angle is a right-angle.

2) Polygons

a) Make sure that you label polygons correctly. The letters must follow round the figure, either clockwise or anti-clockwise. They must not jump across the diagonal.

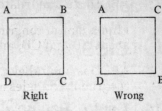

Fig 13.68

b) If you are told that a polygon is a rectangle, then do not assume that it is not a square. It might be. Similarly, a parallelogram may be a rhombus, and an isosceles triangle may be equilateral.

c) Do not assume a polygon is regular unless you are told so. The interior angles of a hexagon add up to 720°. But we can only say that each interior angle is 120° if the hexagon is *regular*.

3) Congruence and similarity

a) If $\triangle ABC$ is congruent to $\triangle DEF$, then the triangles are congruent in that order. So $\hat{A} = \hat{D}$ etc., AB = DE etc. It would then be wrong to state that $\triangle ABC$ is congruent to $\triangle EFD$.

b) If you are told that L is on AB, with AL:LB = 1:3, then the ratio AL:AB is 1:4. Do not think that AL:AB is also 1:3.

c) If figures are similar, with their sides in the ratio $a:b$, then the areas are in the ratio $a^2:b^2$, not $a:b$.

Chapter 14

Circles

14.1 Centre, Radius, Circumference

A *circle* consists of all the points which are the same distance from a fixed point.

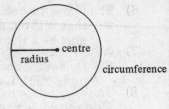

Fig 14.1

The fixed point is the *centre*.

The distance is the *radius*.

The greatest distance across the circle is the *diameter*. The diameter is twice the radius.

The length round the circle is the *circumference*. The ratio of the circumference to the diameter is the same for every circle. The ratio is called π (pronounced *pie*). π can never be written down exactly. It is sometimes approximated by $\frac{22}{7}$ or by 3.14.

☐ **14.1.1 Examples.** *Foundation, level 5*

1) The radius of a big-wheel at a fair is 3 m. How far does a point on the rim travel during one revolution? (Take π to be 3.14).

   Solution The diameter of the wheel is twice the radius, i.e. 6 m. diameter.

   Distance travelled is 3.14 × 6 = 18.8 m.

Fig 14.2

2) A and B are points on a circle with centre O. AÔB = 40°. Find the other angles of the triangle AOB.

   Solution OA and OB are both radii of the circle. Hence they are equal.

   The angles OAB and OBA are equal. Together they must come to 180° − 40° = 140°.

   $$O\hat{A}B = O\hat{B}A = \frac{140°}{2} = 70°$$

☐ **14.1.2 Exercises.** *Foundation, level 5*

Throughout these exercises either use the π button on your calculator or take π to be 3.14.

1) On the circles below mark   a) the centre b) a radius   c) a diameter.

   a)   b)   c)

Fig 14.4

2) A circular running track has radius 50 m. What is the diameter? If an athlete runs round once, how far has she gone?

3) The minute hand of a clock is 4 cm. long. How far does the tip of the hand travel during one hour?

4) The hour hand of a clock is $2\frac{1}{2}$ cm. long. How far does the tip travel in 6 hours?

5) A coin of diameter 3 cm. rolls along the ground. How far has the coin travelled when it has made a complete revolution?

6) The driving wheel of a car is 6 inches in radius. How far does a point on the rim move when the wheel goes through a quarter turn?

7) A running track is to be made in the form of a circle with circumference 400 metres. What should the diameter be? What should the radius be?

8) The circumference of a circle is 20 cm. What is the radius?

9) A square is drawn inside a circle of radius 5 cm. What is the length of the circle between two adjacent corners of the square?

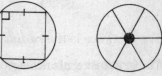

Fig 14.5   Fig 14.6

10) A regular pentagon is drawn inside a circle of radius 8 cm. What is the length of the circle between adjacent corners of the pentagon?

11) A record of diameter 12 inches rotates at 33 revolutions per minute. How far does a point on the rim travel in one minute? How far does it travel in one second?

12) The diameter of a cartwheel is 120 cm., and the boss in the centre of the wheel is of diameter 15 cm. How long are the spokes of the wheel? If thecart travels 100 m., how many times has the wheel turned? Fig 14.6.

13) Find the unknown angles in the following diagrams:

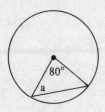
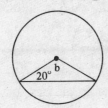
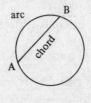

Fig 14.7

14.2 Chords, Arcs, Sectors

The straight line joining two points A and B on a circle is a *chord*.

The part of the circle between A and B is an *arc*.

The region of a circle between two radii is a *sector*.

A diameter splits a circle into two *semi-circles*.

If AB is a diameter of a circle and C a point on the circumference, then $A\hat{C}B = 90°$.

A chord is perpendicular to the radius through its centre.

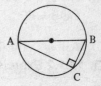

Fig 14.8

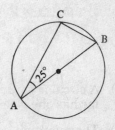

□ **14.2.1 Examples.** *Higher, level 10*

1) AB is a diameter of the circle in Fig 14.9. If $C\hat{A}B = 25°$ find the other angles.

   Solution   Because AB is a diameter it follows that $A\hat{C}B$ is a right-angle.
   The third angle can be found by subtraction:

$$A\hat{C}B = 90° \text{ and } A\hat{B}C = 180° - 90° - 25° = 65°$$

Fig 14.9

2) The chord AB of the circle in Fig 14.10 is bisected by the diameter XY. Show that AX = BX and that AY = BY. What sort of figure is XAYB?

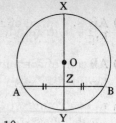

Fig 14.10

Solution The radius OY of the circle goes through the midpoint of AB. Hence XY and AB are perpendicular. Hence $A\hat{Z}X = B\hat{Z}X$. It follows that $\triangle AXZ$ is congruent to $\triangle BXZ$.

This implies that AX = BX.

By similar reasoning AY = BY. The quadrilateral XAYB has two pairs of adjacent sides equal.

XAYB is a kite

☐ **14.2.2 Exercises.** *Higher, level 10*

1) On the circles below mark   a) the chord AB   b) the arc AB.

a)   b)

Fig 14.11

2) Find the unknown angles in the diagrams below.

a)   b) 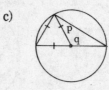   c)

Fig 14.12

Fig 14.13

3) AB and CD are diameters of a circle. Show that ACBD is a rectangle.

4) AB is a diameter of a circle with centre O. C is a point of the circumference. $C\hat{O}B = 40°$. Find the angles $O\hat{C}B$, $O\hat{C}A$. What is angle $C\hat{A}B$?

5) AB is a chord of a circle centre O and radius 5, and $A\hat{O}B = 30°$. What is the length of the arc AB?

6) An athlete runs 100 m. round a running track, which is a circle centre O and radius 70 m. If A is the starting point and B the finishing point, what is the angle between OA and OB?

7) Find the ratio of the arcs AB, BC, CA in Fig 14.14.

8) AB is a chord of a circle centre O, and C is the midpoint of AB. What can you say about $A\hat{C}O$? What is the relationships between the triangles OCA and OCB? If $O\hat{A}B = 50°$ find $C\hat{O}B$.

9) AB and AC are equal chords of a circle centre O. Draw a diagram of the circle. What is the line of symmetry of the figure? If $O\hat{B}A = 20°$ find $B\hat{A}C$.

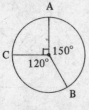

Fig 14.14

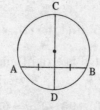

Fig 14.15

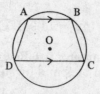

Fig 14.16

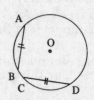

Fig 14.17

10) AB is a chord of a circle, and CD is a diameter perpendicular to AB. Name 3 pairs of equal sides in Fig 14.15. Name a pair of congruent triangles. What sort of quadrilateral is ACBD?

11) AB and CD are parallel chords of the circle centre O in Fig 14.16. Draw the line of symmetry of the figure. If $\hat{A}$ = 80° find the other three angles of the quadrilateral.

12) In Fig 14.17 AB and CD are equal chords of the circle centre O. What is the relationship between ΔOAB and ΔOCD? Draw the perpendiculars from O to the two chords. What can you say about the two perpendicular distances?

14.3 Circle Theorems

A chord divides a circle into two *segments*.

A quadrilateral inscribed inside a circle is *cyclic*.

Fig 14.18

I.   Let AB be a chord of a circle with centre O, and let C be a point on the circumference in the same segment as O. Then.

$$A\hat{O}B = 2 \times A\hat{C}B$$
Fig 14.19

(The angle at the centre is twice the angle at the circumference.)

II. Let AB be a chord of a circle, and let C and D be points in the same segment. Then:

$$A\hat{C}B = A\hat{D}B$$

(Angles in the same segment are equal.)

Fig 14.20

III. The opposite angles of a cyclic quadrilateral add up to 180°.

$$\hat{A} + \hat{C} = 180°$$

IV. The exterior angle of a cyclic quadrilateral is equal to the opposite angle.

$$\hat{A} = B\hat{C}E$$

Fig 14.21

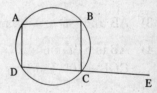

❑ **14.3.1 Examples.** *Higher, level 10*

1)   O is the centre of the circle in Fig 14.22, and $A\hat{O}B$ = 80°. Find C and D.

Solution   C is in the same segment as O. Hence Theorem I applies.

$$\hat{C} = \frac{1}{2} A\hat{O}B = 40°$$

ACBD is a cyclic quadrilateral. Hence Theorem III applies.

$$\hat{D} = 180° - \hat{C} = 140°$$

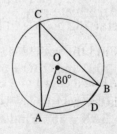
Fig 14.22

2)   The chords AB and CD of a circle cross at X. Show that ΔAXC is similar to ΔDXB.

Solution   C and B are in the same segment of the circle defined by the chord AD. Theorem II applies.

$$A\hat{C}D = A\hat{B}D.$$

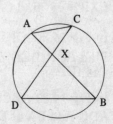
Fig 14.23

A and D are in the same segment of the circle defined by the chord BC. Theorem II applies.

$$C\hat{D}B = C\hat{A}B.$$

The angles $A\hat{X}C$ and $B\hat{X}D$ must also be equal.

ΔAXC is similar to ΔDXB

☐ **14.3.2 Exercises. Higher, level 10**

1) O is the centre of the circle of Fig 14.24. The diagram is not to scale.

   a) If $A\hat{C}B = 25°$ find $A\hat{O}B$.

   b) If $A\hat{O}B = 122°$ find $A\hat{C}B$.

2) The diagram of Fig 14.25 is not to scale.

   a) If $A\hat{D}B = 43°$ find $A\hat{C}B$.

   b) Find an angle equal to $C\hat{A}B$.

   c) Find two pairs of similar triangles.

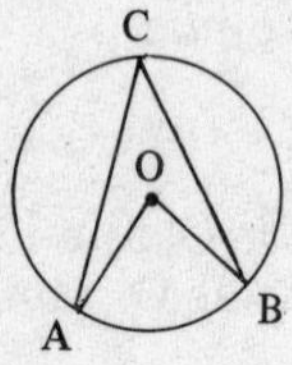

Fig 14.24 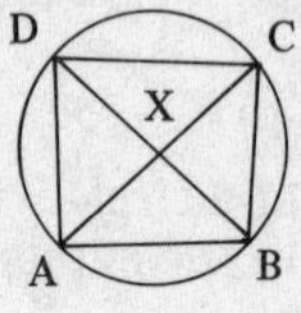   Fig 14.25 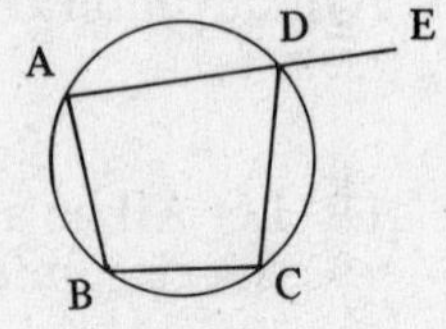   Fig 14.26

3) The diagram of Fig 14.26 is not to scale. ADE is a straight line.

   a) If $\hat{A} = 56°$ find $\hat{C}$.

   b) If $\hat{B} = 74°$ find $A\hat{D}C$ and $C\hat{D}E$.

4) Find the unknown angles in the following figures. Throughout O denotes the centre of the circle.

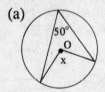

(a) (b) (c) (d) (e)

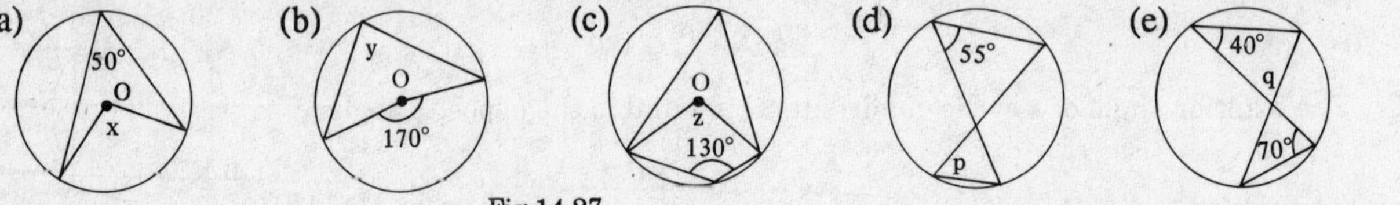

Fig 14.27

5) AB is a chord of a circle centre O. C is a point on the circumference, so that O and C lie in different segments of the circle. Show that $A\hat{C}B$ is equal to half the reflex angle $A\hat{O}B$.

6) In Fig 14.28 O is the centre of the circle. $A\hat{O}C = 108°$ and $C\hat{B}O = 56°$. Find $A\hat{O}B$.

7) In Fig 14.29 the chords AB and DC are parallel. $B\hat{A}X = 35°$. Find $A\hat{X}D$, $C\hat{X}B$, $A\hat{B}D$.

8) In Fig 14.29 the chords AB and DC are parallel. If $C\hat{X}B = 88°$ find $C\hat{A}B$ and $A\hat{C}D$.

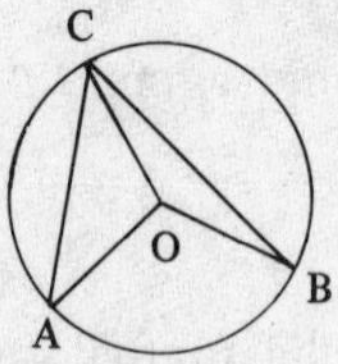

9) In Fig 14.30 ABC and FED are straight lines. $\hat{A} = 110°$. Find $B\hat{E}F$, $B\hat{E}D$, $\hat{C}$. What can you say about the lines AF and CD?

10) In Fig 14.31 ABCD is a parallelogram. The circle through ADC cuts BA at E. Prove that ΔCBE is isosceles.

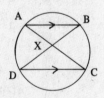

Fig 14.28   Fig 14.29

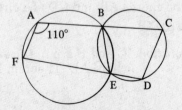

Fig 14.30

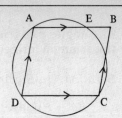

Fig 14.31

11) ABCD is a cyclic parallelogram. Draw a diagram of ABCD and the circle. What else can be said about ABCD?

12) Draw pairs of circles which meet:

   a) At two points   b) at one point   c) at no points.

14.4 Tangents

A line which meets a circle at one point only is a *tangent* to the circle.

A tangent is perpendicular to the radius at that point.

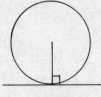

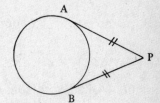

If P lies outside a circle, then the two tangents from P to the circle are equal in length.

$$PA = PB.$$

Fig 14.32   Fig 14.33

Let a chord AB meet a tangent AX. Let C be in the opposite (alternate) segment. The *Alternate Segment* theorem states:

The angle between a chord and a tangent is equal to the angle in the opposite segment.

$$\hat{XAB} = \hat{ACB}$$

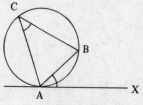

Fig 14.34

☐ **14.4.1 Examples.** *Higher, level 10*

1) In Fig 14.35 XP is a tangent to the circle with centre O. OX cuts the circle at Q. If $P\hat{X}O = 24°$ find $P\hat{O}Q$ and $O\hat{P}Q$.

    Solution   OP is a radius, and PX is a tangent. Hence they are perpendicular.

$$P\hat{O}Q = 180° - 90° - 24° = 66°.$$

    OP and OQ are equal radii. Hence $O\hat{P}Q = O\hat{Q}P$.

$$O\hat{P}Q = \frac{(180° - 66°)}{2} = 57°.$$

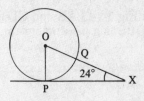

Fig 14.35

2) Two balls touch each other. Show that the two centres and the point of contact lie in a straight line.

    Solution   As the balls touch, they have a common tangent, shown by the dotted line in Fig 14.36.

    Both the radii to the point of contact are perpendicular to the tangent. OTO' = 90° + 90° = 180°.

               OTO' is a straight line.

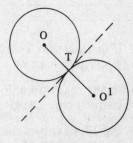

Fig 14.36

3) XA and XB are tangents to a circle. If $A\hat{X}B = 28°$, find $X\hat{A}B$.

    Solution   As XA and XB are tangents, they are equal. Hence △AXB is isosceles. It follows that:

$$X\hat{A}B = \frac{1}{2}(180° - 28°) = 76°$$

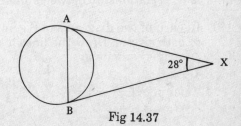

Fig 14.37

4) The angles of a triangle are 48°, 68°, 64°. A circle is inscribed in the triangle. What are the angles of the triangle formed by the three points of contact?

Solution   In Fig 14.38, △APR is isosceles. It follows that $A\hat{P}R = 66°$.

AP is a tangent to the circle, and PR is a chord. By the Alternate Segment theorem:

$$P\hat{Q}R = A\hat{P}R = 66°$$

By similar calculations:

$$P\hat{R}Q = 56° \text{ and } R\hat{P}Q = 58°$$

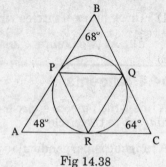

Fig 14.38

❏ **14.4.2 Exercises.** *Higher, level 10*

1) Find the unknown angles in the diagrams below.

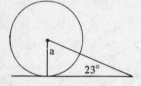 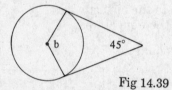 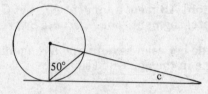

Fig 14.39

2) In Fig 14.40 AT is a diameter of a circle and TB is a tangent. If $\hat{A} = 55°$ find $\hat{B}$.

3) In Fig 14.41 the small circle touches the large circle. Show that the two centres and the point of contact are in a straight line.

4) In Fig 14.42 TA and TB are tangents to the circle with centre O. If $\hat{T} = 72°$ find $\hat{O}$.

5) In Fig 14.43 AB and CD are diameters, and the tangents at A, B, C, D meet at P, Q, R, S. Show that PQRS is a parallelogram.

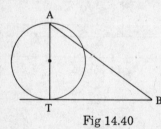 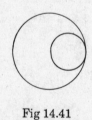 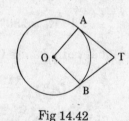 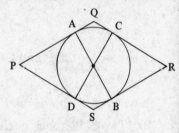

Fig 14.40   Fig 14.41   Fig 14.42   Fig 14.43

6) If in Question 5 we know that AB and CD are perpendicular, what more can we say about PQRS?

7) In the diagram of Fig 14.44 mark in pairs of equal sides and pairs of equal angles.

Fig 14.44

8) Fig 14.45 shows the tangents AXC and BXD to two circles. Mark in pairs of equal sides and pairs of equal angles. What is the relationship between △AXB and △CXD?

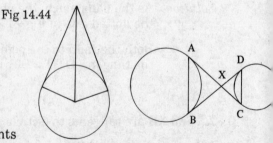

Fig 14.45

9) In each of the diagrams of Fig 14.46 draw the common tangents to both circles. Draw the lines of symmetry of each diagram.

(a)   (b)   (c)   (d)   (e)

Fig 14.46

10) In Fig 14.47 a quadrilateral ABCD is drawn round a circle. Show that AB + CD = BC + AD.

11) Fig 14.48 shows three circles, each of radius 10 cm, touching each other at A, B, C. Find the side of the triangle ABC.

12) Find the unknown angles in the diagrams of Fig 14.49.

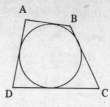

Fig 14.47

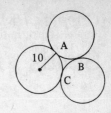

Fig 14.48

   Fig 14.49

13) A circle is drawn inside a triangle with angles 60°, 66°, 54°. Find the angles of the triangle made by the points of contact.

14) A circle is drawn inside a triangle, and the triangle made by the points of contact has angles 55°, 63°, 62°. Find the angles of the original triangle.

15) In Fig 14.50 AB is a diameter of the circle, and CT is the tangent at C. If BĈT = 70°, find AB̂C.

16) The quadrilateral ABCD has angles 100°, 95°, 70°, 95° at A, B, C, D. A circle is drawn inside the quadrilateral, touching AB, BC, CD, DA at P, Q, R, S. Find the angles of PQRS.

17) In Fig 14.51 O is the centre and TA and TB are tangents. AT̂B = 40° and TB̂Y = 25°. Find AÔB, AX̂B, YÂT.

18) Find the unknown angles in the diagrams of Fig 14.52.

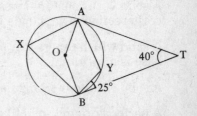

Fig 14.50

Fig 14.51

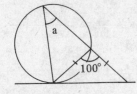 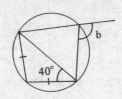 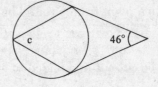

Fig 14.52

Common errors

1) **Circumference and radius**

a) The value of π is only *approximately* 3.14 or $\frac{22}{7}$. If you calculate using these values for π, it is misleading to give your answer to more than 3 significant figures.

b) The circumference of circle is π times the diameter. So if you know the circumference, in order to find the diameter you must *divide* by π. Do not multiply.

c) Be careful not to confuse the radius and the diameter. Read the question carefully, to see which you are given or which you are required to find.

2) **Chords**

a) Do not assume that a chord is a diameter unless you are told so.

b) Do not assume that two chords cross in the centre of a circle unless you are told so.

These two mistakes are often made when a picture like Fig 14.53 is used. It is often wrongly assumed that:

AD is a diameter.   X is the centre of the circle.   AB is parallel to CD.

The last error leads one to say that Â = D̂, instead of Â = Ĉ.

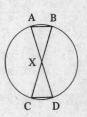

Fig 14.53

Chapter 15

Solids

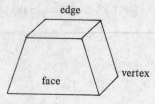
Fig 15.1

15.1 Solids and Nets

The flat side of a solid is a *face*.

The line on a solid where two faces meet is an *edge*.

The point on a solid where three or more edges meet is a *vertex* or *corner*.

A solid with six square faces is a *cube*.

A solid with six rectangular faces is a *cuboid*.

A solid with constant cross-section is a *prism*. If the cross-section is a triangle it is a *triangular* prism.

A solid which tapers to a point from a rectangular base is a *pyramid*.

A solid with four triangular faces is a *tetrahedron*.

A perfectly round solid like a ball is a *sphere*.

A solid whose cross-section is a circle is a *cylinder*.

A solid which tapers to a point from a round base is a *cone*.

A diagram which can be cut out and folded to make a solid is a *net*.

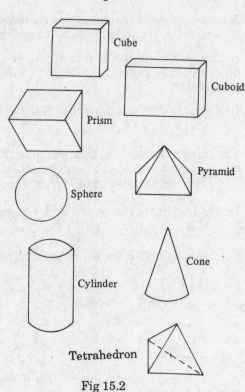
Fig 15.2

☐ **15.1.1 Examples.** *Foundation, levels 4, 5*

1) Fig 15.3 shows a cube. How many faces, edges, vertices does it have? Name a pair of parallel edges and a pair of parallel faces.

   Solution   Looking at the diagram, there are 2 horizontal faces and 4 vertical faces.

   There are 4 vertical edges and 8 horizontal edges.

   There are 4 vertices on the top face and 4 more on the bottom face.

   There are 6 faces, 12 edges and 8 vertices.

   Any two vertical edges are parallel. The two horizontal faces are parallel.

   AE and CG are parallel: ABCD and EFGH are parallel.

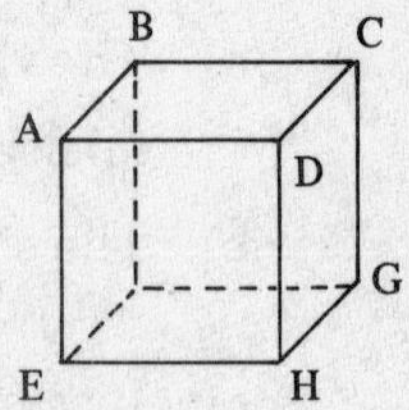

   Fig 15.3

2) Fig 15.4 shows the net of a solid. What is the name of the solid? Which point will F join?

   Solution   When the solid is assembled, there will be 2 triangular faces CAE and GBD at each end.

   The solid is a triangular prism.

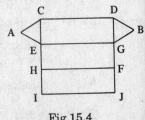
Fig 15.4

IJ is glued to CD, JF is glued to DB.

F will join B.

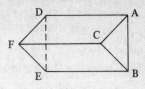

Fig 15.5

☐ **15.1.2 Exercises.** *Foundation, levels 4, 5*

1) Find the number of faces, edges and vertices of the pyramid of Fig 15.2.

2) Fig 15.5 shows a prism with a triangular cross-section. How many faces, edges and vertices does it have? Name a pair of parallel faces, and two pairs of parallel edges.

3) What are the mathematical names for the following common solids?

   a) A matchbox   b) Dice   c) A wedge of cheese

   d) A tin can   e) A ping-pong ball   f) A six-sided pencil (unsharpened)

   g) The tip of a sharpened pencil   h) A bell-tent.

4) Fig 15.6 shows a number of cubes arranged to form another solid. How many cubes are there? How many more cubes would be needed to make the solid a cuboid?

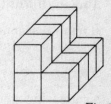

Fig 15.6

5) Fig 15.7 shows a house which is 3 storeys high, 3 rooms wide and 3 rooms deep. In each room, there is a window in every wall that faces the outside.

   How many rooms are there?

   How many rooms have no windows?

   How many rooms have 1 window?

   How many rooms have 2 windows?

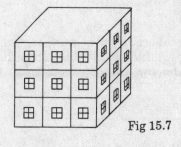

Fig 15.7

Suppose that skylights are put in to the ceilings of all the rooms on the top floor. How many rooms have 3 windows?

6) Fig 15.8 shows a cuboid formed by 24 smaller cubes. The cuboid is dipped into paint.

   a) How many cubes have one painted face?

   b) How many cubes have 2 painted faces?

   c) How many cubes have 3 painted faces?

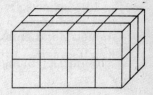

Fig 15.8

7) Fig 15.9 shows a net. What solid is formed from the net? What point will join B? What point will join N?

8) A die is marked so that the number of dots on opposite faces always adds up to 7. Fig 15.10 shows the net of a die. Fill in the number of dots on the blank faces.

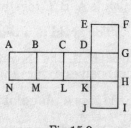

Fig 15.9

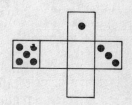

Fig 15.10

9) Fig 15.11 shows several possible nets. Which of them will be the net of a cube?

  (a)   (b)   (c)   (d)   (e)

 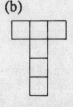 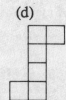

Fig 15.11

10) Fig 15.12 shows a square sheet of cardboard. The shaded bits are removed, and the rest is folded along the dotted lines. What sort of object is made?

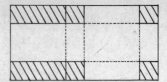

Fig 15.12

11) Describe the solids formed from the following nets:

(a)   (b)   (c)   (d)

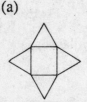 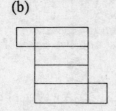 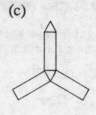 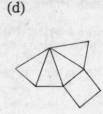

Fig 15.13

15.2 Symmetry

If a solid is identical on both sides of a plane, then it is *symmetrical* about the plane.

Fig 15.14

If a solid occupies the same region of space after being rotated about a line, then the line is an *axis of symmetry*.

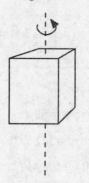

Fig 15.15

☐ **15.2.1 Example.** *Foundation, level 5*

Fig 15.16 shows an equilateral triangle, with LMN at the midpoints of its sides. What solid is made when it is folded along the dotted lines? Find a plane of symmetry and an axis of symmetry.

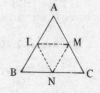

Fig 15.16

Solution   Bring the points A, B, C together. Fig 15.17.

The solid is a tetrahedron

A plane through AL and the midpoint X of MN will cut the solid into identical halves. Hence this plane is a plane of symmetry.

ALX is a plane of symmetry

Take the line through A and the centre G of △LMN. If the solid is rotated through 120° about this line it will return to the same region of space.

AG is an axis of symmetry

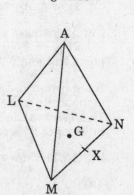

Fig 15.17

☐ **15.2.2 Exercises.** *Foundation, level 5*

1)   For each of the following figures draw a plane of symmetry and an axis of symmetry.

a) A cuboid, 3 by 4 by 5. b) A cuboid, 3 by 3 by 5.

c) A cube, 3 by 3 by 3. d) A cylinder.

e) A pyramid with square base. f) A sphere.

g) A cone. h) A prism, with cross-section an equilateral triangle

(a) 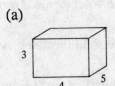 (b) (c) (d)

(e) (f) (g) (h)

Fig 15.18

2) The net of Fig 15.19 is folded to make a solid.

   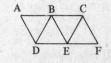

   a) What is the name of the solid? Which point is joined to F? What point is joined with A?

   b) Mark on the net the triangle which is opposite to A.

   c) How many faces, edges and vertices does the solid have?

   Fig 15.19

3) Draw nets for the following solids. Your diagrams should be clearly labelled, but they need not be to scale.

   a) A cuboid 3 by 4 by 5. b) A cuboid 3 by 3 by 5

   c) A wedge. d) A prism, whose cross-section is a right-angled isosceles triangle.

(a) 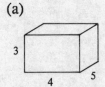 (b) (c) (d)

Fig 15.20

4) Fig 15.21 shows 3 edges of a cube. Complete the diagram.

5) Fig 15.22 shows part of the net for a cube. Complete the net.

6) Fig 15.23 shows a cylinder. The curved side is cut along the dotted line. Draw the side when it has been laid flat.

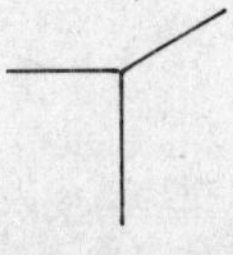

Fig 15.21 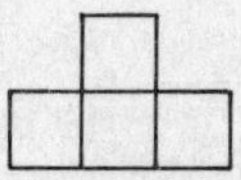 Fig 15.22 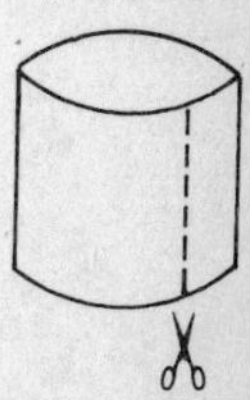 Fig 15.23

7) Find out how many planes of symmetry there are for each of the solids in Question 1.

8) Find out how many axes of symmetry there are for each of the solids in question 1.

9) The curved surface of the cone of Fig 15.24 is made out of paper. It is cut along the dotted line. Make a sketch (not to scale) of the paper when it is laid flat.

10) Fig 15.25 shows a square based pyramid. X and Y are the midpoints of AD and BC.

Fig 15.24 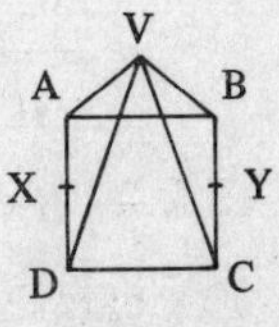 Fig 15.25 143

a) The pyramid is cut in half by the plane through VXY. What is the name of the solid formed by one of the halves? How many faces, edges, vertices does it have?

b) The pyramid is cut in half by the plane through VAC. What is the name of the solid formed by one of the halves? How many faces, edges, vertices does it have?

11) Fig 15.26 shows a cube. W, X, Y, Z are the midpoints of AB, CD, GH, EF.

a) The cube is cut in half by the plane through WXYZ. What is the name of the solid formed by one of the halves? How many faces, edges, vertices does it have?

b) The cube is cut in half by the plane through ACGE. What is the name of the solid formed by one of the halves? How many faces, edges, vertices does it have?

c) The cube is cut in two by the plane through BED. What is the name of the smaller part? How many faces, edges, vertices does it have?

12) The longest distance within the cube of Fig 15.26 is AG. Name 3 other lines in the cube which have the same length.

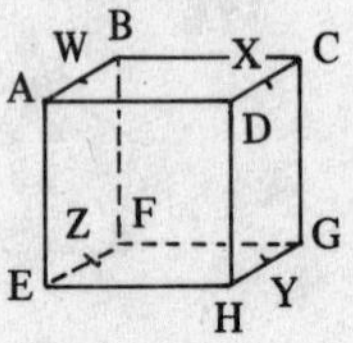

13) Make a copy of the cube of Fig 15.27. On your diagram mark the midpoints of each of the 6 faces. Join the midpoints together, except when they are directly opposite each other. You should now have a diagram of another solid, called an *octahedron*.

Fig 15.26

a) How many faces, edges, vertices does this solid have?

b) Draw the net for this solid.

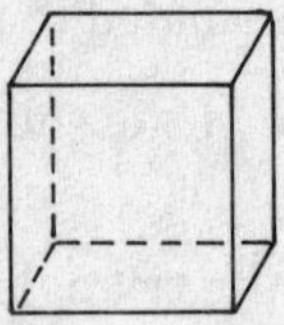

Common Errors

1) **Diagrams**

Fig 15.27

Angles and lines get distorted in drawings of solids. Right-angles may not seem to be right-angles, and equal sides may look different. Be careful that you do not misunderstand a diagram.

2) **Names of solids**

If a solid is described as a cuboid, then do not assume that it is not a cube. A cube is a special case of a cuboid.

Chapter 16

Constructions and Drawings

16.1 Constructions and Measurement

The main instruments used in geometrical drawings are the *ruler* (to measure distances), the *protractor* (to measure angles), and *compasses* (to draw circles).

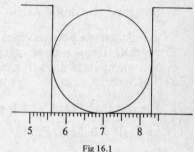
Fig 16.1

❑ **16.1.1 Examples.** *Foundation, levels 4, 5, 6*

1) Fig 16.1 shows a cylinder held on top of a ruler. What is the diameter of the cylinder?

   Solution   The left hand edge is at 5.6 cm. The right hand edge is at 8.3. Subtract one measurement from the other.

   The diameter is 8.3 - 5.6 = 2.7 cm.

2) The triangle ABC has AB = 2 cm, BC = 3 cm, CA = 4 cm. Make an accurate drawing of the triangle, and measure the angle at A.

   Solution   The steps are shown in Fig 16.2.

   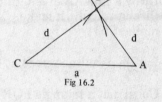
   Fig 16.2

   a)   Use a ruler to draw the line CA of length 4.

   b)   Separate compasses to a width of 3 cm, and with the point at C draw an arc.

   c)   Separate the compasses to 2 cm, and with the point at A draw an arc.

   d)   Where the arcs intersect is B. Join up the sides.

   Lay the protractor on AC so that the central point is at A. Notice that AB makes an angle of about 47°.

   BÂC = 47°

   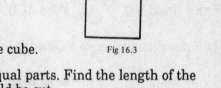
   Fig 16.3

❑ **16.1.2 Exercises.** *Foundation, levels 4, 5, 6*

1) Fig 16.3 shows a ruler laid on top of a cube. Find the side of the cube.

2) Fig 16.4 shows a length of wood which is to be divided into 5 equal parts. Find the length of the wood. Divide that length by 5. Mark on the wood where it should be cut.

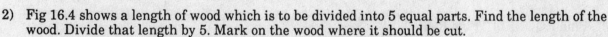
Fig 16.4

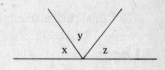
Fig 16.5

3) Measure the angles *x*, *y*, *z* of Fig 16.5. Find *x* + *y* + *z*. What do you notice?

4) Draw a line AB of length 2 inches. Use your protractor to make an angle of 40° at A.

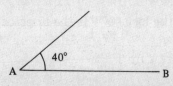

Fig 16.6

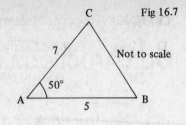

Fig 16.7

5) Draw a line AB of length 5 cm. Use your protractor and ruler to draw a line AC such that BÂC = 50° and AC is of length 7 cm. Find the length BC and the angle ACB.

6) The dots of Fig 16.8 are 1 cm apart.

   a) Measure AC.   Fig 16.8

   b) Find the angle AĈB.

   c) Separate your compasses to the length DC. Draw a circle with centre at D. How many dots does this circle go through?

7) Draw a line PQ of length 6 cm. Use your compasses and ruler to find a point R such that PR = 7 cm and QR = $8\frac{1}{2}$ cm. Find the angles of the triangle PQR.

8) Construct a triangle XYZ such that XY = 10 cm, YZ = 11 cm, XZ = 12 cm. Find the angle XŶZ.

9) Construct a triangle ABC such that AB = 5 cm, AC = 12 cm, BC = 13 cm. What is Â?

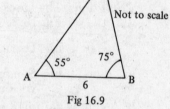

Not to scale

10) Draw a line AB of length 6 cm. Use your protractor to draw a line through A which makes 55° with AB. Draw a line through B which makes 75° with BA.

Let the two lines cross at C. Check that AĈB = 50°. Measure AC and CB.

Fig 16.9

11) Construct a triangle PQR such that PQ = 7 cm, RP̂Q = 82° and RQ̂P = 72°. What are RQ and RP?

12) Construct a triangle XYZ such that XY = 4 in., ZX̂Y = 35°, ZŶX = 47°. Find ZX and ZY.

13) Draw a line AB of length 7 cm. Draw a line through A which makes 46° with AB. Let C be on this line, with AC = $5\frac{1}{2}$ cm.

   a) Find BC.   b) Find AĈB.

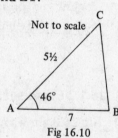

Not to scale

Fig 16.10

14) Construct a triangle PQR with PQ = 7 cm, RP̂Q = 65°, RP = $6\frac{1}{4}$ cm.

   a) Find RQ   b) Find RQ̂P.

15) Construct a triangle XYZ with XY = XZ = 11 cm and YX̂Z = 50°. Find YZ.

Locus

Suppose a point moves according to a rule. The path that the point follows is its *locus*.

If P is equidistant from two points A and B, the locus of P is the perpendicular bisector of the line segment AB.

If P is equidistant from two lines L and M, the locus of P is the bisector of the angle between L and M.

□ **16.1.3 Examples.** *Intermediate, level 7*

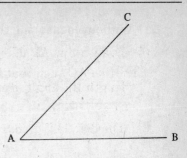

Fig 16.11

1) Make an accurate copy of the lines AB and AC of Fig 16.11.

Draw the bisector of the angle B$\hat{A}$C.

Construct the perpendicular bisector of the line AB.

Let these two new lines meet at D. Measure AD.

Solution The bisector of B$\hat{A}$C is constructed as follows.

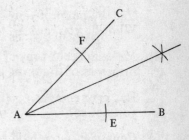

Fig 16.12

    a) Put the point of your compasses on A, and draw an arc cutting AB and AC at E and F.

    b) Draw further arcs with the compass point at E and at F. Join up A with the meet of these further arcs.

The perpendicular bisector of AB is constructed as follows.

    c) Draw equal arcs with the point of your compasses at A and at B, crossing at H and G. Join HG.

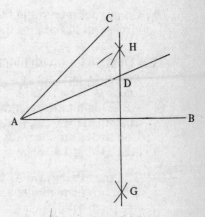

Fig 16.13

Draw further arcs from G and H, and join up their meeting points.

Finally measure AD.

AD = 2.4 cm.

2) A ship is sailing due East at 10 m.p.h. An enemy gun site with a range of 3 miles is 5 miles East and 2 miles South.

Draw a scale diagram, using 1 cm to represent 1 mile, of the ship's course and the region which is within the range of the gun. For how long is the ship in danger?

Solution The positions of ship and gun-site are shown in Fig 16.14.

The ship's course is the horizontal line AB.

The region within range of the gun is the circle of radius 3 cm. with centre at G.

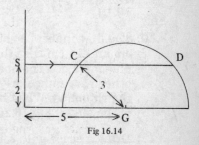

Fig 16.14

The ship is in danger while it is within the circle. The length of the chord CD is 4.5 cm, which represents 4.5 miles.

The ship is in danger for $\frac{4.5}{10}$ = 0.45 hours

□ **16.1.4 Exercises.** *Intermediate, level 7*

1) Construct, using compasses and ruler, the bisectors of the angles below.

(a)   (b)   (c)

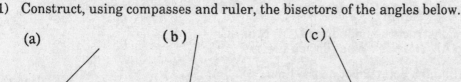

Fig 16.15

2) Without measuring the lengths of the lines below construct their perpendicular bisectors.

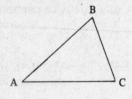

Fig 16.16

3) Make a copy of the triangle ABC of Fig 16.17. Construct as accurately as you can the perpendicular bisectors of the three sides. The three bisectors should all go through the same point G.

   Put the point of your compasses on G and draw a circle radius AG. What do you notice?

Fig 16.17

4) Make another copy of the triangle ABC of Fig 16.17. Construct as accurately as you can the bisectors of the three angles. What do you notice about the intersection points?

   Put your compass point on the intersection point and draw a circle which just touches AB. What do you notice?

5) Construct a triangle ABC with sides 5 cm, 6 cm, 7 cm. By the method of Question 3 construct the circle which goes through the points A, B, C. What is the radius of this circle?

6) Construct a triangle PQR with sides 7 cm, 9 cm, 13 cm. By the method of Question 4 construct the circle which touches the three sides of the triangle. What is the radius of this circle?

7) Construct a quadrilateral ABCD with AB = 5 cm, $\hat{A}$ = 100°, DA = 7 cm, $\hat{B}$ = 120°, BC = 12 cm. Measure DC and $\hat{C}$.

8) ABCD is a parallelogram with AB = 10 cm, AD = 8 cm, and A = 75°. Construct an accurate diagram of ABCD. Fig 16.18

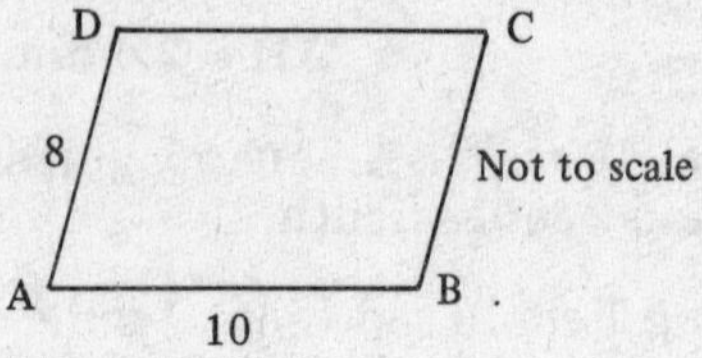

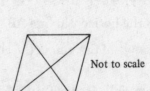

Fig 16.18

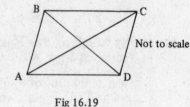

Fig 16.19

9) The parallelogram ABCD has AB = 5 cm, and the diagonals AC and BD are 6 cm and 11 cm respectively. Use the fact that the diagonals of a parallelogram bisect each other to construct ABCD. What is AD? Fig 16.19

10) The diagonals of a rhombus are 10 cm and 14 cm. Use the fact that the diagonals of a rhombus bisect each other perpendicularly to construct the rhombus. What are the lengths of the sides? Fig 16.20.

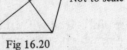

Fig 16.20

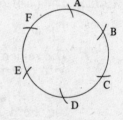

Fig 16.21

11) Draw a circle with radius 5 cm. With your compasses at the same separation mark out points A, B, C, D, E, F on the circumference, as in Fig 16.21. What is the figure ABCDEF? What is the figure ACF? Measure AC. Fig 16.21

12) Draw any triangle ABC. Construct the midpoints L, M, N of BC, CA, AB respectively. Join AL, BM, CN. What do you notice?

13) Fig 16.22 shows a line AB and a point P not on the line.

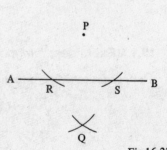

Fig 16.22

   Put the point of your compasses on P and draw an arc to cut AB at two points R and S.

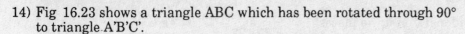

Put the point of your compasses on R and draw an arc below AB. Do the same with S. Let the two arcs meet at Q. Join PQ. What can you say about the lines AB and PQ?

14) Fig 16.23 shows a triangle ABC which has been rotated through 90° to triangle A'B'C'.

The *centre of rotation* must be the same distance from A as from A'.

Construct the perpendicular bisectors of AA' and of BB'. Let them cross at G.

Measure the distances GC and GC'. What do you notice?

Fig 16.23

15) Draw a square ABCD with side 8 cm.

a) Shade the region of those points which are less than 3 cm from the side AB.

b) Shade the region of those points which are less than 2 cm from the point.

c) Mark the point X which is 4 cm from AD and 3 cm from AB.

d) Mark the point Y which is 2 cm from BC and 3 cm from B.

16) Draw a rectangle PQRS with PQ = 10 cm and PS = 8 cm.

a) Shade the region of those points which are less than 2 cm from P.

b) Mark the point Z which is 7 cm from P and 5 cm from Q.

c) Mark the points which are within 2 cm of the diagonal PR and within 4 cm from R.

17) Fig 16.24 shows a scale drawing of a room, in which 1 cm represents 1 m. X marks the only power point. A lamp has a cable of 1.5 m. Make a copy of the diagram, and shade the region corresponding to the places where the lamp could be put.

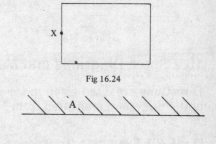

Fig 16.24

18) Fig 16.25 shows a map of two countries, A and B, with sea in between. The scale is 1 cm to 50 km. Make a copy of the diagram.

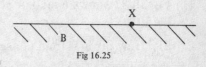

a) Country A does not allow boats from B to fish within 20 km of its coast. Shade the region on the map where B's boats may not fish.

b) A fishing smack can only travel 100 km from its port at X. Shade the region within which it can fish.

Fig 16.25

19) Fig 16.26 shows the diagram of a field, in the scale of 1 cm to 2 m. A goat is tethered to the point of intersection of AC and BD. Make a copy of the diagram.

a) Mark the point where the goat is tethered.

b) The goat is on a 3 m. lead. Shade the region within which it can graze.

c) A fierce dog is kennelled at A, on a leash of 2 m. Shade the region within which the goat is in danger from the dog.

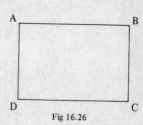

Fig 16.26

20) An old manuscript tells of buried treasure near the walls of an abbey.

The directions are:

The treasure is buried 3 feet from the South wall, and is 5 feet from the oak tree.

Make a copy of the map, and mark on it the places where the treasure might be.

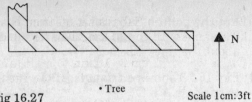

Fig 16.27

• Tree

Scale 1cm: 3ft

21) In the diagrams below, the wheel A moves clockwise at 60 r.p.m. Predict the motion of the wheel B.

a)   b)   c)

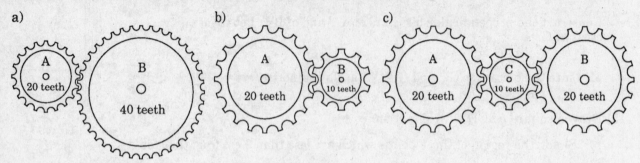

Fig 16.28

22) In the diagrams below, the top pulleys are fixed but free to rotate. The rope at X is pulled downwards at 1 m s⁻¹. Predict the motion of the pulley block Y.

a)   b)   c)

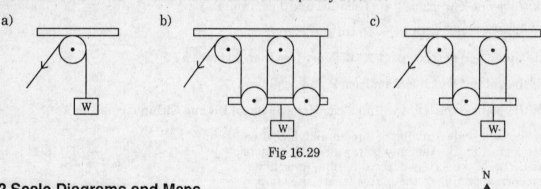

Fig 16.29

16.2 Scale Diagrams and Maps

Directions are often given in the form of a *bearing*. This is the angle of the direction, measured from North in a clockwise direction.

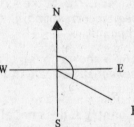

Fig 16.30

☐ **16.2.1 Examples.** *Foundation, levels 4, 5, 6*

1) Fig 16.31 shows a map of Scotland. The scale is 1 cm for 50 miles.

   a) How far is it from Glasgow to Aberdeen?

   b) What is the bearing of Perth from Stranraer?

   Solution   a)   Measure the distance on the map. It comes to 2.2 cm. Multiply this by 50.

   The distance from Glasgow to Aberdeen is 110 miles.

   b)   The line from Stranraer to Perth makes 28° with the North direction.

   The bearing of Perth from Stranraer is 028°.

Fig 16.31

2) Explorers in the Sahara set off from base camp. They travel South for 50 km, then SE for 20 km, then for 30 km on a bearing of 035°.

Use a scale of 1 cm to 10 km to make an accurate drawing of the journey. How far are they from the base camp and what is its bearing?

Solution The first leg of the journey is the 5 cm line AB.

The second stage BC makes 135° with this line.

The third stage CD makes 35° with the North South line.

Measure DA:

They are 50 km from base

Measure the angle DA makes with North South.

$$D\hat{A}B = 38°.$$

Subtract from 360°:

The bearing is 322°

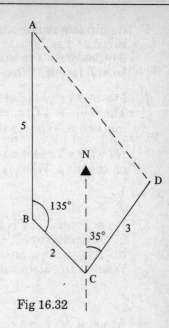

Fig 16.32

□ 16.2.2 Exercises. *Foundation, levels 4, 5, 6*

1) What bearings are the following equivalent to?

a) East   b) South   c) North

d) North East   e) South West.

(2) Express the following bearings in terms of compass directions:

a) 090°   b) 270°   c) 135°   d) 315°.

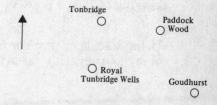

Fig 16.33

3) Fig 16.33 shows part of a map of Kent. The scale is 1 cm per 6 km.

a)   Find the distance from Tonbridge to Goudhurst.

b)   What is the bearing of Paddock Wood from Tunbridge Wells?

4) The scale of Fig 16.34 is 1 cm per 10 km.

a)   What is the distance of Blantyre from Zomba?

b)   What is the bearing of Blantyre from Zomba?

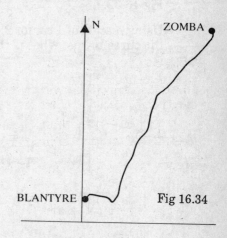

Fig 16.34

5) Fig 16.35 shows a map of Southern England. The scale is 1 cm to 60 km.

a)   What is the distance and the bearing of London from Southampton?

b)   Which town lies on a bearing of 270° from London?

c)   Which town is 50 km from Ipswich?

Fig 16.35

6) A ship leaves harbour and sails North for 20 miles. It then turns and sails East for 10 miles. Using a scale of 1 cm per 5 miles draw a map of the ship's voyage. How far is the ship from harbour? What is the bearing of the ship from the harbour?

7) Fig 16.36 shows part of a map of the Lleyn peninsula. The scale is 1 inch per mile. Gareth starts from Botwnnog, marked A. What is the bearing of the peak at point B?

He walks 2 miles on a bearing of 340°. Mark his journey on the map. What is the distance and bearing of the peak at B?

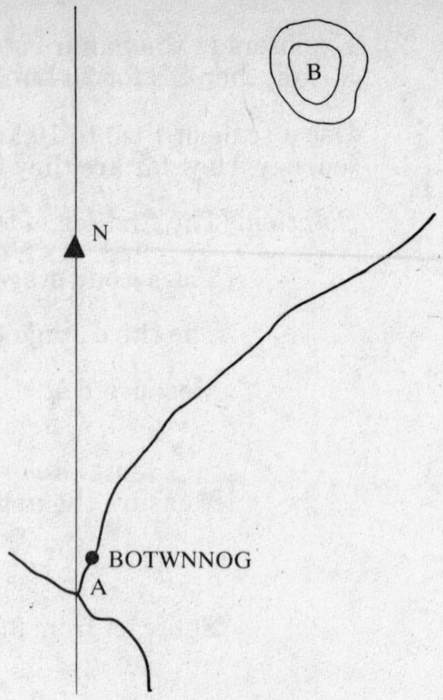

8) From point X the bearing of a target Z is 028°. From point Y, 100 m East of X, the bearing is 340°. Use a scale of 1 cm to 10 m to draw an accurate diagram of the triangle XYZ. What is the distance of Z from X? Fig 16.37.

9) At point A I see a church spire on a bearing of 350°. I walk due North for 200 m to point B and the spire is now on a bearing of 260°. Fig 16.38.

Using a scale of 1 cm per 50 m. make an accurate diagram of my journey. How far is the spire from point B?

10) Aytown is 10 km North of Beetown. The bearing of Ceetown from Beetown is 030° and its bearing from Aytown is 100°. Fig 16.39.

Using a scale of 1 cm for 2 km make an accurate diagram of the three towns. What is the distance of Ceetown from Beetown?

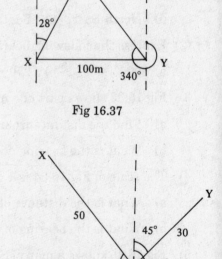

Fig 16.36

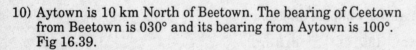

Fig 16.37

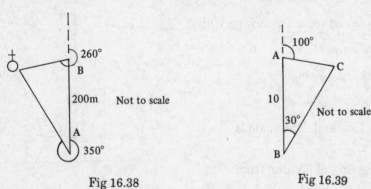

Fig 16.38

Fig 16.39

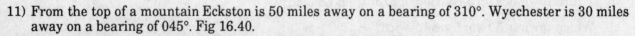

Fig 16.40

11) From the top of a mountain Eckston is 50 miles away on a bearing of 310°. Wyechester is 30 miles away on a bearing of 045°. Fig 16.40.

Using a scale of 1 inch per 10 miles make an accurate map of the mountain and the two towns. What is the distance between the towns? What is the bearing of Wyechester from Eckston?

Common Errors

1) Use of instruments

a) Make sure that you get the best out of your instruments. You cannot make accurate diagrams with a blunt pencil or with a wobbly compass.

b) After making a construction, do not rub anything out. The lines and arcs are part of your answer, and the examiner will want to see them.

c) Do not give your answers to too high a degree of accuracy. With a ruler, you can measure distances to about 0.5 mm. It is misleading to give your answers to a greater degree of accuracy than that.

2) Bearings

Make sure that you measure bearings *clockwise* round from North. Due West, for example, is on a bearing of 270°.

Chapter 17

Areas and Volumes

17.1 Lengths and Areas

The *area* of a rectangle is the product of the length and the breadth.

$$A = l \times b$$

The *perimeter* of a rectangle is the sum of twice the length and twice the breadth.

$$P = 2l + 2b$$

The area of a triangle is half the product of the base and the height.

$$A = \frac{1}{2}b \times h$$

The perimeter or *circumference* of a circle is 2π times the radius.

$$C = 2\pi \times r$$

The area of a circle is π times the square of the radius.

$$A = \pi r^2$$

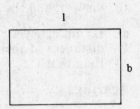

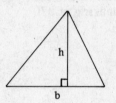

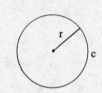

Fig 17.1

☐ **17.1.1 Examples.** *Foundation, levels 4, 5*

1) A rectangular lawn measures 5 metres by 16 metres. What is its area and its perimeter?

   Solution   For the area multiply the length by the breadth.

   $$A = 5 \times 16 = 80 \text{ m}^2.$$

   For the perimeter add twice the length to twice the breadth.

   $$L = 2 \times 5 + 2 \times 16 = 42 \text{ m}.$$

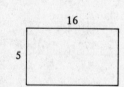

Fig 17.2

2) The dots of Fig 17.3 are 1 cm apart. Find the area of the triangle shown.

   Solution   The base of the triangle is 3 cm. Its height is 2 cm.

   $$A = \frac{1}{2} \times 3 \times 2 = 3 \text{ cm}^2$$

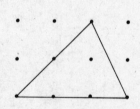

Fig 17.3

3) A circular flower-bed has radius 2 m. Find its area. Take π to be $\frac{22}{7}$.

   Solution   Fig 17.4 shows a diagram of the flower bed.

   $$\text{Area} = \pi 2^2 = 12.6 \text{ m}^2.$$

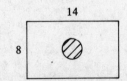

Fig 17.4

4) A string of length 15 inches just goes round a circular post. Find the radius of the post. (Take $\pi = 3.142$).

Solution The circumference of the post is 15 inches. This is 2π times the radius. Divide 15 by 2π:

Radius = 15 ÷ 2π = 2.387 inches.

❏ **17.1.2 Exercises.** *Foundation, levels 4, 5*

1) Find the areas of the rectangles below.

a)
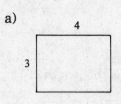

b)
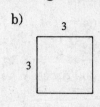

c)
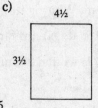

d)
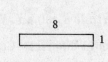

Fig 17.5

2) Find the perimeters of the rectangles in Question 1.

3) A lawn measures 30 feet by 63 feet.

   a) What is its area and its perimeter?

   b) What would it cost to spread it with fertiliser at $\frac{1}{2}$p per square foot?

   c) What would it cost to surround it with a fence costing 35p per foot?

4) A kitchen floor measures $3\frac{1}{2}$ metres by 2 metres. What would it cost to cover it with linoleum at £2.30 per square metre?

5) A bathroom wall is 2 m high and 3 m long. What is its area? What would it cost to tile it with 10 by 10 cm tiles costing 16 p each?

6) One side of a rectangle is 20 cm, and its area is 600 cm^2. What is the length of the other side?

7) The perimeter of a square is 100 m. What is its area?

8) Find the areas of the triangles shown.

a)

b)

c)
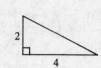

d)
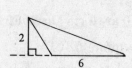

Fig 17.6

9) The dots of Fig 17.7 are 1 cm. apart. Find the areas of the triangles and rectangles.

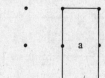
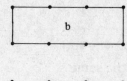

Fig 17.7

10) The base of a triangle is 3 ft., and its area is 18 ft^2. What is its height?

11) The height of a triangle is 5 cm., and its area is 25 cm^2. What is its base?

12) A rectangle on graph paper has its vertices at (1,1), (3,1), (3,4), (1,4). What is its area and its perimeter?

13) A triangle on graph paper has its vertices at (1,1), (5,1), (2,4). What is its area?

14) A triangle on graph paper has its vertices at (2,1), (6,3), (2,4). What is its area?

15) The radius of a running track is 50 m. What is its circumference?

16) The circumference of a race track is 1,200 m. What is its radius?

17) The radius of the Earth is 6,400 km. What is its circumference?

18) A cotton reel is 2.6 cm in diameter. Cotton is wound round it 200 times. What is the length of cotton?

19) Find the areas of the circles with:

   a) Radius 3 cm.   b) Diameter 4 ft.   c) Radius 12 m.   d) Diameter $\frac{1}{2}$ in.

20) Find the radii of the circles with:

   a) Diameter 6 cm.   b) Circumference 12 in.   c) Area 9 m².   d) Area 5 ft².

21) A long-playing record has diameter 12 inches, and the playing surface ends 2 inches from the centre. What is the area of the playing surface?

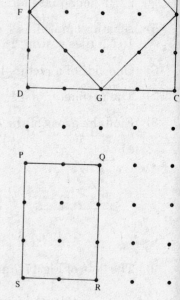

22) A square carpet of side 2 metres is on the floor of a room $2\frac{1}{2}$ metres by 4 metres. What is the area of the uncarpeted floor?

23) A garden is 40 ft by 20 ft. The flower beds are 2 ft wide on all sides. What area of the garden is covered by lawn? Fig 17.8.

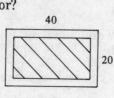

Fig 17.8

24) A picture which is 25 cm by 15 cm is framed within a border 4 cm wide. What is the total area of picture and border?

25) Fig 17.9 shows a semicircle. What is the curved length?

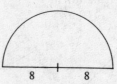

Fig 17.9

26) The dots of Fig 17.10 are 1 cm apart. Find:

   a) The area of the square ABCD.

   b) The area of the triangle AEF.

   c) The area of the square EFGH.

27) a) Find the area of PQRS.

   b) Draw a triangle with equal area to PQRS.

   c) Draw a rectangle with twice the area of PQRS.

   d) Draw a rectangle, with one side of length 1 cm, with the same perimeter as PQRS.

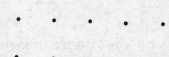

Parallelograms and Trapezia

The area of a parallelogram is the product of the base and the height.

Fig 17.10

$$A = b \times h$$

The area of a trapezium is the product of the height and the average of the parallel sides.

$$A = h \times \frac{(d + e)}{2}$$

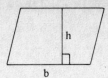
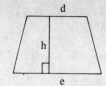

Fig 17.11

❑ **17.1.3 Examples.** *Intermediate, level 7*

1) The dots of Fig 17.12 are 1 cm apart. Find the area of the parallelogram shown.

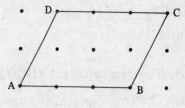

Solution   The base AB is 3 cm. long, and the height is 2 cm.

The area is $3 \times 2 = 6$ cm^2.

Fig 17.12

2) A garden shed is 2 m. from back to front: the back wall is $1\frac{1}{2}$ m. high and the roof slopes up uniformly to the front wall which is $2\frac{1}{2}$ metres high. Find the area of a side wall.

Solution   Fig 17.13 shows a diagram of the shed. The side wall is a trapezium, in which the parallel sides are of length $1\frac{1}{2}$ and $2\frac{1}{2}$. The height is 2.

The area is $2 \times \dfrac{(1\frac{1}{2} + 2\frac{1}{2})}{2} = 4$ m^2

Fig 17.13

❑ **17.1.4 Exercises.** *Intermediate, level 7*

1) Find the area of the parallelograms shown.

(a)
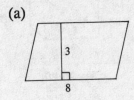

(b)
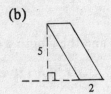

(c)
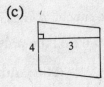

Fig 17.14

2) Find the areas of the trapezia shown.

(a)
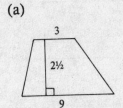

(b)
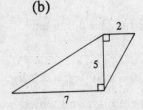

(c)
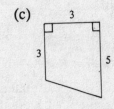

Fig 17.15

157

3) The dots of Fig 17.16 are 1 cm apart. Find the areas of the parallelograms and trapezia in the figure.

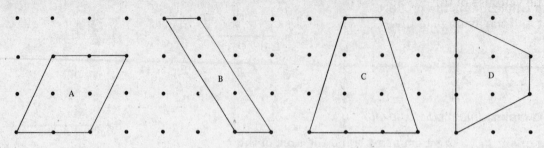

Fig 17.16

4) The parallelogram ABCD has its vertices at (1,1), (2,2), (5,2), (4,1). Find its area.

5) The trapezium ABCD has its vertices at (4,2), (4,7), (3,6), (3,4). Find its area.

6) The kite of Fig 17.17 has diagonals of length 4 cm. and 7 cm. Find its area.

Fig 17.17

17.2 Volumes and Surface Areas

The *volume* of a cuboid is the product of its length, breadth and height.

$$V = l \times b \times h$$

The volume of a cylinder is the product of the height and the area of the base circle.

$$V = \pi r^2 h$$

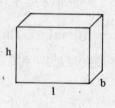

Fig 17.18

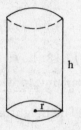

Fig 17.19

The *density* of a solid is its mass divided by its volume.

☐ **17.2.1 Examples.** *Foundation, levels 4, 5*

1) Find the volume of the cuboid of Fig 17.20.

Solution The height is 6 cm., the width 5 cm. and the length 4 cm.

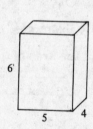

The volume is 4 × 5 × 6 = 120 cm³.

Fig 17.20

2) The surface of a pond is a rectangle 50 cm. by 30 cm. During a cold spell it froze to a depth of 2 cm. Find the mass of ice, given that the density of ice is 0.9 g/cm³.

Solution The ice may be thought of as a cuboid, 50 cm long, 30 cm wide and 2 cm deep. Its volume in cm³ is:

$$50 \times 30 \times 2 = 3{,}000 \text{ cm}^3.$$

Its mass is:

0.9 x 3,000 = 2,700 g. = 2.7 kg.

3) Find the volume of a soup can which is 20 cm high and which has base radius 3 cm.

Solution The can is a cylinder, with $h = 20$ and $r = 3$.

The volume is $\pi 3^2 \times 20 = 565$ cm³.

☐ **17.2.2 Exercises.** *Foundation, levels 4, 5*

1) Find the volumes of the cuboids in Fig 17.21.

(a)   (b)   (c)

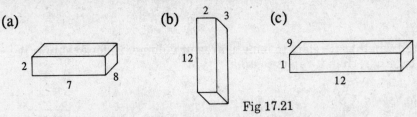

Fig 17.21

2) Squash balls are sold in cubes of side 4 cm. How many of these cubes can be packed in a carton which is 20 cm. by 24 cm. by 16 cm.?

3) A field is in the shape of a rectangle 50 m. by 40 m. Snow falls on the field to a depth of $1\frac{1}{2}$ m. What is the weight of snow when it melts, given that the density of snow is 50 kg/m^3.?

4) A garden is a rectangle 6 m wide by 13 m long. 0.04 m of rain falls: what is the volume of water?

5) A cuboid has a square base of side 3 cm. What is its height, given that its volume is 36 cm^3.?

6) A water tank is a cuboid, with a base of 1.2 m. by 0.8 m. How deep is the water when the tank contains 0.384 m^3 of water?

7) A classroom is 5 m. by 6 m. by 3 m. Health regulations require that each pupil must have at least 5 m^3 of air. How many pupils can the classroom legally hold?

8) Sugar cubes are cubes of side $1\frac{1}{2}$ cm. How many fit in a box which is 24 cm. by 6 cm. by 12 cm.? If the density of sugar is 1.5 g cm^{-3}, what is the mass of one sugar cube?

9) The dots of Fig 17.22 are 1 cm apart. (a) is the net of a cube and (b) is the net of a cuboid. Find their surface areas and their volumes.

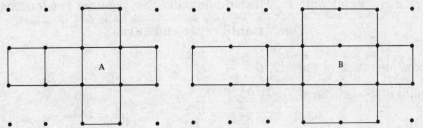

Fig 17.22

10) Find the volumes of the cylinders of Fig 17.23, giving your answers to 3 significant figures.

(a)   (b)   (c)

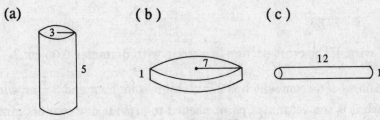

Fig 17.23

11) A tin of shoe-polish is 2 cm. high, and the diameter of its base is 7 cm. Find its volume.

Prisms and Dimensions

The volume of a prism is the product of the height and the area of the cross-section.

$$V = A \times h$$

By looking at a formula we can tell whether it refers to length, area or volume. If, for example, it involves two lengths multiplied together, then it refers to area.

Fig 17.24

☐ **17.2.3 Examples.** *Intermediate, levels 7, 8*

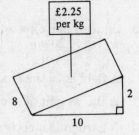

1) Fig 17.25 shows a wedge of cheese, with the dimensions as shown. Find its price, given that the density of the cheese is 1.5 g/cm³ and that it costs £2.25 per kilogram.

   Solution The wedge is a triangular prism. The cross-sectional area is a right-angled triangle, with area $\frac{1}{2} \times 2 \times 10 = 10$ cm².

   $$\text{Volume} = 8 \times 10 = 80 \text{ cm}^3.$$

   $$\text{Weight} = 1.5 \times 80 = 120 \text{ grams} = 0.12 \text{ kg}.$$

   Price = 0.12 × 2.25 = £0.27 = 27 p.

   Fig 17.25

2) A room is 2 m. high, 4 m. long and 3 m. wide. Find the surface area of the walls and ceiling.

   Solution The ceiling is 4 by 3. There are two walls which are 2 by 4, and two walls which are 2 by 3. The total area is therefore:

   $$3 \times 4 + 2 \times 2 \times 4 + 2 \times 2 \times 3 = 40 \text{ m}^2$$

3) A formula connected with a cylinder is $2\pi r(r + h)$, where r and h represent lengths. Does this formula represent length, area or volume?

   Solution 2 and π are plain numbers. The formula multiplies together two lengths.

   The formula represents area

☐ **17.2.4 Exercises.** *Intermediate, levels 7, 8*

1) Find the volumes of the prisms of Fig 17.26.

   (a) 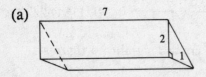 (b)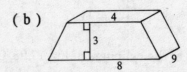

   Fig 17.26

2) What is the volume of 1,000 cm. of wire, if the cross-section is a circle with diameter 0.05 cm.?

3) What is the area of the walls and ceiling of a room which is $2\frac{1}{2}$ m. high, 4 m. long and $3\frac{1}{2}$ m. wide?

   If a coat of paint is 0.01 cm thick, what is the volume of paint needed to provide one coat of paint? (Give your answer in cubic centimetres.)

4) What is the surface area of a cube of side 3 cm.?

5) A tank is 1 m. high and its base is a rectangle which is 0.8 by 0.6 m. What is the surface area of the sides and base of the tank?

6) Fig 17.27 shows the cross section of a copper pipe. The inside radius is 2 cm and the outside radius is $2\frac{1}{2}$ cm.

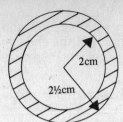

Fig 17.27

   a) Find the shaded area.

   b) Find the volume of copper in 40 m. of the pipe.

7) An electrical cable consists of a copper cylinder surrounded by rubber. The diameter of the copper is $1\frac{1}{2}$ cm, and the rubber is $\frac{1}{4}$ cm. thick. The cable is 100 m. long.

   a) Find the volume of the copper.

   b) Find the total volume of the cable.

   c) Find the volume of the rubber.

8) The base radius of a cylinder is 2 cm. If its volume is 25 cm^3, what is its height?

9) What length of wire with radius of cross-section 0.5 mm. can be made from 20 cm^3 of copper?

10) A cylinder which holds 1 litre has height 20 cm. What is the area of its cross-section?

11) A measuring jar is a cylinder with radius of cross-section 5 cm. If 500 c.c. of water is poured into it, how deep will the water be?

12) Fig 17.28 shows a shed which is 3 m. long, $1\frac{1}{2}$ m. deep, and 2 m. high at the back and $1\frac{3}{4}$ m. high at the front.

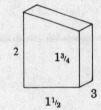

Fig 17.28

   a) Find the area of a side wall.

   b) Find the volume of the shed.

13) Wire is made with a cross-sectional area of 0.1 cm^2. What length of this wire can be made from 100 cm^3 of metal? How many times will the wire wrap round a cylinder of radius 4 cm?

14) x and y represent lengths: A represents area: V represents volume. Which do the following represent?

   a) xy b) xA c) $\dfrac{V}{A}$

   d) $4\pi x^2 y$ e) $\dfrac{A}{(x+y)}$ f) $\dfrac{A^2}{y}$

Spheres, Cones, Pyramids

The volume V and the surface area A of a sphere with radius r are given by:

$$V = \frac{4}{3}\pi r^3$$

$$A = 4\pi r^2.$$

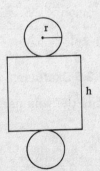

The surface area of a cylinder, with height h and base radius r, is the sum of the areas of the two circles at the ends and the curved surface.

$$A = 2\pi r^2 + 2\pi rh$$

The volume V of a cone with base radius r and height h is given by:

$$V = \frac{1}{3}\pi r^2 h$$

Fig 17.29

The volume V of a pyramid with base area A and height h is given by:

$$V = \frac{1}{3}Ah$$

□ **17.2.5 Examples.** *Higher, level 9*

1) A hollow rubber ball has outside radius 6 cm. and inside radius $5\frac{1}{2}$ cm. Find the volume of rubber.

   Solution The volume is found by subtracting the volume of the hollow sphere inside from the volume of the whole ball.

   $$\textbf{Volume} = 4\pi\frac{6^3}{3} - 4\pi\frac{(5\frac{1}{2})^3}{3} = \textbf{208 cm}^3$$

2) A cone with height 6 cm. and base radius 3 cm. is immersed in a cylinder of water with base radius 7 cm. By how much does the water rise?

   Solution See Fig 17.30. The extra volume introduced is the volume of the cone:

   $$\frac{1}{3}\pi 3^2 \times 6 = 18\pi \text{ cm}^3$$

   This extra volume adds an extra cylinder of water. The base radius of the extra cylinder is still 7 cm. The height of the extra cylinder is found by dividing the extra volume by the base area.

   Fig 17.30

   $$\textbf{Rise} = 18\pi \div (\pi 7^2) = 18 \div 49 = \textbf{0.367 cm.}$$

3) A pyramid is of height 12 in. and its base is a square of side 9 in. It is cut by a plane parallel to the base and 4 inches from the peak. What is the volume of the lower part?

   Solution Fig 17.31 shows the solid. The top part is similar to the original pyramid. It is a third of the height of the original. The side of its base must be a third of the original side.

   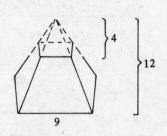

   Subtract the volume of the small pyramid from the volume of the original pyramid.

   Fig 17.31

   $$\textbf{Volume of lower part} = 12 \times \frac{9^2}{3} - 4 \times \frac{3^2}{3} = \textbf{312 in}^3$$

□ **17.2.6 Exercises.** *Higher, level 9*

1) Find the volumes of the spheres with radii:

   a) 5 cm.   b) 2 m.   c) $7\frac{1}{2}$ in.

2) Find the surface areas of the spheres in Question 1.

3) Find the surface areas of the cylinders in Question 10 of 17.2.2.

4) A cube has volume 24 cm³. What is its side?

5) A cylinder of height 5 in. has volume 64 cm³. What is the radius of its base?

6) Find the radii of the spheres with volumes:

   a) 46 cm³   b) 12 in³   c) 5.56 mm³

7) A sphere has volume 3.44 cm³. What is its surface area?

8) A sphere has surface area 66 mm². What is its volume?

9) The great pyramid of Gizah has a square base of side 230 m. and is 148 m. high. What is its volume?

10) A cone has base radius 6 cm. and is 7 cm. high. What is its volume?

11) A sphere of radius 20 cm. is painted. A coat of paint is 0.01 cm thick. Evaluate the volume of paint by the following two methods:

   a) Multiply the surface area of the sphere by the thickness of the coat of paint.

   b) Subtract the volume of the sphere before painting from the volume of the sphere after painting.

12) A cylindrical jar with a base radius of 5 cm contains water to a depth of 10 cm. Find the rise in water level if the following are immersed:

   a) A sphere of radius 2 cm.

   b) A cube of side $1\frac{1}{2}$ cm.

   c) A cone with base radius 3 cm and height 4 cm.

13) A hemispherical bowl of radius 7 cm. is full of water. It is then emptied into a cylindrical jar of base radius 5 cm. How deep is the water in the cylinder?

14) Paper which is 0.03 cm thick is rolled round a cylindrical core of radius 8 cm. The completed roll of paper is a cylinder with radius 20 cm. Find the length of the paper.

15) A certain sort of ball is a sphere with radius $2\frac{1}{2}$ cm. It is made by winding rubber string round a spherical core of radius 1 cm.

   a) Find the volume of rubber.

   b) If the rubber string has area of cross-section 0.02 cm² find the length of rubber.

16) A cylindrical pipe delivers 2 litres of water per second. If the radius of cross-section is $1\frac{1}{2}$ cm find the speed of water in the pipe.

17) A hose-pipe delivers 3 litres of water per second, at a speed of 17 m/sec. What is the radius of cross-section of the pipe?

18) Fig 17.32 shows a cone which has been cut in two by a plane parallel to its base, halfway from the base. The original height was 24 cm and the base radius is 4 cm.

   a) Find the volume of the original cone.

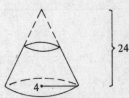

Fig 17.32

b)   Find the radius of the cut circle.

c)   Find the volume of the cone which has been removed.

d)   Find the volume of the remaining solid.

Show that the volume of the remaining solid is $\frac{7}{8}$ of the original volume.

19) A cone is cut in two by a plane parallel to its base and $\frac{1}{4}$ of the height down from the apex.

a)   Show that the upper cone is similar to the original cone.

b)   Show that the upper cone has volume $\frac{1}{64}$ of the original volume.

c)   What is the ratio of the volumes of the two parts?

20) A yoghourt carton consists of the lower half of a cone after the top has been cut off. The height of the carton is 10 cm, and the radii at top and bottom are 2 cm and 4 cm respectively.

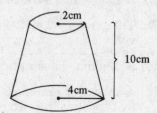

a)   Find the height of the original cone.

b)   Find the volume of the carton.

Fig 17.33

21) If a circle is rotated about a diameter, the solid region it passes through is a sphere.

a)   If an isosceles triangle is rotated about its axis of symmetry, what solid region does it pass through?

b)   What plane figure would you rotate in order to pass through a cylindrical region of space?

17.3 Combinations of Figures

❏ **17.3.1 Examples.** *Intermediate, level 7*

1)   Find the area of the shape in Fig 17.34.

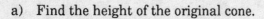

Solution   Divide the shape into two rectangles, one 10 by 2, the other 6 by 1.

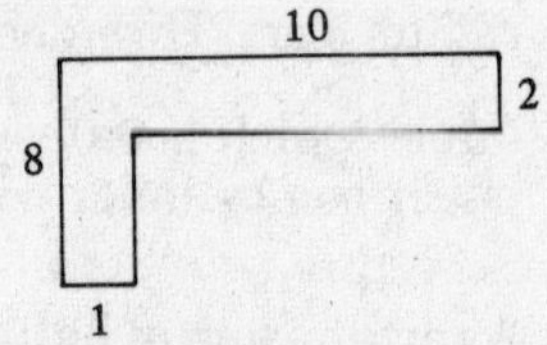

The area is $10 \times 2 + 6 \times 1 = 26$

Fig 17.34

2)   A barn is 12 m. long, 3 m. deep and $2\frac{1}{2}$ m. high. The roof consists of half a cylinder. Find the total volume.

Solution   The shape is shown in Fig 17.35. The side walls are in the shape of a semi-circle of diameter 3 m. over a 3 by $2\frac{1}{2}$ rectangle.

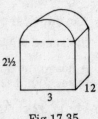

$$\text{Surface area of end wall} = \pi\,\frac{(1\frac{1}{2})^2}{2} + 3 \times 2\frac{1}{2} = 11 \text{ m}^2.$$

Fig 17.35

Volume = length × surface area of end = $12 \times 11 = 132$ m^3

❑ **17.3.2 Exercises.** *Intermediate, level 7*

1) Find the areas of the shapes in Fig 17.36.

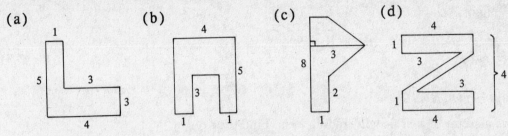

(a) (b) (c) (d)

Fig 17.36

2) The dots of Fig 17.37 are 1 cm apart. Find the areas of the shapes shown.

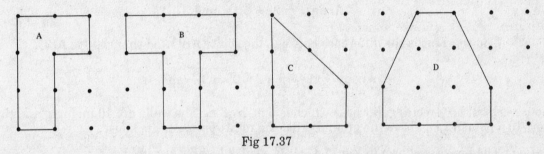

Fig 17.37

3) A running track consists of two parallel straight sections of length 50 m, which are 20 m apart, and semi-circular ends. Find the total length and the area enclosed by the track. Fig 17.38.

4) A steel girder has the cross-section shown in Fig 17.39. Find the area of cross section and the volume of 6 m. of the girder.

5) The end wall of a hut is shown in Fig 17.40. Find the area of the wall. If the hut is $2\frac{1}{2}$ m. long find its volume.

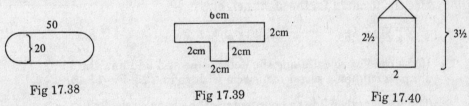

Fig 17.38   Fig 17.39   Fig 17.40

6) A window consist of a semi-circle on top of a rectangle: the rectangle is 60 cm high and 40 cm wide. Find the area of the window.

7) A box is made out of wood 1 cm thick, so that its interior dimensions are 20 by 15 by 12, and its exterior dimensions are 22 by 17 by 14. By subtracting the interior volume from the exterior volume find the volume of the wood.

Cones, Sectors and Arcs

Suppose a cone has radius r and height h. Then its *slant height* is the distance from the rim to the top. It is given by the formula:

$$l = \sqrt{r^2 + h^2}$$

The curved surface area of the cone is:

$$A = \pi r l.$$

Suppose an arc subtends θ at the centre of a circle of radius *r*.

$$\text{Arc length} = 2\pi r \times \frac{\theta}{360}$$

$$\text{Sector area} = \pi r^2 \times \frac{\theta}{360}$$

Fig 17.41

❐ 17.3.3 Examples. *Higher, level 9*

1) Fig 17.42 shows a sector of a circle with radius 2 cm. Find the area of the sector ABC and of the shaded segment.

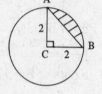

Solution   The area of the sector is a quarter of the area of the circle.

$$\textbf{Area} = \frac{1}{4}\pi 2^2 = \textbf{3.14 cm}^2$$

Fig 17.42

The segment is the difference between the sector ABC and the triangle ABC.

$$\textbf{Area} = \frac{1}{4}\pi 2^2 - \frac{1}{2} \times 2 \times 2 = \textbf{1.14 cm}^2$$

2) A house is built on a rectangular base which is 6 m. by 8 m. The walls are 10 m. high, and the roof tapers to a point 12 m. above the ground. Find the total volume of the house.

Solution   The house is shown in Fig 17.43. It is a cuboid with a pyramid on top. The total volume is:

$$\textbf{6} \times \textbf{8} \times \textbf{10} + \textbf{6} \times \textbf{8} \times \frac{\textbf{2}}{\textbf{3}} = \textbf{512 m}^3$$

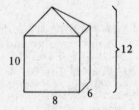

Fig 17.43

3) A cone is 24 cm high and its base radius is 7 cm. Find its slant height. If the curved area were laid out flat, what would be the angle of the sector obtained?

Solution   Use the formula for the slant height:

$$l = \sqrt{7^2 + 24^2} = 25.$$

If the cone were cut along the dotted line and laid flat, the shape would be a sector of a circle with radius 25. Fig 17.45.

The arc length AB is the perimeter of the base circle, $2\pi 7$.

Reduce 360° in the ratio $2\pi 7 : 2\pi 25$

$$\textbf{Angle of sector} = \textbf{360}° \times \frac{\textbf{7}}{\textbf{25}} = \textbf{100.8}°$$

Fig 17.44

❐ 17.3.4 Exercises. *Higher, level 9*

1) Find the areas of the sectors in Fig 17.46.

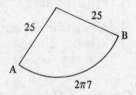

Fig 17.45

(a)   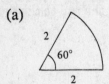   (b)   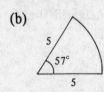   (c)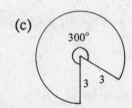

Fig 17.46

2) Find the arc lengths of the sectors in Question 1.

3) A square is drawn inside a circle of radius 5 cm. Find the area of the square, and hence find the area between the circle and the square. Fig 17.47.

4) Four cog wheels of radius 2 cm. have their centres at the corners of a square of side 5 cm. A chain passes tightly round the wheels: how long is the chain? Fig 17.48.

5) In Fig 17.49 part of a circle of radius 1 has been drawn from the centre (1,1), so that it touches the x-axis and the y-axis. The shape between the axes and the circle is shaded. Find the area and the perimeter of the shaded region.

6) In Fig 17.50 a square of side 8 cm. has had one corner cut by a circle radius 4 cm. The centre of the circle is at the centre of the square. Find the area and the perimeter of the shape.

Fig 17.47

Fig 17.48

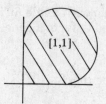
Fig 17.49

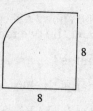
Fig 17.50

7) A test-tube consists of a cylinder with a hemisphere at one end. The length of the cylinder is 20 cm, and its radius of cross-section is 1 cm. Find the volume of the test-tube and its surface area. Fig 17.51.

8) Before it is sharpened, a pencil is a cylinder 15 cm long with diameter 1 cm. Find its volume.

It is now sharpened, so that the last 2 cm of the pencil forms a cone. How much volume has been lost? Fig 17.52.

9) An cornet ice-cream is a cone surmounted by a hemisphere. The total height is 5 inches and the radius of the hemisphere is 1 inch. Find the total volume. Fig 17.53.

10) A peg-top consists of a hemisphere on top of a cone. The hemisphere has radius 2 cm. How high should the cone be if it is to have the same volume as the hemisphere? Fig 17.54.

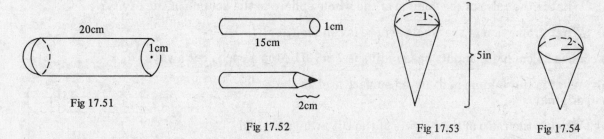
Fig 17.51   Fig 17.52   Fig 17.53   Fig 17.54

11) A boiler consist of a cylinder, with base radius 20 cm and length 50 cm, with hemispheres at both ends. Find its volume and its surface area. Fig 17.55.

12) A sphere of radius 3 cm fits exactly inside a cylinder of base radius 3 cm and height 6 cm. Show that their surface areas are equal. Find the ratio of their volumes. Fig 17.56.

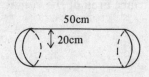
Fig 17.55

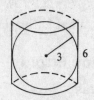
Fig 17.56

13) A child makes a model of a 'Dalek' as follows. He takes a cone with base radius 5 cm and height 15 cm and cuts off the top 9 cm. The top now has radius 3 cm. Fig 17.57.

The cone he has cut off is replaced by a hemisphere with the same radius. Find the volume of the 'Dalek'.

14) Find the curved surface area of a cone with height 4 cm and base radius 3 cm.

Fig 17.57

15) Find the angle of the sector obtained when the surface of Question 14 is laid flat.

16) A tent is cone shaped, with height 8 ft and base radius 6 feet. Find its volume, and the area of canvas of its sides.

17) Fig 17.58 shows a sector of a circle with angle 60° and radius 8 cm. It is folded and the two radii are joined to form a cone.

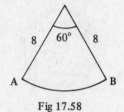

a)   What is the arc length AB?

b)   What is the base radius of the cone?

c)   What is the height of the cone?

Fig 17.58

d)   What is the volume of the cone?

18) One sphere has twice the radius of another. What is the ratio of their surface areas? What is the ratio of their volumes?

19) A cylinder has its base radius halved and its height tripled. What has the volume been multiplied by?

20) A cone has its height halved and its base radius multiplied by 4. What has the volume been multiplied by?

21) Two golf balls differ by 1% in their radii. Find the ratio of their radii. What is the ratio of their areas? By what percentage do they differ in area? By what percentage do they differ in volume?

22) A sphere of radius 10 cm is hollow, the cavity being a sphere of radius 5 cm.

a)   What is the ratio of the volume of the whole sphere to the volume of the cavity?

b)   What proportion of the whole sphere is solid matter?

23) A cone is 24 cm high, and its base radius is 7 cm. The top 4 cm. is cut away.

a)   What is the volume and curved surface area of the original cone?

b)   What is the ratio of the heights of the cut away cone and the original cone?

c)   What is the ratio of the surface areas of the cut away cone and the original cone?

d)   What is the curved surface area of the remaining solid?

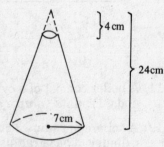

Fig 17.59

Common Errors

1) **Areas**

a)   Be careful that you do not confuse the radius and the diameter of a circle. Read the question carefully to see which is being used.

b) The area of a triangle or a parallelogram is found by the product of the base and the *height*. It is not found by the product of adjacent sides, unless the triangle or parallelogram happens to be right-angled.

c) Mistakes are common with questions in which the units are mixed. It is safest to express all the measurements in the same units.

2) Volumes

The formulas for the volumes of cones and pyramid refer to the height of the solid. This means the *vertical* height, not the slant height.

3) Combinations

If you have two similar solids, then the ratio of their lengths is not the same as the ratio of their areas or of their volumes.

If the length ratio is $a{:}b$, then the area ratio is $a^2{:}b^2$ and the volume ratio is $a^3{:}b^3$.

Chapter 18

Trigonometry

The equations of *trigonometry* connect together the sides and angles of a triangle.

18.1 Right-angled Triangles

The right-angled triangle ABC of Fig 18.1 is labelled as shown.

The ratios between the sides are functions of the angle P°.

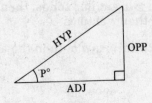

Fig 18.1

$\sin P° = \dfrac{OPP}{HYP}$ (The *sine* of P°)

$\cos P° = \dfrac{ADJ}{HYP}$ (The *cosine* of P°)

$\tan P° = \dfrac{OPP}{ADJ}$ (The *tangent* of P°)

These can be remembered by the "word":

OHSAHCOAT

(**O**pposite over **H**ypoteneuse equals **S**ine, **A**djacent over **H**ypoteneuse equals **C**osine, **O**pposite over **A**djacent equals **T**angent)

or SOHCAHTOA

(**S**ine is **O**pp over **H**yp, **C**os is **A**dj over **H**yp, **T**an is **O**pp over **A**dj).

These equations give the ratios in terms of the angle. The *inverse functions* give the angle in terms of the ratios.

$$P° = \sin^{-1}\frac{OPP}{HYP} = \cos^{-1}\frac{ADJ}{HYP} = \tan^{-1}\frac{OPP}{ADJ}.$$

When using a calculator to find ratios, the angle is pressed first. To find sin 43°, press the following buttons:

| 4 | 3 | sin |

The answer 0.682 will appear.

When using a calculator to find angles, the ratio is pressed first. To find cos⁻¹0.3, press the following buttons:

| . | 3 | inv | cos |

The answer 72.5° will appear.

❑ **18.1.1 Examples.** *Intermediate, level 8*

1) In the triangle ABC of Fig 18.2, $\hat{A} = 90°$, $\hat{B} = 53°$, and BC = 7 cm. Find AC, to three significant figures.

    Solution   Comparing with Fig 18.1, the HYP side is BC, OPP is AC, ADJ is AB. Using the equations above:

$$\sin 53° = \frac{\text{OPP}}{7}$$

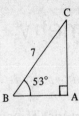

Fig 18.2

Use tables or a calculator to find that:

$$\sin 53° = 0.7986$$

This gives:

$$0.7986 = \frac{\text{OPP}}{7}$$

AC = 7 × 0.7986 = 5.59

2) In the triangle DEF of Fig 18.3 $\hat{E} = 90°$, $\hat{F} = 43°$, DF = 9 metres. Find EF, giving the answer to three significant figures.

Solution In this figure the triangle is not in the same position as in Fig 18.1. The sides are labelled as follows:

DF = HYP

(Because it is farthest from the right-angle)

ED = OPP

(Because it is farthest from the angle of 43°)

EF = ADJ

(Because it is next to the angle of 43°)

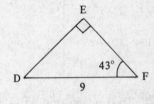

Fig 18.3

EF = 9 × cos 43° = 9 × 0.7314 = 6.58 m

3) In triangle GHI, $\hat{H} = 90°$, HG = 3, HI = 5. Find the angle $\hat{I}$, to the nearest degree.

Solution Here OPP = GH = 3, and ADJ = HI = 5. This gives the equation:

$$\tan I = \frac{\text{OPP}}{\text{ADJ}} = \frac{3}{5}.$$

Use the inverse tan button:

I = tan⁻¹ 0.6 = 31°

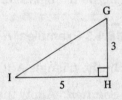

Fig 18.4

❑ **18.1.2 Exercises.** *Intermediate, level 8*

1) Use your calculator to find the following, giving the answers to 3 decimal places:

a) sin 23°   b) cos 47°   c) tan 75°   d) sin 64.34°   e) cos 0.934°

2) Use your calculator to find the following, giving the answers to the nearest tenth of a degree:

a) sin⁻¹0.8   b) cos⁻¹0.24   d) tan⁻¹0.63   e) tan⁻¹4.9

3) In the triangles shown, find the unknown sides.

Fig 18.5

171

4) In the triangles shown, find the unknown sides.

  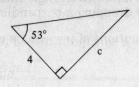

Fig 18.6

5) In the triangles shown, find the angles.

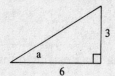

Fig 18.7

6) In the triangles shown, find the angles.

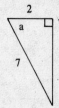 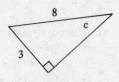

Fig 18.8

7) In the triangle ABC, $\hat{A}$ = 90°, $\hat{B}$ = 27° and BC = 8 cm. Find AB and CA.

8) In the triangle PQR, $\hat{P}$ = 90°, QR = 57 metres and QP = 23 metres. Find the angles at Q and at R.

9) ABC is a triangle in which $\hat{A}$ = 90°, AB = 17 cm and AC = 12 cm. Find the other angles.

10) In the triangle LMN, $\hat{L}$ = 90°, $\hat{M}$ = 43° and MN = 83 m. Find LM and LN.

☐ **18.1.3 Examples.** *Higher, levels 8, 9*

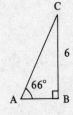

1) In the triangle ABC of Fig 18.9, $\hat{B}$ = 90°, $\hat{A}$ = 66°, BC = 6 cm. Find BA.

Solution Apply the basic formula:

$$\tan 66° = \frac{6}{BA}$$

Multiply both sides by BA:

$$BA \times \tan 66° = 6$$

Now *divide* both sides by tan 66°:

$$BA = \frac{6}{\tan 66°}$$

$$BA = \frac{6}{2.2460} = 2.67$$

Fig 18.9

2) In the triangle ABC of Fig 18.10, AB = AC = 8cm, BC = 4 cm. Find the angles of the triangle.

Solution The trigonometric ratios are only defined for a right-angled triangle.

Convert the isosceles triangle into two right-angled triangles by dropping a perpendicular. Fig 18.11.

Now the equations of trig can be used.

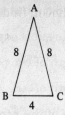
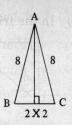

$$\sin B\hat{A}X = \frac{2}{8}$$

$$B\hat{A}X = \sin^{-1}\frac{1}{4} = 14.5°.$$

Fig 18.10   Fig 18.11

$$\hat{A} = 2 \times 14.5 = 29° \text{ and } \hat{B} = \hat{C} = 75.5°$$

☐ **18.1.4 Exercises.** *Higher. levels 8 and 9*

1) Find the sides of the triangles shown.

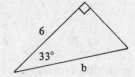

Fig 18.12

2) Find the angles of the triangles shown.

Fig 18.13

3) In the quadrilateral ABCD, $A\hat{C}B = D\hat{A}C = 90°$, AB = 10 cm, $A\hat{B}C = 50°$ and $D\hat{C}A = 40°$. Find AC and DC.

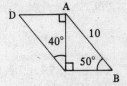

Fig 18.14

4) For the triangle ABC, BC = 3 and $\hat{B} = \hat{C} = 57°$. Find AB.

5) For the triangle ABC, AB = AC = 7, BC = 4. Find $\hat{A}$.

6) The diagonals of a rhombus are 8 cm and 4 cm in length. Find the angles of the rhombus.

7) A pair of compasses is 3 inches in length, and can be separated to 54°. Find the greatest separation of the points.

8) When the neck of a giraffe is at 20° to the vertical, it can reach leaves which are $4\frac{1}{2}$ metres above the ground. Assuming that the shoulders of the giraffe are 2 metres high, how long is its neck?

9) Pincers whose blades are 20 cm long can just grasp a pipe of radius 5 cm. Find the angle of separation of the blades.

10) A line joins the point (1,5) and (6,6). Find the angle this line makes with the *x*-axis.

11) Find the angle that the line $y = 2x + 3$ makes with the *x*-axis. Find the angle between this line and $y = \frac{1}{2}x - 4$.

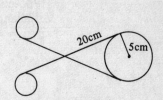

Fig 18.15

12) In a circle of radius 12 cm a chord AB of length 7 cm is drawn. Find the angle this chord subtends at the centre of the circle.

13) A circle of radius 6 cm has a regular pentagon ABCDE inscribed in it.

    a) Find the angle subtended at the centre by AB.

    b) Find the length of the perpendicular from the centre to AB.

    c) Find the length of AB.

    d) Find the area of the pentagon.

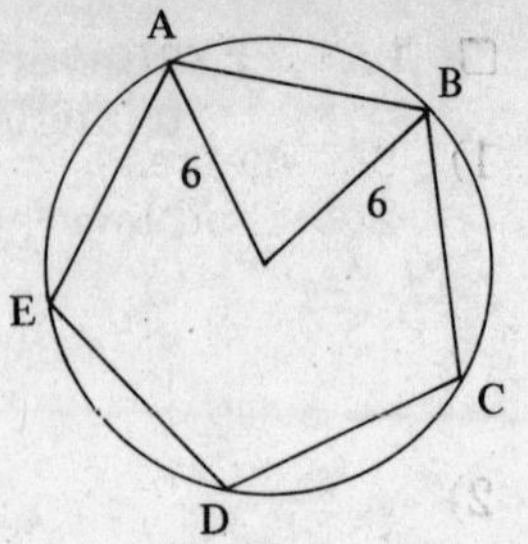

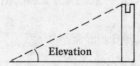

Fig 18.16

18.2 Elevation, Depression, Bearings

When we look up at something, the angle our line of vision makes with the horizontal is called the *angle of elevation*.

When we look down on something, the angle between the horizontal and our line of vision is called the *angle of depression*.

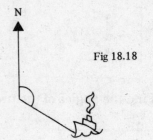

Fig 18.17

A direction can be described by the angle its path makes with due North, measured clockwise. This angle is called its *bearing*. Fig 18.18.

Fig 18.18

☐ **18.2.1 Examples.** *Intermediate, level 8*

1) From the top of a cliff, the angle of depression of a boat is 5°. If the boat is 2000 m. out to sea, how high is the cliff?

    Solution Let *d* be the height of the cliff. From the picture:

$$\tan 5° = \frac{d}{2000}$$

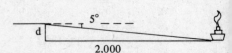

$$d = 2000 \times \tan 5° = 175 \text{ m.}$$

Fig 18.19

2) A kite on 75 m of string is 55 m. high. What angle does it make with the horizontal?

    Solution Let P° be the angle. From the picture:

$$\sin P° = \frac{55}{75}$$

$$P° = \sin^{-1}\frac{55}{75} = 47°.$$

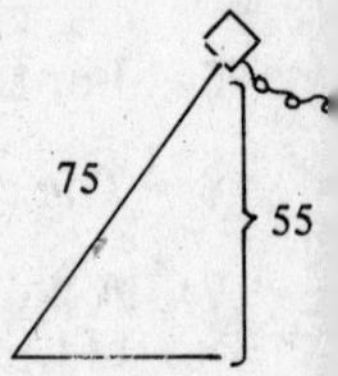

Fig 18.20

3) A ship sails 50 km on a bearing of 070°. How far North and how far East has it gone?

    Solution In the triangle as shown, the actual path of the ship is the Hypoteneuse of the triangle. The North distance N is the Adjacent side and the Easterly distance E is the Opposite side.

$$\cos 70° = \frac{N}{50}$$

$$\sin 70° = \frac{E}{50}$$

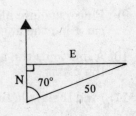

Fig 18.21

N = 17.1 km and E = 47 km.

☐ **18.2.2 Exercises.** *Intermediate, level 8*

1) 200 metres from the base of a tower the angle of elevation of the top is 24°. Find the height of the tower.

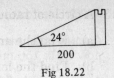

200

Fig 18.22

2) A cliff is 75 metres high, and a boat is 3000 metres out to sea. Find the angle of depression of the boat from the cliff.

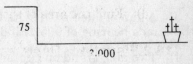

Fig 18.23

3) A ladder of length 2 metres leans against a wall at an angle of 35° to the vertical. Find how far up the wall it reaches.

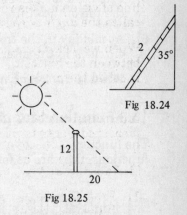

Fig 18.24

4) A flagpole 12 feet high throws a shadow of length 20 feet. What is the angle of elevation of the sun?

5) A girl's eyes are 5 feet above the ground. When she is 30 feet from the base of a tree, the angle of elevation of the top is 35°. Find the height of the tree.

Fig 18.25

6) Two minarets of a mosque are 40 m. and 55 m. in height. Their bases are 30 m. from each other. What is the angle of depression of the lower minaret from the higher?

7) A terrace of seats at a football ground is 30 metres long and rises 4 metres from top to bottom. Find the angle of slope of the terrace.

8) A girder is 15 metres long. One end is on the ground. When the other end of the girder is 7 metres above the ground, what angle does it make with the vertical?

9) A rocket is fired at an elevation of 50°. If it travels at 40 metres per second, how high is it after 5 seconds?

10) I stand on the bank of a river, and see a tree directly opposite me on the other side. I now walk 20 metres upstream, and the line from me to the tree makes 83° with the bank. How wide is the river?

11) 100 m. from a church the base of the spire has angle of elevation 12°, and the top of the spire has elevation 23°. What is the height of the spire?

12) The gradient of a hill is 1 in 10. (This means that we go 1 unit vertically for every 10 units horizontally.) Find   a)   the angle of slope   b)   How far we have climbed after walking 100 metres.

13) A plane flies at 6,000 metres above the ground. I spot it directly above me, and 10 seconds later its angle of elevation is 72°. How fast is the plane flying?

14) A line is drawn from the origin to the point (5,2). What angle does this line make with the *x*-axis?

15) A plane flies 400 km at 057°. How far North and how far East is it from the starting point?

16) I walk 5 miles North and then 3 miles West. What is the bearing of my starting point?

17) A ship is 30 km due South of a port, and is sailing on a bearing of 048°. It sails until it is due East of the port. How far East will it then be?

☐ **18.2.3 Exercises.** *Higher, levels 8, 9*

1) A spire is 50 metres high, and the angle of elevation is 15°. How far away am I from the base? If I now walk 150 metres towards the tower, what is the new angle of elevation?

2) A ship travels 50 km on a bearing of 050°, then for 60 km on a bearing of 140°. What is the bearing of the ship from its starting point?

3) A gun is due South of its target. From an observation post 50 m. due West of the gun, the target is on a bearing of 005°. Find the distance of the target from the gun.

4) Two men stand on opposite sides of a flagpole, which is 12 metres taller than they are. The angles of elevation of the top of the flagpole from the two men are 10° and 15°. How far apart are the men?

5) From the top of a 50 m cliff, the angle of depression of a ship is 3°. How far away is the ship from the top of the cliff?

6) At midday in the tropics the sun is vertically overhead. What is the angle of elevation of the sun at 4 o'clock in the afternoon?

   What is the length of the shadow of a 6 foot man at this time?

18.3 Functions for all Angles

The functions sin, cos and tan can be found for angles bigger than 90°. When we plot the graphs of these functions, they are as follows:

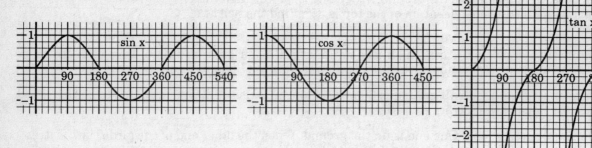

Fig 18.26

Notice that the graphs of sin and cos follow a wave pattern, repeating itself every 360°. The tan function is not defined at 90°, 270° etc. The tan graph repeats itself every 180°.

☐ **18.3.1 Example.** *Higher, level 9*

Solve the equations   a) $\tan x = 1$ and   b) $\tan y = -1$, for angles x and y bigger than 90°.

Solution   a)   We know that tan 45° = 1. As tan repeats itself after 180°, another solution is obtained by adding 180° to 45°.

$$x = 225°$$

b)   Looking at the graph of tan, we see that it is negative between 90° and 180°. By the symmetry of the graph, a solution to the equation is found by subtracting 45° from 180°.

$$y = 135°$$

18.3.2 Exercises. *Higher, level 9*

1) Use your calculator to find the following:

   a)   sin 150°   b)   cos 200°   c)   tan 160°

d) sin 270°   e)   cos 300°   f)   tan 280°

2) Draw an accurate graph of $y = \sin x$ for x between 0° and 360°. Use the graph to find two solutions for each of the following equations:

  a) $\sin x = 0.5$ b) $\sin x = 0.4$ c) $\sin x = -0.3$

3) Draw an accurate graph of $y = \cos x$ for x between 0° and 360°. Use the graph to find two solutions for each of the following equations:

  a) $\cos x = 0.5$ b) $\cos x = -0.8$ c) $\cos x = -0.5$

4) Draw an accurate graph of $y = \tan x$ for x between 0° and 720°. Use the graph to find two solutions for each of the following equations:

  a) $\tan x = 2$ b) $\tan x = -1$ c) $\tan x = -0.7$

5) Find the solutions of the following inequalities in the range $0° \leq x \leq 360°$:

  a) $\sin x > 0.5$ b) $\cos x < 0$ c) $\tan x > 0$

18.4 The Sine and Cosine Rules

The *sine rule* and the *cos rule* extend trigonometry for triangles which are not necessarily right-angled.

When labelling the sides and angles of the triangle ABC, the convention is that each side is labelled with the same letter as the opposite angle. So side c is opposite C etc.

The *sine* rule is:

$$\frac{\sin \hat{A}}{a} = \frac{\sin \hat{B}}{b} = \frac{\sin \hat{A}}{c}$$

Fig 18.27

The rule can be written the other way up:

$$\frac{a}{\sin \hat{A}} = \frac{b}{\sin \hat{B}} = \frac{c}{\sin \hat{C}}$$

The *cosine* rule is:

$$c^2 = a^2 + b^2 - 2ab \cos C.$$

This rule gives a side in terms of the other sides and the opposite angle. The rule can be re-arranged to give the angles in terms of the sides.

$$\cos \hat{C} = \frac{a^2 + b^2 - c^2}{2ab}$$

❑ **18.4.1 Examples.** *Higher, level 10*

1) In Fig 18.28, $\hat{Q} = 68°$, PQ = 4 and PR = 5. Find $\hat{R}$.

  Solution   The first version of the sine rule is used.

$$\frac{\sin \hat{R}}{4} = \frac{\sin 68°}{5}$$

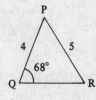

Fig 18.28

    Multiply across to obtain:

  $\sin R = \sin 68 \times \dfrac{4}{5}$

$$\hat{R} = \sin^{-1}0.7417 = 48°$$

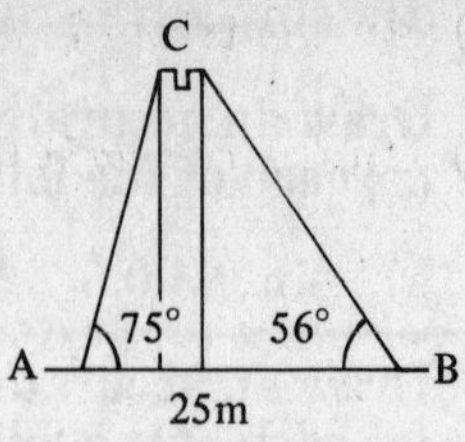

2) Two men are 25 m. apart. One is North of a tower, and one is South. They measure the angles of elevation of the top of the tower as 75° and 56°. Find the height of the tower.

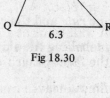

Fig 18.29

Solution   Let C be the top of the tower, and A and B the two men.
AB = 25, $\hat{C} = 180° - 56° - 75° = 49°$.

Use of the sine rule gives:

$$AC = \sin 56° \times \frac{25}{\sin 49°} = 27.46.$$

Trigonometry now gives:

Height = AC × sin 75° = 27 metres.

3) In the triangle PQR, PQ = 5, QR = 6.3, RP = 7.4. Find $\hat{P}$.

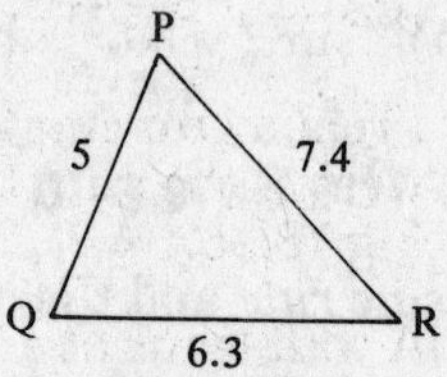

Solution   The cosine rule gives, on putting $\hat{C} = \hat{P}$, $c = 6.3$, $a = 5$, $b = 7.4$:

$$\cos \hat{P} = \frac{5^2 + 7.4^2 - 6.3^2}{2 \times 5 \times 7.4}$$

$$\cos \hat{P} = 0.5415$$

$$\hat{P} = 57.2°$$

Fig 18.30

4) The hour hand of a clock is 15 cm long, and the minute hand is 20 cm. long. How far apart are the tips of the hands at five o'clock?

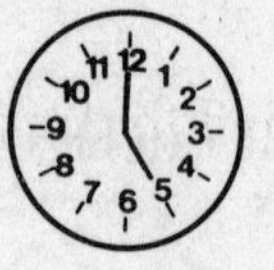

Solution   At five o'clock the angle between the hands is $360 \times \frac{5}{12} = 150°$.

Let d be the distance between the tips. Use the cosine rule:

$$d^2 = 15^2 + 20^2 - 2 \times 15 \times 20 \times \cos 150°$$

d = 33.8 cm.

Fig 18.31

❏ **18.4.2 Exercises.** *Higher, level 10*

Sine rule exercises

1) Find the unknown sides in the following triangles:

Fig 18.32

2) Find the unknown angles in the following triangles:

Fig 18.33

3) In the triangle ABC, $\hat{A}$ = 23°, $\hat{B}$ = 68°, AB = 5.6. Find the sides AC and CB.

4) In the triangle DEF, DE = 10, EF = 13, $\hat{D}$ = 62°. Find the angles F and E.

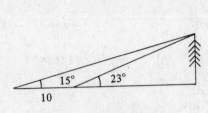

Fig 18.34

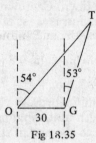

Fig 18.35

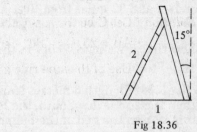

Fig 18.36

5) I measure the angle of elevation of a tree as 15°. I then walk 10 metres towards the tower, and the angle of elevation is now 23°. Find my distance from the top of the tree, and hence find the height of the tree. Fig 18.34.

6) From a gun emplacement, the bearing of its target is 053°. From an observation point 30 metres west of the gun, the bearing of the target is 054°. What is the distance from the gun to its target? Fig 18.35.

7) A wall leans at 15° to the vertical. A ladder of length 2 metres is placed so that one end of the ladder is 1 metre from the base of the wall. What angle is the ladder leaning at? How high up the wall does the ladder reach? Fig 18.36.

8) Ship A leaves from port C at a bearing of 123°. Ship B leaves from port D, 200 miles south of C, on a bearing of 046°. How far from C do their paths cross?

Cosine rule exercises

9) Find the unknown sides in the following triangles.

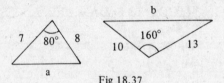

Fig 18.37

10) Find the unknown angles in the following triangles.

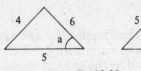

Fig 18.38

11) In the triangle ABC, AB = 17, BC = 23, $A\hat{B}C$ = 43°. Find AC.

12) In the triangle DEF, DE = 45, EF = 52, DF = 27. Find $\hat{B}$.

13) A pilot flies his plane for 50 miles, then turns through 20° and flies a further 73 miles. How far is he from home? Fig 18.39.

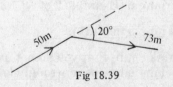

Fig 18.39

14) The lower jaw of a crocodile is 58 cm, and the upper jaw is 52 cm. It can open its jaws to an angle of 43°. What is the greatest width of object that it can grasp in its jaws? Fig 18.40.

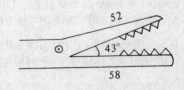

Fig 18.40

15) A parallelogram has sides of length 12 and 7 cm, and the longer diagonal is 15 cm. Find the angles of the parallelogram. Find the length of the shorter diagonal. Fig 18.41.

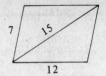

Fig 18.41

16) Aytown is 50 miles due north of Beetown; Ceetown is 27 miles from Aytown and 32 miles from Beetown, and is west of them. Find the bearings of Ceetown from Aytown and from Beetown. Fig 18.42.

17) Show that if we put C = 90° in the cosine rule, we obtain Pythagoras' theorem.

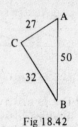

18) A flagpole of length 5 metres leans due North at 17° to the vertical. When the sun is due South, the length of its shadow is 6 metres. How far is it from the end of the flagpole to the end of the shadow?

Fig 18.42

Miscellaneous exercises

19) Find the unknown sides of the following triangles.

Fig 18.43

20) Find the unknown angles of the following triangles.

Fig 18.44

21) A quadrilateral ABCD has sides given by AB = 12, BC = 14, CD = 13.5, DA = 11.2. $\hat{A}$ = 58°. Find BD and $\hat{C}$.

22) In triangle ABC, AB = 4.5, BC = 5.3, $\hat{B}$ = 39°. Find $\hat{A}$.

Common Errors

1) Right-angled triangles

Be careful that you do not apply trigonometry to the wrong sort of triangle. The ratios are only defined for those triangles which are *right-angled*. If there is not any right-angle in the triangle, then more work has to be done so that there is one. This was done in the case of an isosceles triangle (Example 2 of 18.1.3)

2) The three ratios

Make sure that you do not use the wrong ratio for trig. Usually this error occurs because the triangle has been wrongly labelled.

It does not matter which way up the triangle is, the longest side, (the side farthest from the right-angle) is always the HYP. The side which is next to the angle we are dealing with is the ADJ, and the side opposite the angle we are dealing with is the OPP.

If you have any difficulty with these, make sure that the triangle is labelled with HYP, ADJ and OPP before you start working out the ratios.

3) Depression

The angle of depression is the angle which the line of vision makes with the *horizontal*, not with the vertical.

4) Algebra

Mistakes in algebra are easy to make. When the unknown term is on the bottom of the ratio instead of the top, it is very common to get the wrong answer. For example, from the equation $\sin 67° = \dfrac{6}{h}$, the answer is $h = \dfrac{6}{\sin 67°} = 6.518$.

If you obtain $h = 5.523$, it means that you must have multiplied by $\sin 67°$ instead of dividing.

There is always a quick check to make sure that you have done the correct thing. The hypoteneuse is always the largest side of the triangle. So if you find that your value for HYP is smaller than your value for OPP or ADJ, then you must have multiplied instead of divided.

5) Use of calculator

a) Your calculator works "backwards" for the sin, cos and tan ratios. Do not press the sin button before the angle button.

b) Do not press the = button after the function button.

c) Be careful that you give the answer to the right degree of accuracy. If you are asked to give an answer to 3 decimal places, then it is a good idea to work to more than 3 decimal places.

For example, to find $53 \times \sin 26°$:

$53 \times \sin 26° = 53 \times 0.43837 = 23.23361 = 23.234$ (to 3 decimal places).

6) Sine and Cosine rules

a) When using either the sine rule or the cosine rule, be sure that you have labelled the sides and angles correctly. Side a is opposite angle A, b is opposite B, c is opposite C.

b) Calculator errors are very common, especially when using the cosine rule for finding an angle from the sides.

For example, if you have the formula:

$$\cos \hat{C} = \frac{6^2 + 9^2 - 7^2}{2 \times 6 \times 9}$$

The answer you should obtain is $\cos \hat{C} = 0.6296$

If you obtain $\cos \hat{C} = 1836$ it means that you have multiplied by 6 and by 9 instead of dividing by them.

Chapter 19

Pythagoras' Theorem

19.1 Pythagoras in two Dimensions.

Pythagoras' Theorem gives a relationship between the sides of a right-angled triangle.

The square on the hypoteneuse is equal to the sum of the squares on the other two sides.

In Fig 19.1, the side farthest from the right-angle is the hypoteneuse. The theorem says:

$$c^2 = a^2 + b^2.$$

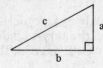

Fig 19.1

☐ 19.1.1 Examples. *Intermediate, level 7*

1) In the triangle of Fig 19.2, $\hat{C} = 90°$, CB = 3 and AC = 5. Find BA.

   Solution Comparing with Fig 19.1, BA = c, AC = b, BC = a. The theorem now gives:

   $$c^2 = 3^2 + 5^2$$
   $$c^2 = 9 + 25 = 34$$
   $$c = \sqrt{34} = 5.831$$

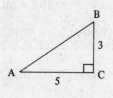

Fig 19.2

2) A ladder of length $2\frac{1}{2}$ metres leans against a wall, so that the foot of the ladder is 1 metre from the base of the wall. How far up the wall does the ladder reach?

   Solution Here the length of the ladder is c, and b is the distance to the base of the wall. The height up the wall is a. The theorem gives:

   $$2\left(\frac{1}{2}\right)^2 = a^2 + 1^2$$

   $$a^2 = 6\frac{1}{4} - 1 = 5\frac{1}{4}$$

   $$a = \sqrt{5\frac{1}{4}} = 2.291 \text{ m}$$

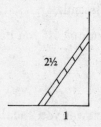

Fig 19.3

☐ 19.1.2 Exercises. *Intermediate, level 7*

1) For the following triangles, find the unknown sides.

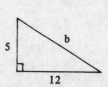

Fig 19.4

2) I travel 3 miles North and 5 miles East. How far am I from my starting point? Fig 19.5.

3) A rectangle is 7 cm by 8 cm. How long are the diagonals? Fig 19.6.

4) A kite is flying so that it is 55 feet high, and is above a point 75 feet from the flyer. How long is the string of the kite? Fig 19.7.

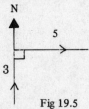

Fig 19.5

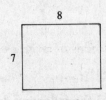

Fig 19.6

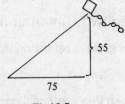

Fig 19.7

5) The diagonals of a rhombus are 10 cm and 13 cm. What is the length of the side of the rhombus? Fig 19.8.

6) P is a point 6 cm from the centre of a circle of radius 2 cm. A tangent is drawn from P, touching the circle at T. How long is the tangent PT? Fig 19.9.

7) A chord of length 5 is drawn in a circle of radius 20. What is the shortest distance from the centre of the circle to the chord? Fig 19.10.

8) A cylindrical horizontal drain has radius 4 inches. There is water in it to a depth of 1 inch. How far below the centre of the drain is the water? How wide is the water in the drain? Fig 19.11.

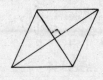

Fig 19.8

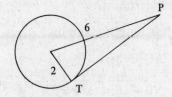

Fig 19.9

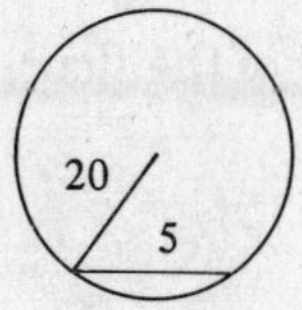

Fig 19.10

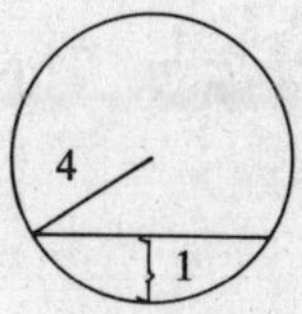

Fig 19.11

9) A road has a gradient of 1 in 20. (Which means that it rises 1 metre for every 20 metres horizontally.) If I walk so that I have risen $2\frac{1}{2}$ metres, how far have I travelled horizontally? How far have I walked along the road?

10) A stick of length 5 feet is held at an angle, so that when the sun is vertically above the stick the shadow is 3 feet long. How high is the end of the stick above the ground?

11) A pendulum 2 metres long is pulled aside so that the bob has moved 10 cm to the left. How far is the bob below the support? By how much has the bob risen?

12) The sides of a triangle are 7, 11 and 12. Is this a right-angled triangle?

13) The sides of a triangle are 15, 8 and 17. Is this a right-angled triangle?

Fig 19.12

14) An isosceles triangle ABC has sides AB = AC = 7 cm and BC = 6 cm. Find the length of the perpendicular from A to BC.

15) PQR is an equilateral triangle of side 4 cm. Find the length of the perpendicular from P to QR. Find the area of the triangle.

16) The dots of the figure below are 1 cm. apart. Find the lengths of:

a) AB   b) AC   c) CD.

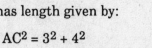

Fig 19.13

17) A graph is drawn to a scale of 1 cm per unit. Find the following distances:

a) (0,0) to (1,2)   b) (0,0) to (5,-3)

c) (1,1) to (5,7)   d) (1,3) to (6,1).

19.2 Pythagoras and Trigonometry in Three Dimensions

☐ **19.2.1 Examples.** *Higher, level 9*

1) A room ABCDEFGH is a cuboid, $2\frac{1}{2}$ metres high, 4 metres long, 3 metres wide. What is the length of the 'space diagonal' AG? What angle does this line make with the horizontal?

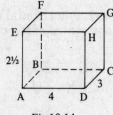

Fig 19.14

Solution   The diagonal of the floor, AC, has length given by:

$$AC^2 = 3^2 + 4^2$$

$$AC = \sqrt{9 + 16} = \sqrt{25} = 5.$$

In the triangle AGC, ACG is a right angle. This then gives:

$$AG^2 = AC^2 + CG^2$$

$$AG = \sqrt{25 + 6\frac{1}{4}} = \sqrt{31.25} = 5.59 \text{ metres}$$

For the angle, trigonometry gives:

$$G\hat{A}C = \tan^{-1}\frac{2\frac{1}{2}}{5} = 26.6°.$$

AG makes 26.6° with the horizontal.

2) A section of a hillside is shown in Fig 19.15.

ABFE is a horizontal plane; AB, EF and CD are horizontal lines of length 50 metres. BF is 12 metres and FD and CE are vertical lines of length 5 metres. Find:

a) BD   b) CB   c) The slope of DB   d) The slope of CB.

Solution   BDF is a right-angled triangle. This then gives:

$$BD^2 = FD^2 + BF^2$$

$$BD = \sqrt{169} = 13$$

BCD is a right-angled triangle. This gives:

$$BC^2 = CD^2 + DB^2$$

$$BC = \sqrt{2500 + 169} = \sqrt{2669} = 52 \text{ metres}$$

Using trigonometry in triangle BFD:

$$\tan F\hat{B}D = \frac{5}{12}$$

BD makes $\tan^{-1}\dfrac{5}{12} = 22.6°$ **with the horizontal**

Using trigonometry in triangle BCE:

$$\sin C\hat{B}E = \frac{5}{CB}$$

CB makes $\sin^{-1}\dfrac{5}{52} = 6°$ **with the horizontal**

☐ **19.2.2 Exercises.** *Higher, level 9*

1) A cuboid ABCDEFGH is shown in Fig 19.16. Find the diagonal FD and the angles $F\hat{D}B$, $E\hat{H}B$, $F\hat{D}C$.

2) A wedge ABCDEF is shown in Fig 19.17. Find the lengths AF, AC, AD and the angles $A\hat{C}E$, $B\hat{C}A$.

3) In Fig 19.18 the pyramid VABCD has a square base ABCD of side 4 cm. VA = VB = VC = VD = 6 cm. Find AC, and the height of V above the plane ABCD. Find the angles $V\hat{A}C$ and $V\hat{A}D$.

4) A prism is of height 10 cm and its cross-section is an equilateral triangle of side 3 cm. (Fig 19.19). Find the length AE. Find the angle $B\hat{A}E$.

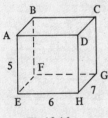

Fig 19.16

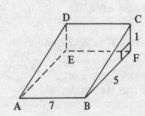

Fig 19.17

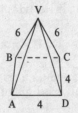

Fig 19.18

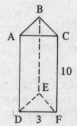

Fig 19.19

5) A cardboard carton is a cuboid with sides 2 ft, $1\frac{1}{2}$ ft, $2\frac{1}{2}$ ft. Find the length of the longest stick which can be put in the carton. Find the angles that the stick now makes with the sides of the carton.

6) In Fig 19.20 the pyramid VABCD is on a square base ABCD of side 100 metres, and V is 50 metres above the ground. Find:

   a) The diagonal AC of the base.

   b) The length of VA.

   c) The distance from V to the midpoint X of AB.

   d) The angle $V\hat{A}C$.

   e) The inclination of VX to the horizontal.

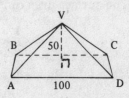

Fig 19.20

185

7) In Fig 19.21 ABCD is a regular tetrahedron of side 8 cm. It is placed with BCD horizontal, so that A is vertically above the centre O of triangle BCD. X is the midpoint of BD. Find:

a) AX and XC.

b) The angle $A\hat{X}C$.

c) XO and OC.

d) The angle $A\hat{C}O$.

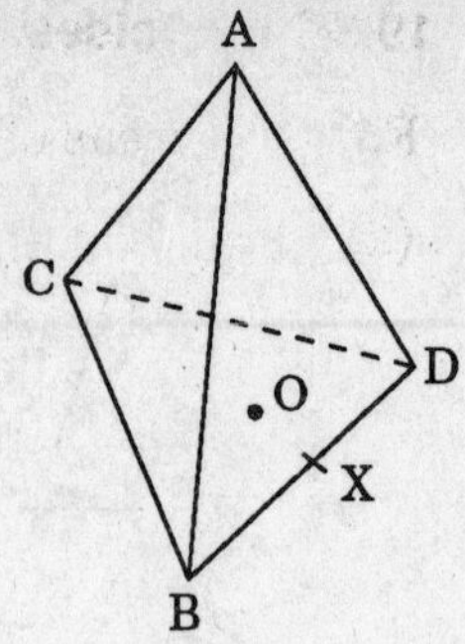

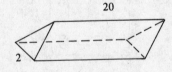

Fig 19.21

19.3 Problems of Areas and Volumes

A useful formula for the area R of the triangle ABC is:

$$R = \tfrac{1}{2}AB \times AC \times \sin \hat{A}$$

□ **19.3.1 Examples.** *Higher, level 9*

1) A regular octagon is inscribed inside a circle of radius 10 cm. Find the area of the octagon.

Solution Join up the points of the octagon to the centre of the circle as shown in Fig 19.22.

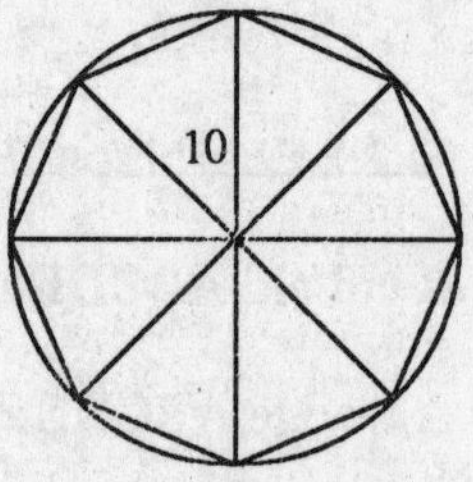

The angle at the centre of each of the triangles is $\dfrac{360}{8} = 45°$.

Using the formula, the area of each little triangle is:

$$\tfrac{1}{2} \times 10 \times 10 \times \sin 45°.$$

There are eight little triangles, so the total area is:

$$8 \times \tfrac{1}{2} \times 10 \times 10 \times \sin 45° = 283 \text{ cm}^2$$

Fig 19.22

2) A stick of chocolate is shaped as a triangular prism 20 cm long. Each end is an equilateral triangle of side 2 cm. Find its volume and its surface area.

Solution Use the formula to find the area of each triangle.

$$\tfrac{1}{2} \times 2 \times 2 \times \sin 60°.$$

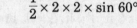

Fig 19.23

The volume is:

$$20 \times \tfrac{1}{2} \times 2 \times 2 \times \sin 60° = 34.6 \text{ cm}^3.$$

The surface area consists of the two ends and the three sides.

$$2 \times \tfrac{1}{2} \times 2 \times 2 \times \sin 60° + 3 \times 20 \times 2 = 123.5 \text{ cm}^2$$

◻ **19.3.2 Exercises.** *Higher, level 9*

1) Find the areas of the triangles shown below:

(a) (b) 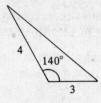 (c)

Fig 19.24

2) For the triangle LMN, ML = 3.4, NL = 5.6, $\hat{L}$ = 66°. Find the area of the triangle.

3) A regular hexagon is inscribed inside a circle of radius 8 cm. Find the area of the hexagon.

4) A regular pentagon of side 12 cm is inscribed inside a circle. Find the radius of the circle. Find the area of the pentagon. Fig 19.25.

5) A circle of radius 13 cm is inscribed inside a regular 10 sided figure. Find the side of the figure. Find the area of the figure. Fig 19.26.

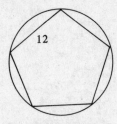
Fig 19.25

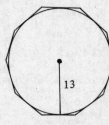
Fig 19.26

6) The dots of the figure below are 1 cm apart. Find the perimeter of the triangle ABC. Make a copy of the diagram. Mark the points D, E such that BCDE is a square, and find its area.

Fig 19.27

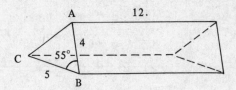
Fig 19.28

7) The prism of Fig 19.28 is 12 cm long, and its cross section is a triangle ABC with AB = 4 cm, BC = 5 cm, $A\hat{B}C$ = 55°. Find the area of △ABC and hence find the volume of the prism.

8) A prism is 10 cm long, and its cross section is an isosceles triangle with sides of length 7 cm, 7 cm, 8 cm. Find the area of the triangle, and hence find the volume of the prism.

Common Errors

1) **Right-angled triangles**

  a) Do not use Pythagoras for a triangle which is not right- angled. The theorem only holds if the triangle does have a right-angle.

  b) Do not use the wrong sides for the theorem. The longest side is always the hypoteneuse. Make sure that you have labelled the sides correctly before using the equation $c^2 = a^2 + b^2$.

2) **Errors in algebra**

  When using the theorem, the a^2 and b^2 terms must be added *before* the square root is taken. Do not take the square root first.

$$\sqrt{a^2 + b^2} \neq a + b.$$

3) Three-dimensional problems

It is very easy to get the wrong triangle, or to label the sides incorrectly. Any three-dimensional diagram is in perspective, so a right-angled triangle may not look like one on the diagram.

It is often a good idea to make a separate two-dimensional picture of the triangle you are using.

Chapter 20

Transformations

20.1 Single Transformations

A *transformation* changes the positions of points and shapes in the plane. The standard transformations are as follows:

A *translation* shifts all points in a fixed direction.

A *rotation* turns all points through a fixed angle.

A *reflection* moves each point to its mirror image on the other side of a fixed line.

An *enlargement* increases the size of shapes by a constant number. The number is the *scale factor*.

If a shape is unchanged after reflection in a line, then it has *symmetry* about that line.

If a shape is unchanged after rotation round a point, then it has *symmetry* about that point. The *order of symmetry* is the number of turns before it returns to the original position.

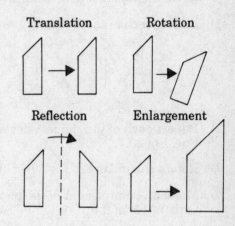

Fig 20.1

❑ **20.1.1 Examples.** *Foundation, levels 4, 5, 6*

1) The letter F is shown on the grid of Fig 20.2. Show its position after it has been:

   a) Reflected in the dotted line.

   b) Translated 1 up and 2 to the right.

   c) Rotated through 90° clockwise.

   d) Enlarged by a factor of 2.

Fig 20.2

Solution The effect of these operations is shown in Fig 20.3.

a) 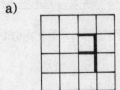   b) 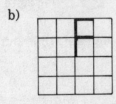   c) 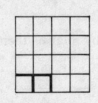   d)

Fig 20.3

2) Describe the symmetries of the letter H and the three-spoked wheel of Fig 20.4.

   Solution The letter H will remain unchanged if it is reflected about either of the dotted lines shown in Fig 20.5 (overleaf). It will also be unchanged if it is rotated about the central point through half a turn.

   Hence there are two lines of symmetry, and the letter has rotational symmetry of order 2.

Fig 20.4

The wheel will remain unchanged if it is reflected about any of the spokes. It will also remain unchanged if it is rotated through a third of a circle about the centre.

Hence there are three lines of symmetry, and the symbol has rotational symmetry of order 3.

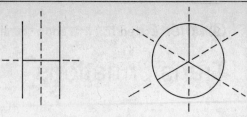

Fig 20.5

☐ **20.1.2 Exercises.** *Foundation, levels 4, 5, 6*

1) Reflect each of the shapes below in the horizontal dotted line.

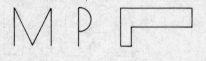

Fig 20.6

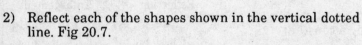

2) Reflect each of the shapes shown in the vertical dotted line. Fig 20.7.

3) Rotate the picture of Fig 20.8 through 90°, 180°, 270°.

4) Make an enlargement of the picture in Fig 20.8, by a scale factor of 2.

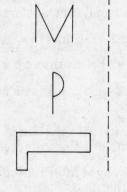

Fig 20.7

Fig 20.8

5) Complete each of the shapes in Fig 20.9 by reflecting them in the dotted line.

Fig 20.9

6) Complete each of the shapes in Fig 20.10 by rotating them through 180°, about the point X.

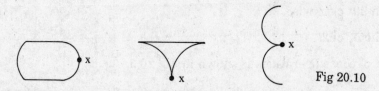

Fig 20.10

7) Complete each of the shapes below by rotating them through 90°, 180° and 270°, about the point X.

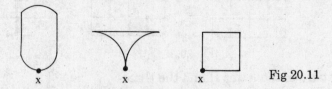

Fig 20.11

8) Make copies of the figures below, enlarged by a scale factor of 3.

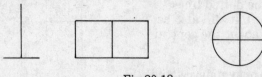

Fig 20.12

9) Draw dotted lines to show the lines of symmetry of the following figures:

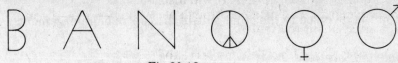

Fig 20.13

10) Do the following figures have any rotational symmetry? If they do, give the order of symmetry.

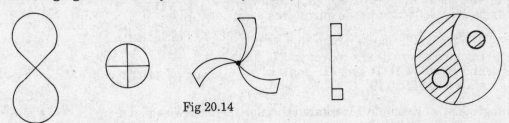

Fig 20.14

11) The year 1881 was the last year which was symmetrical about a vertical line through the centre. When will be the next such year?

12) The year 1961 was the last year which had rotational symmetry. When will be the next such year?

13) Fig 20.15 shows a 4 by 6 grid, lettered A to X.

A	B	C	D	E	F
G	H	I	J	K	L
M	N	O	P	Q	R
S	T	U	V	W	X

Fig 20.15

   a) What translation will take G to P?

   b) What translation will take N to E?

   c) A translation takes M to J. Where will this translation take G? Where will it take N?

   d) Describe the transformation which takes the shape made from the blocks AGMN to the blocks EKQP.

   e) Describe the transformation which takes the AGMN shape to QKED.

   f) Describe the transformation which takes the AGMN shape to LKJP.

   g) Describe the transformation which takes the MST block to AGMSTUV.

14) A wallpaper pattern is obtained by reflection, rotation and translation of a simple shape. Complete the following patterns:

a)   b)

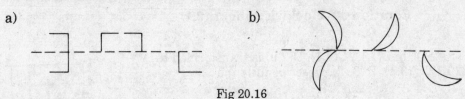

Fig 20.16

15) The triangles A, B, C, D, E of Fig 20.17 are all congruent. Describe the transformations which take:

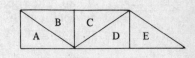

   a) A to E   b) B to C.   c) A to B.

Fig 20.17

Centres of rotation and enlargement

In a rotation, one point remains fixed. This point is the *centre of rotation*.

In an enlargement, one point remains fixed. This point is the *centre of enlargement*. Distances from the centre of enlargement are multiplied by the scale factor λ.

If λ is positive but less than 1, the transformation is called a *reduction*.

If λ is negative, each point is reversed to the other side of the centre of enlargement.

☐ **20.1.3 Examples.** *Intermediate, level 7*

1) A triangle is drawn on the grid of Fig 20.18 by joining up the points A(1,3), B(1,4), C(3,3). Draw this triangle after reflection in the mirror line y = x.

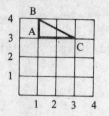

Fig 20.18

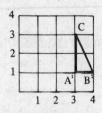

Fig 20.19

Solution   The point C is on the mirror line, so it remains where it is. A moves to A'(3,1), and B to B'(4,1). The new triangle is shown in Fig 20.19.

2) The triangle ABC of Example 1 is rotated through 90° clockwise about A. Draw the new triangle.

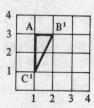

Fig 20.20

Solution   Because A is the centre of rotation, it remains fixed. The vertical line AB moves to a horizontal line AB', where B' is at (2,3). The horizontal line AC moves to a vertical line AC', where C' is at (1,1). The new triangle is shown in Fig 20.20.

3) The triangle ABC is given a translation (1,-2). Find the new co-ordinates, and draw the new triangle.

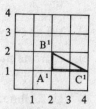

Fig 20.21

Solution   A translation (1,-2) means that 1 is added to the x co-ordinate and 2 is subtracted from the y co-ordinate. The new positions are A'(2,1), B'(2,2), C'(4,1). The triangle is shown in Fig 20.21.

4) The rectangle with vertices at A(1,1), B(1,3), C(2,3), D(2,1) is transformed to the rectangle with vertices at A'(2,2), B'(4,2), C'(4,1), D'(2,1). Describe fully the transformation.

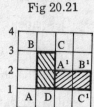

Fig 20.22

Solution   The rectangle before and after the transformation is shown in Fig 20.22. The rectangle has been rotated clockwise. AB is vertical and A'B' is horizontal, hence the rotation is through 90°. D is fixed, so D is the centre of rotation.

Rotation 90° clockwise about (2,1)

5) The square with vertices A(1,0), B(1,1), C(2,1), D(2,0) is transformed to the square A'(0,-2), B'(0,1), C'(3,1), D'(3,-2). Describe fully the transformation.

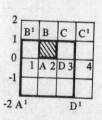

Fig 20.23

Solution   The square before and after transformation is shown in Fig 20.20. The square has been enlarged by a factor of 3. The midpoint of BC is unchanged, hence this must be the centre of enlargement.

An enlargement of scale factor 3 from the point $(1\frac{1}{2},1)$

☐ **20.1.4 Exercises.** *Intermediate, level 7*

1) Draw axes on a sheet of graph paper. Mark the axes from -5 to 10, allowing 1 cm per unit. On the paper draw the triangle with vertices A(1,1), B(3,1), C(2,5).

Draw the triangles obtained by applying the following transformations to △ ABC. Write down the co-ordinates of the vertices:

a)   A reflection in the *x*-axis.

b)   A reflection in the *y*-axis.

c)   A reflection in the line *y = x*.

d)   A reflection in the line *y = -x*.

2)   Draw axes as in Question 1. Draw the triangle with vertices D(1,2), E(1,3), F(5,2). Apply the following transformations to △ DEF: draw the transformed triangles and write down their vertices.

a)   A rotation of 180° about D.

b)   A rotation of 180° about E.

c)   A rotation of 180° about the origin (0,0).

d)   A rotation of 90° clockwise about F.

e)   A rotation of 90° anti-clockwise about (1,1).

3)   The triangle with vertices A(1,1), B(1,3), C(4,3) is given a transformation so that it moves to A'(-1,1), B'(-1,3), C'(-4,3). Draw both triangles on graph paper, and describe the transformation.

4)   The triangle ABC of Question 3 is transformed to A"(1,1), B"(3,1), C"(3,4). Draw the new triangle on graph paper, and describe the transformation.

5)   The rectangle with vertices A(1,1), B(1,3), C(2,3), D(2,1) is transformed to the rectangle with vertices A'(1,-1), B'(3,-1), C'(3,-2), D'(1,-2). Draw both rectangles on graph paper and describe the transformation.

6)   The rectangle ABCD of Question 5 is transformed to A"(-1,-1), B"(-1,-3), C"(-2,-3), D"(-2,-1). Draw this rectangle and describe the transformation.

7)   A translation of (1,-2) is applied to the triangle ABC of Question 3. Find the new co-ordinates, and draw the translated triangle.

8)   A translation takes the point (3,4) to (6,2). Where would the same translation take the triangle with vertices at A(1,1), B(2,3), C(0,4)?

9)   The triangle ABC of Question 3 is transformed to A‴(1,1), B‴(1,5), C‴(7,5). Describe the transformation.

10) A triangle has vertices at A(0,2), B(2,3), C(2,2). After a transformation it has moved to A'(3,1), B'(2,3), C'(3,3). Describe fully the transformation.

11) The triangle of Question 10 is transformed to A"(0,2), B"(-2,1), C"(-2,2). Describe fully the transformation.

12) The rectangle PQRS with vertices P(0,0), Q(0,2), R(1,2), S(1,0) is transformed to P'(1,-1), Q'(3,-1), R'(3,-2), S(1,-2). Describe the transformation.

13) The rectangle of Question 12 is transformed to P"(-2,-2), Q"(-2,-4), R"(-3,-4), S"(-3,-2). Describe the transformation.

14) The triangle of Question 10 is transformed to A‴(0,0), B‴(2,-1), C‴(2,0). Describe the transformation.

15) The rectangle of Question 12 is transformed to P‴(-4,0), Q‴(-4,2), R‴(-5,2), S‴(-5,0). Describe the transformation.

16) The triangle of Question 10 is transformed to A""(-6,-1), B""(2,3), C""(2,-1). Describe the transformation.

17) The rectangle of Question 12 is transformed to P""(0,0), Q""(0,-4), R""(-2,-4), S""(-2,0). Describe the transformation.

18) The triangle of Question 10 is reflected in the line $x = 3$. Give the co-ordinates of the new vertices.

19) The triangle of Question 10 is reflected in the line $y = -1$. Give the co-ordinates of the new vertices.

20) On graph paper draw the rectangle PQRS of Question 12. Rotate the figure through 40° anti-clockwise about the origin (0,0). Write down as accurately as you can the co-ordinates of the new vertices.

21) On graph paper draw the triangle ABC of Question 10. Rotate the figure through 60° clockwise about A. Write down as accurately as you can the co-ordinates of the new vertices.

22) The triangle ABC of Question 10 is enlarged by a scale factor of 2 about the point (1,1). Find the new coordinates.

23) The triangle ABC of Question 10 is reduced by a scale factor of $\frac{1}{2}$ about the point (2,0). Find the new coordinates.

24) The rectangle PQRS of Question 12 is enlarged by a scale factor of -3 about the point (2,1). Find the new coordinates.

☐ **20.2 Combined Transformations.** *Higher, level 10*

If a transformation P is followed by a transformation Q, the *combined transformation* is written QP. In particular, if P is done twice then the transformation is written P^2.

If a transformation P takes A to A', the *inverse* transformation which takes A' back to A is written P^{-1}.

☐ **20.2.1 Examples.** *Higher, level 10*

1) The triangle T with vertices at A(1,1), B(1,3), C(2,3) is reflected in the line $x = 2$, giving the triangle T'. T' is reflected in the line $x = -1$, to give the triangle T".

   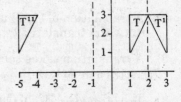

   Draw the three triangles on graph paper. Describe the transformations which take T to T" and which take T" to T.

   Fig 20.24

   Solution   The triangles are shown in Fig 20.24. T is taken to T" by a translation, of 6 units to the left. T" is taken to T by a translation of 6 units to the right.

2) The triangle T of example 1 is rotated through a quarter turn clockwise about (1,1), then through a further quarter turn clockwise about (0,0). To what single transformation is this equivalent?

   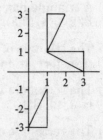

   Solution   The three triangles are shown in Fig 20.25. The two rotations of 90° give a rotation of 180°. The centre of rotation is (1,0).

   Fig 20.25

☐ **20.2.2 Exercises.** *Higher, level 10*

1) The triangle S with vertices A(1,1), B(2,2), C(4,1) is reflected in the line $x = 1$ and then in the line $x = 2$. Find the single transformation equivalent to the two reflections.

2) The triangle S of Question 1 is reflected in $y = 1$ and then in $y = -1$. Find the single transformation equivalent to the two reflections.

3) The triangle S of Question 1 is reflected in $x = 1$ and then in $y = 1$. Describe fully the single transformation.

4) The triangle S of Question 1 is reflected in $y = -1$ and then in $x = 2$. Describe fully the single transformation.

5) LMN is the triangle with vertices L(0,2), M(3,2), N(0,1). Describe the single transformation equivalent to the following successive transformations:

a) Rotation of 90° clockwise about L.

b) Rotation of 90° anti-clockwise about (0,0).

6) With $\triangle$LMN as in Question 5, describe the single transformation equivalent to the following successive transformations:

a) Rotation of 90° clockwise about (0,0).

b) Rotation of 90° clockwise about (2,2).

7) With $\triangle$LMN as in Question 5, describe the single transformation equivalent to the following successive transformations:

a) Rotation of 180° about (2,3).

b) Rotation of 180° about (1,0).

8) Let R be the rectangle with vertices at X(1,1), Y(1,3), Z(2,3), W(2,1). Let P be the operation of rotation through 90° anti-clockwise about (0,0), and let Q be the operation of rotation through 90° clockwise about (1,1). Describe the effect on R of the following combined transformations:

a) P^2 b) P^{-1} c) PQ d) QP e) $(PQ)^{-1}$ f) $(QP)^{-1}$ g) $P^{-1}Q^{-1}$.

What can you conclude about inverses of combined transformations?

9) Let T be the triangle with vertices at J(-1,1), K(0,3), L(2,1). Let F be the operation of reflection in the line $x = y$, and G the operation of reflection in $y = 0$. Describe the effect on T of the following transformations:

a) F^{-1} b) G^2 c) FG d) GF.

20.3 Matrix Transformations

A *matrix* is a rectangular block of numbers. A 2 by 2 matrix can represent a transformation in the plane.

To apply the matrix $\begin{pmatrix} 1 & 2 \\ 5 & 7 \end{pmatrix}$ to the point (3,4), write the coordinates vertically.

$$\begin{pmatrix} 1 & 2 \\ 5 & 7 \end{pmatrix}\begin{pmatrix} 3 \\ 4 \end{pmatrix} = \begin{pmatrix} 1 \times 3 + 2 \times 4 \\ 5 \times 3 + 7 \times 4 \end{pmatrix} = \begin{pmatrix} 11 \\ 43 \end{pmatrix}$$

So the new point will be at (11,43).

Matrices can be multiplied to give a combined transformation. The

transformation obtained by applying $\begin{pmatrix} 1 & 2 \\ 5 & 7 \end{pmatrix}$ and then $\begin{pmatrix} 2 & 3 \\ 4 & 8 \end{pmatrix}$ is:

$$\begin{pmatrix} 2 & 3 \\ 4 & 8 \end{pmatrix}\begin{pmatrix} 1 & 2 \\ 5 & 7 \end{pmatrix} = \begin{pmatrix} 2 \times 1 + 3 \times 5 & 2 \times 2 + 3 \times 7 \\ 4 \times 1 + 8 \times 5 & 4 \times 2 + 8 \times 7 \end{pmatrix} = \begin{pmatrix} 17 & 25 \\ 44 & 64 \end{pmatrix}$$

☐ **20.3.1 Examples.** *Higher, level 10*

1) The points of the triangle T are A(1,1), B(3,1), C(1,2). Draw the triangle T on graph paper.

The points ABC are transformed to A'B'C', forming the triangle T', by the matrix M where:

$$M = \begin{pmatrix} 0 & -1 \\ 1 & 0 \end{pmatrix}$$

Draw the triangle T' on the same graph paper.

T' is reflected in the *x*-axis, to obtain T". Draw T" on the graph paper. What single matrix would send T to T"?

Solution Plot A, B, C on the graph. Apply the matrix M to the points.

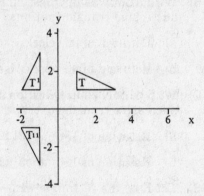

$$\begin{pmatrix} 0 & -1 \\ 1 & 0 \end{pmatrix} \begin{pmatrix} A & B & C \\ 1 & 3 & 1 \\ 1 & 1 & 2 \end{pmatrix} \begin{pmatrix} A' & B' & C' \\ -1 & -1 & -2 \\ 1 & 3 & 1 \end{pmatrix}$$

Plot the points A', B', C'.

Reflect the triangle T' in the *x* axis. Notice that the *y* coordinates have all been multiplied by -1. Hence the matrix of this transformation is:

$$N = \begin{pmatrix} 1 & 0 \\ 0 & -1 \end{pmatrix}$$

Fig 20.26

The matrix which will go from T to T" is the product of M and N.

$$\begin{pmatrix} 1 & 0 \\ 0 & -1 \end{pmatrix} \begin{pmatrix} 0 & -1 \\ 1 & 0 \end{pmatrix} = \begin{pmatrix} 0 & -1 \\ -1 & 0 \end{pmatrix}$$

2) Find the matrix which takes (1,0) to (3,5) and (0,1) to (4,-2)

Solution Notice that for a general matrix:

$$\begin{pmatrix} a & b \\ c & d \end{pmatrix} \begin{pmatrix} 1 \\ 0 \end{pmatrix} = \begin{pmatrix} a \\ c \end{pmatrix}$$

So (1,0) always goes to the left column of the matrix. Similarly (0,1) is sent to the right column of the matrix. The required matrix can now be written down:

$$\begin{pmatrix} 3 & 4 \\ 5 & -2 \end{pmatrix}$$

☐ **20.3.2 Exercises.** *Higher, level 10*

1) Let T be the triangle with vertices L(1,2), M(3,2), N(2,1). Find the image of T after action by the following matrices:

$$A = \begin{pmatrix} 1 & 0 \\ 0 & -1 \end{pmatrix} \quad B = \begin{pmatrix} -1 & 0 \\ 0 & 1 \end{pmatrix} \quad C = \begin{pmatrix} 0 & 1 \\ -1 & 0 \end{pmatrix} \quad D = \begin{pmatrix} 0 & -1 \\ 1 & 0 \end{pmatrix} \quad E = \begin{pmatrix} 0 & -1 \\ -1 & 0 \end{pmatrix}$$

2) Repeat Question 1, for the rectangle with vertices at P(1,1), Q(1,3), R(2,3), S(2,1).

3) Describe the action of the matrices A, B, C, D, E defined in Question 1.

4) Let S be the unit square with vertices at (0,0), (1,0). (1,1), (0,1). Find the image of S after action by the following matrices:

$$F = \begin{pmatrix} 3 & 0 \\ 0 & 3 \end{pmatrix} \quad G = \begin{pmatrix} \frac{1}{2} & 0 \\ 0 & \frac{1}{2} \end{pmatrix} \quad H = \begin{pmatrix} \frac{1}{2} & 0 \\ 0 & 1 \end{pmatrix} \quad J = \begin{pmatrix} 1 & 0 \\ 0 & 2 \end{pmatrix} \quad K = \begin{pmatrix} -2 & 0 \\ 0 & -2 \end{pmatrix}$$

5) Describe the action of the matrices F, G, H, J, K defined in Question 4.

6) On squared paper plot the triangle T with vertices (1,2), (1,–1), (2,2). T is taken to U by the matrix:

$$A = \begin{pmatrix} 0 & 1 \\ -1 & 0 \end{pmatrix}$$

Plot U on the same squared paper. U is now taken to V by the matrix:

$$B = \begin{pmatrix} -1 & 0 \\ 0 & 1 \end{pmatrix}$$

g Plot V on the same paper. What single matrix will take T to V??

7) Let S be the square with corners at (1,1), (1,2), (2,2), (2,1). Plot S on squared paper. S is sent to S' by the matrix:

$$C = \begin{pmatrix} 1 & 2 \\ 0 & 1 \end{pmatrix}$$

Plot S' on the paper. S' is taken to S" by the matrix:

$$D = \begin{pmatrix} 1 & 0 \\ 0 & -1 \end{pmatrix}$$

Plot S" on the same paper. What single matrix will take S to S"?

8) On squared paper plot the triangle with vertices at (1,4), (5,4), (1,3). Draw the image of the triangle after action by A, A^2, A^3, A^4, where:

$$A = \begin{pmatrix} 0 & 1 \\ -1 & 0 \end{pmatrix}$$

Write down A^{17}.

9) Let M be the rectangle with corners at (1,1), (1,5), (2,5), (2,1). Plot this rectangle on squared paper, and its image after action by B, C and BC where:

$$B = \begin{pmatrix} 1 & 0 \\ 0 & -1 \end{pmatrix} \quad C = \begin{pmatrix} -1 & 0 \\ 0 & 1 \end{pmatrix}$$

10) Find the matrix which takes (1,0) to (4,-2) and (0,1) to (2,-1).

11) Find the matrix which takes (2,0) to (4,6) and (0,3) to (9,6).

12) A certain matrix takes the square with vertices (0,0), (3,0), (3,3), (0,3) to the parallelogram with vertices (0,0), (2,5), (6,7), (4,2). Find the matrix.

Common Errors

1) **Single transformations**

a) Be careful that you do not confuse the different types of transformation. It is common to mistake reflections for rotations and vice-versa. A quick check is as follows: A reflection reverses the order of lettering of a triangle, so that if it is lettered clockwise before the reflection then it will be anti-clockwise after. A rotation preserves the order.

b) It is very easy to confuse transformations when a square has been transformed. The square of Fig 20.27 has been translated to (i), rotated to (j), and reflected to (k).

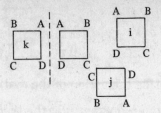

Fig 20.27

2) Combined transformations

The order in which transformations are done is important. PQ is not the same as QP. Be sure to remember that PQ means that Q is done first and then P.

3) Transformation matrices

a) Usually, the co-ordinates of a point may be written either horizontally or vertically. But when a matrix is applied to a point, its co-ordinates must be vertical.

b) Be careful when multiplying matrices to give a combined transformation. The product AB is a transformation which does B first and then A. It is easy to get this the wrong way round.

Chapter 21

Vectors

Vectors are quantities which have direction as well as magnitude.

21.1 Vectors and Coordinates

Translations in the plane can be represented by *vectors*. If a translation moves x units to the right and y upwards then it is given by:

$$\mathbf{v} = \begin{pmatrix} x \\ y \end{pmatrix}$$

Vectors can be added, or multiplied by ordinary numbers.

$$\begin{pmatrix} a \\ b \end{pmatrix} + \begin{pmatrix} c \\ d \end{pmatrix} = \begin{pmatrix} a+c \\ b+d \end{pmatrix} \qquad m \times \begin{pmatrix} a \\ b \end{pmatrix} = \begin{pmatrix} ma \\ mb \end{pmatrix}$$

The length or *modulus* of a vector is given by:

$$\left| \begin{pmatrix} a \\ b \end{pmatrix} \right| = \sqrt{a^2 + b^2}$$

Vectors **v** and **u** can be added by completing the third side of the triangle as in Fig 21.1.

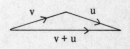

Fig 21.1

☐ **21.1.1 Examples.** *Intermediate, level 8*

1) For the vectors **a, b, c** as defined below find   a) **a + b**,   b) 3**a** + 2**b**,   c) |**c**|
   d) numbers x and y such that $x\mathbf{a} + y\mathbf{b} = \mathbf{c}$.

$$\mathbf{a} = \begin{pmatrix} 3 \\ 2 \end{pmatrix} \qquad \mathbf{b} = \begin{pmatrix} 5 \\ -3 \end{pmatrix} \qquad \mathbf{c} = \begin{pmatrix} 4 \\ 9 \end{pmatrix}$$

Solution   a)   Add the terms of the vectors:

$$\mathbf{a} + \mathbf{b} = \begin{pmatrix} 3 \\ 2 \end{pmatrix} + \begin{pmatrix} 5 \\ -3 \end{pmatrix} = \begin{pmatrix} 8 \\ -1 \end{pmatrix}$$

b)   Multiply the terms of the vectors and add:

$$3\mathbf{a} + 2\mathbf{b} = 3 \times \begin{pmatrix} 3 \\ 2 \end{pmatrix} + 2 \times \begin{pmatrix} 5 \\ -3 \end{pmatrix} = \begin{pmatrix} 9 \\ 6 \end{pmatrix} + \begin{pmatrix} 10 \\ -6 \end{pmatrix} = \begin{pmatrix} 19 \\ 0 \end{pmatrix}$$

c)   Use the formula for modulus:

$$|\mathbf{c}| = \left| \begin{pmatrix} 4 \\ 9 \end{pmatrix} \right| = \sqrt{4^2 + 9^2} = \sqrt{16 + 81} = \sqrt{97} = 9.85$$

d)   Re-write the equation $x\mathbf{a} + y\mathbf{b} = \mathbf{c}$ as:

$$x \times \begin{pmatrix} 3 \\ 2 \end{pmatrix} + y \times \begin{pmatrix} 5 \\ -3 \end{pmatrix} = \begin{pmatrix} 4 \\ 9 \end{pmatrix}$$

$$\begin{pmatrix} 3x + 5y \\ 2x - 3y \end{pmatrix} = \begin{pmatrix} 4 \\ 9 \end{pmatrix}$$

This gives the simultaneous equations:

$$3x + 5y = 4$$
$$2x - 3y = 9$$

Solve these to obtain:

$$x = 3 \text{ and } y = -1$$

2) Illustrate on a graph the vectors **a, b, a + b**, where **a** and **b** are as in question 1.

Solution For **a** start from the origin and draw a
line 3 units to the right and 2 units up.
Draw **b** similarly.

For **a + b** start at the end of **a** and draw a
line 5 units to the right and 3 downwards.
Join the end of this line to the origin to
show **a + b**.

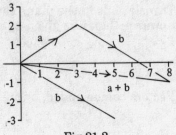

Fig 21.2

□ **21.1.2 Exercises.** *Intermediate, level 8*

1) For the vectors **a, b, c** defined below find   a) **a + b**,   b) **2a + 5b**,
    c) **2a - 3b**,   d) **a - b**.

$$a = \begin{pmatrix} 2 \\ -7 \end{pmatrix} \quad b = \begin{pmatrix} 5 \\ 1 \end{pmatrix}$$

2) With the vectors **a** and **b** as in question 1, evaluate:

    a) |**a**| b) |**b**| c) |**a + b**| d) |**2a + 3b**| e) |**a − b**|

3) Find x and y from the following vector equation:

$$\begin{pmatrix} x \\ 2 \end{pmatrix} + \begin{pmatrix} 3 \\ y \end{pmatrix} = \begin{pmatrix} 7 \\ 9 \end{pmatrix}$$

4) Find x and y from the following vector equation:

$$\begin{pmatrix} 2 \\ 1 \end{pmatrix} + \begin{pmatrix} -1 \\ y \end{pmatrix} = \begin{pmatrix} x \\ 3 \end{pmatrix}$$

5) With **a** and **b** as in Question 1, solve the following vector equations:

    a) **v + a = b**   b) **v − b = 2a**

    c) $x\mathbf{a} + y\mathbf{b} = \begin{pmatrix} 22 \\ -3 \end{pmatrix}$ d) $\lambda\mathbf{a} + \mu\mathbf{b} = \begin{pmatrix} 9 \\ -13 \end{pmatrix}$

6) On squared paper draw the vectors **a, b** of Question 1. Shift the tail of **b**
    to the head of **a**, and hence draw **a + b**.

7) On squared paper draw **c, d, c + d, c − d**, where:

$$c = \begin{pmatrix} 2 \\ 1 \end{pmatrix} \qquad d = \begin{pmatrix} 3 \\ 5 \end{pmatrix}$$

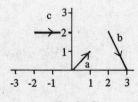

8) Write down in terms of co-ordinates the vectors shown
    in Fig 21.3.

Fig 21.3

9) Let **a** and **b** be as in Question 1. Write down the co-ordinates of A(1,1) after translation by

    a) **a**   b) **b**   c) **b + b**.

10) Under a translation, A(3,4) is moved to B(4,2). Write down the vector of the translation. What translation takes B to A?

21.2 Vector Geometry

If A and B are two points in the plane the vector from A to B is written <u>AB</u>.

The vector which goes from the origin to A is called the *position vector* of A.

Two vectors are parallel when one is a multiple of the other.

☐ **21.2.1 Examples.** *Intermediate, level 8*

1) A quadrilateral has vertices A(1,1), B(2,3), C(3,4), D(2,2). Show that ABCD is a parallelogram but is not a rhombus.

   Solution The vectors of the sides of the quadrilateral are found by subtracting the co-ordinates.

   $$\underline{AB} = \begin{pmatrix} 2-1 \\ 3-1 \end{pmatrix} = \begin{pmatrix} 1 \\ 2 \end{pmatrix} : \quad \underline{DC} = \begin{pmatrix} 3-2 \\ 4-2 \end{pmatrix} = \begin{pmatrix} 1 \\ 2 \end{pmatrix}$$

   Since <u>AB</u> = <u>DC</u>, opposite sides are equal and parallel. Hence ABCD is a parallelogram.

   $$|\underline{AB}| = \sqrt{1^2 + 2^2} = \sqrt{5}$$

   $$|\underline{AD}| = \sqrt{1^2 + 1^2} = \sqrt{2}$$

   Since adjacent sides are not equal, ABCD is not a rhombus.

2) ABC is a triangle, and <u>AB</u> = **b**, <u>AC</u> = **c**. X and Y are the midpoints of <u>AB</u> and <u>AC</u> respectively. Express in terms of **b** and **c**:

   a) <u>BC</u>   b) <u>AX</u>   c) <u>AY</u>   d) <u>XY</u>.

   What can you conclude about BC and XY?

   If △ABC has area 10, what is the area of △AXY?

   Solution Go from B to C by way of A. <u>BA</u> is minus <u>AB</u>, which gives:

   a) <u>BC</u> = –**b** + **c** = **c** – **b**.

   <u>AX</u> is halfway along <u>AB</u>. Hence:

   b) $\underline{AX} = \frac{1}{2}\mathbf{b}$ and   c) $\underline{AY} = \frac{1}{2}\mathbf{c}$

   By similar reasoning to part a) :

   d) $\underline{XY} = \frac{1}{2}\mathbf{c} - \frac{1}{2}\mathbf{b} = \frac{1}{2}(\mathbf{c} - \mathbf{b})$.

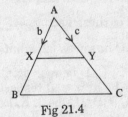

Fig 21.4

   Since $\underline{XY} = \frac{1}{2}\underline{BC}$, conclude that XY is parallel to BC and half its length.

   The two triangles are similar in the ratio 1:2. Hence the areas are in the ratio $1^2:2^2 = 1:4$.

   The area of △AXY is $\frac{10}{4} = 2.5$

❏ **21.2.2 Exercises.** *Intermediate, level 8*

1) Four points in the plane are A(1,2), B(2,4), C(2,-4), D(4,0). Write down the vectors <u>AB</u>, <u>AC</u>, <u>CD</u>, <u>BD</u>, <u>DA</u>. Which of these vectors are parallel to each other?

2) J(-3,1), K(-2,-2), L(-1,6), M(2,-3) are four points in the plane. Write down the vectors <u>JL</u>, <u>JK</u>, <u>ML</u>, <u>MK</u>, <u>MJ</u>. Which of these vectors are parallel to each other?

3) Four points in the plane are A(2,3), B(5,9), C(2,2), D(3,4). Show that <u>AB</u> is parallel to <u>CD</u>. What is the ratio of their lengths? Do the four points form a parallelogram?

4) A quadrilateral ABCD has vertices at A(-1,1), B(1,1), C(4,-1), D(3,-1). Show that ABCD is a trapezium. Is it a parallelogram?

5) Show that the four points P(1,0), Q(3,3), R(4,2), S(2,-1) form a parallelogram. Find the lengths of the sides. Is PQRS a rhombus?

6) A quadrilateral has its vertices at W(1,6), X(1,1), Y(4,5), Z(4,10). Show that WXYZ is a rhombus.

7) J(1,2), K(5,4), L(6,2), M(2,0) are the four vertices of a quadrilateral. Show that they form a parallelogram. By considering the lengths of the diagonals show that they form a rectangle.

8) W(1,2), X(2,4), Y(5,3), Z(4,1) are the four vertices of a quadrilateral. Prove that they form a parallelogram. Do they form a rectangle? Do they form a rhombus?

9) A(-4,4), B(-1,3), C(1,2), D(5,1) are four points in the plane. By considering the vectors between them, find out which three of them lie on a straight line.

10) Which three of the points L(-2,-2), M(0,-1), J(2,1) and K(4,1) lie on a straight line?

11) Given the three points A(1,1), B(2,3), C(1,2) find the point D such that ABCD is a parallelogram.

12) Given L(0,-2), M(-1,3), N(4,4) find the point P so that LMNP is a parallelogram.

13) Find x to ensure that A(1,1), B(2,5), C(x,9) lie on a straight line.

14) ABC is a triangle, and <u>AB</u> = **b**, <u>AC</u> = **c**. D and E lie on AB and AC respectively so that AD = $\frac{1}{4}$AB and AE = $\frac{1}{4}$AC. Express in terms of **b** and **c**:

   <u>BC</u>, <u>AD</u>, <u>AE</u>, <u>DE</u>, <u>DB</u>, <u>DC</u>.

   Which of these vectors are parallel to each other?

   If △ABC has area 32, what is the area of the quadrilateral DBCE?

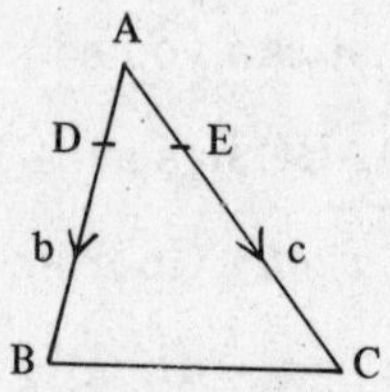

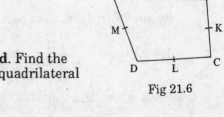

Fig 21.5

15) Let A, B, C, D be four points, and let their position vectors be **a**, **b**, **c**, **d** respectively. Let the midpoints of AB, BC, CD, DA be J, K, L, M respectively.

   Find the position vectors of J, K, L, M in terms of **a**, **b**, **c**, **d**. Find the vectors <u>JK</u>, <u>KL</u>, <u>LM</u>, <u>MJ</u>. What can you deduce about the quadrilateral JKLM?

Fig 21.6

16) Fig 21.7 shows three equilateral triangles arranged in a row. <u>AB</u> = **b**, <u>AC</u> = **c**. Express in terms of **b** and **c**:

   <u>BD</u>, <u>CD</u>, <u>AE</u>, <u>BC</u>

17) OABCDE is a regular hexagon, with the position vectors of A and E relative to O being **a** and **e** respectively. Express in terms of **a** and **e**:

   <u>BC</u>, <u>CD</u>, <u>EB</u>, <u>OB</u>.

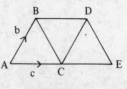

Fig 21.7

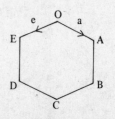

Fig 21.8

18) (For draughts players)

Think of a draughts-board as graph paper, in which the *x*-axis is the base line for White. Each white piece can move one forward diagonally, so its possible movements are described by the vectors:

$$\begin{pmatrix} 1 \\ 1 \end{pmatrix} \quad \text{and} \quad \begin{pmatrix} -1 \\ 1 \end{pmatrix}$$

Write down the possible movements for the black pieces and for the kings.

19) (For chess-players)

Write down the vectors describing the possible movements for chess pieces.

20) Which three of the points with position vectors **a**, **b**, **a** + **b**, **a** - **b** lie on the same straight line?

21) ABCD is a parallelogram: $\underline{AB}$ = **b**, $\underline{AD}$ = **d**. Let X, Y be the midpoints of AB and CD respectively. Express in terms of **b** and **d**:

$\underline{AC}$, $\underline{AX}$, $\underline{AY}$, $\underline{XY}$.

What can you conclude about the line XY?

22) OABC is a parallelogram, in which the position vectors of A, B, C relative to O are **a**, **b**, **c** respectively. Let X and Y be the midpoints of the diagonals OB and AC respectively.

Express **b** in terms of **a** and **c**.

Find expressions for $\underline{AB}$, $\underline{AY}$, $\underline{OX}$, $\underline{OY}$.

What can you say about the points X and Y?

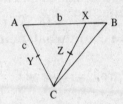

23) In the triangle ABC, $\underline{AB}$ = **b** and $\underline{AC}$ = **c**. X lies on AB so that $AX = \frac{3}{4}AB$. Y is the midpoint of AC, and Z is the midpoint of CX.

Find in terms of **b** and **c** the vectors $\underline{AY}$, $\underline{AX}$, $\underline{CX}$, $\underline{CZ}$, $\underline{AZ}$, $\underline{ZY}$. What can you say about ZY?

Fig 21.9

24) In the triangle OAB, P is the midpoint of AB, and Q lies on OA so that $OQ = \frac{2}{3}OA$. Extend OB to R so that OB = BR.

Let the position vectors of A and B relative to O be **a** and **b**. Find the position vectors of P, Q, R. Find $\underline{PQ}$ and $\underline{PR}$. What can you say about the points P, Q, R?

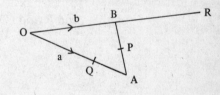

Fig 21.10

21.3 Applications of Vectors

Vectors have direction as well as size. There are many physical quantities, such as velocity and force, which are best described by vectors.

❒ **21.3.1 Examples.** *Higher, level 9*

1) A plane can fly at 200 m.p.h. The pilot wishes to fly 400 miles North, but there is a wind from the East of 40 m.p.h. In which direction should he fly? How long will the journey take him?

Solution   Let **v** and **u** be the wind velocity and the plane's velocity relative to the air. Then the actual velocity, relative to the ground, is **v** + **u**. The pilot wants this actual velocity to be due North. Hence the diagram on the right must be fitted.

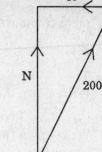

The direction in which the pilot should steer is given by trigonometry. Take $\sin^{-1}$ of $\dfrac{40}{200}$:

He should steer at a bearing of 011.5°

His actual speed, relative to the ground, is given by Pythagoras:

$$\text{Actual speed} = \sqrt{200^2 - 40^2} = 196 \text{ m.p.h.}$$

Fig 21.11

$$\textbf{Time taken} = \frac{400}{196} = \textbf{2.04 hours}$$

2)   A force of 5 N acts due North, and a force of 6 N acts due East. Find a single force which will counterbalance them.

Solution   Draw the forces as shown. The third side of the triangle is the force which will cancel out the original forces. Use Pythagoras to find its length, and trigonometry to find its direction.

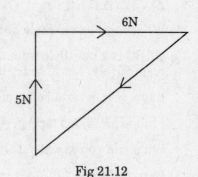

The counterbalancing force is 7.81 N at 230°

3)   Two forces are $\mathbf{F} = \begin{pmatrix} 3 \\ 4 \end{pmatrix}$ and $\mathbf{G} = \begin{pmatrix} 1 \\ 4 \end{pmatrix}$ Find a)   the magnitude of **F**

Fig 21.12

b)   the force which will counterbalance both of them.

Solution   a)   Use Pythagoras to find the magnitude of **F**.

$$|\mathbf{F}| = \sqrt{3^2 + 4^2} = 5$$

b)   Add F and G to obtain $\begin{pmatrix} 4 \\ 8 \end{pmatrix}$. The counterbalancing force will be minus this.

$$\textbf{The counterbalancing force is } \begin{pmatrix} -4 \\ -8 \end{pmatrix}.$$

❏ **21.3.2 Exercises.** *Higher, level 9*

1)   The grid shown represents a river. **w** is the water velocity and **r** is the velocity of a rower relative to the water. Draw the vector representing the velocity of the boat relative to the bank. If the boat starts at X where does it reach the other side?

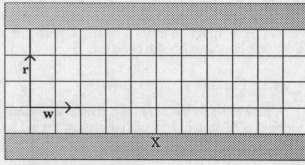

2)   A river flows at 3 km.p.h. A man can row at 5 km.p.h. If he points the boat directly across the river, find his actual speed and direction.

Fig 21.13

3)   If the rower of Question 1 wishes to go directly across the river, in what direction should he steer? If the river is 0.2 km wide, how long will it take him to cross?

4)   A plane can fly at 400 m.p.h. There is a 50 m.p.h wind from North to South. If the plane is pointed East, what bearing and speed will it fly at?

5) If the pilot of Question 4 wishes to fly East, what bearing should be set? What will be the speed of the plane?

6) Two forces are of 5 N due North and 7 N due East. Find the size and direction of the single equivalent force.

7) Two forces are of 6 N North West and of 7 N North East. Find the size and direction of the single force which will counterbalance them.

8) Two forces are given by $A = \begin{pmatrix} 2 \\ 8 \end{pmatrix}$ and $B = \begin{pmatrix} 3 \\ 4 \end{pmatrix}$. Find the single equivalent force C. Find $|A|$, $|B|$, $|C|$.

9) Find in vector form the force needed to counteract $3A + 2B$, where A and B are as in Question 8.

Common Errors

1) **Co-ordinates**

Be careful not to confuse vectors with points. $(3,4)$ is a point, and $\begin{pmatrix} 3 \\ 4 \end{pmatrix}$ is a vector which goes from the origin to that point.

The vector will also go from $(1,1)$ to $(4,5)$, or from $(-1,-3)$ to $(2,1)$ and so on. The vector does not have any position: it can start from anywhere in the plane.

2) **Modulus**

Be careful when working out the modulus of a vector. If v is the vector $\begin{pmatrix} 3 \\ 4 \end{pmatrix}$, then its modulus is given by:

$|v| = \sqrt{3^2 + 4^2} = 5$.

The modulus of v is not equal to $3 + 4$.

Similarly, when adding vectors do not add the moduli.

$|a + b| \neq |a| + |b|$.

Be careful with negative signs. The modulus of $\begin{pmatrix} -3 \\ 4 \end{pmatrix}$ is 5, not $\sqrt{7}$.

3) **Vector Geometry**

a) A vector is only a position vector if it starts from the origin. Otherwise take account of where it does start.

b) The vector from $(3,2)$ to $(5,7)$ is $\begin{pmatrix} 2 \\ 5 \end{pmatrix}$, not $\begin{pmatrix} -2 \\ -5 \end{pmatrix}$,

If A has position vector a, and B has position vector b, then $\underline{AB}$ is $b - a$. It is not b, or $a + b$, or $a - b$.

4) Do not confuse relative velocity (relative to the water or the air) with absolute velocity.

Chapter 22

Statistical Diagrams

A *statistical diagram* makes sense of a whole mass of data by summarizing the important results in a picture.

22.1 Tables, Bar Charts, Pie-Charts

A set of statistical figures can often be divided into several groups. There are several ways of displaying these groups.

A *table* gives the numbers in each of the different groups.

A *bar-chart* shows the numbers in each group by bars, whose lengths are proportional to the numbers.

A *pie-chart* compares the sizes by representing them as the slices of a pie.

◻ **22.1.1 Examples.** *Foundation, levels 4, 5*

1) A group of schoolchildren were asked how they had come to school that morning. The following table gives their answers.

Type of transport	Number of children
Car	7
Bus	12
Train	9
Cycle	5
Walking	17

a) How many walked to school?

b) How many children were there in all?

Solution a) Read from the table, to find that:

17 walked to school

b) Add up all the figures, to find that:

7 + 12 + 9 + 5 + 17 = 50 children were asked

2) Erica's record collection has pop, jazz, classical and country records. She shows how many she has of each by a table and by a bar chart.

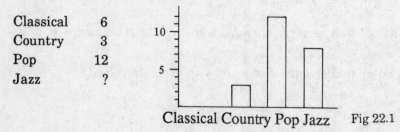

Classical	6
Country	3
Pop	12
Jazz	?

Fig 22.1

a) How many jazz records does she have?

b) Fill in the bar corresponding to her Classical records.

c) How many records does she have in all?

Solution   a)   From the bar chart, the bar for jazz is 8 high.

She has 8 jazz records

b) The bar corresponding to Classical records is 6 high. Fill in the bar chart as shown.

c) Add together all the figures in the table.

Classical Country Pop Jazz

Fig 22.2

She has 12+8+6+3 = 29 records in all

3) All the children in a class of 24 were asked what their favourite sport was. The answers were illustrated in the pie chart shown.

a) How many preferred Soccer?

b) If 4 children liked rugger best, what is the size of the angle in the rugby slice?

Fig 22.3

Solution   a)   There are 120° in the soccer slice, out of a total of 360°. 120° is a third of 360°. Hence the number of children who liked soccer best is one third of the total:

$$\frac{1}{3} \times 24 = 8 \text{ children liked soccer best}$$

b) 4 children out of 24 prefer rugger. 1 in 6 of the children prefer rugger. The slice is one sixth of the whole pie.

$$\text{The angle is } \frac{360°}{6} = 60°$$

❑ **22.1.2 Exercises.** *Foundation, levels 4, 5*

1) A class of children were asked how many brothers or sisters they had. The answers were displayed in the following table:

Number of brothers or sisters	0	1	2	3	4	5	6+
Frequency	5	9	8	4	2	1	0

a) What was the most common number of brothers and sisters? 1

b) How many only children were there? 5

c) How many children were there in the class? 29

2) A survey of the colour of cars was carried out in a car-park. The results were as follows:

Colour	Red	Black	White	Blue	Yellow	Green	Other
Number	4	7	6	8	3	4	8

a) How many red cars were there? 4

b) How many cars were blue or green? 12

c) How many cars were not black? 33

3) The results in an exam were as follows:

Grade	A	B	C	D	E	F	U
Numbers	23	47	124	98	65	43	21

a) How many candidates got grade A? 23

b) How many got below grade D? 129

c) How many took the exam in all? 421

d) If grades A, B and C represent a pass, how many passed the exam? 194

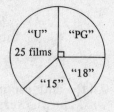

Fig 22.4

4) The cinemas in a town have shown 60 films throughout the year. The category of film is shown by the pie chart in Fig 22.4.

a) How many PG films were there? 15   360 ÷ 24 = 15

b) What is the angle for the U sector of the pie? 150°

60 ÷ 25 = 2·4   360/2·4 = 150

5) A boy makes a pie-chart showing how he spends each day. It is given in Fig 22.5.

a) For how many hours is he at school? 6

b) What is the angle for the meals sector of the chart? 15

c) For how many hours is he awake? 16

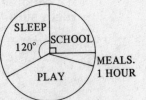

Fig 22.5

6) A girl receives some money for her birthday. She decides to spend it as shown in the pie-chart of Fig 22.6.

a) How much money did she receive in all?

b) How much did she spend on magazines?

Fig 22.6

7) The inflation rate for 5 countries is shown by the bar-chart in Fig 22.7. Fill in the table below to show the same information.

Country	UK	USA	France	Japan	Germany
Inflation	6	3	8	4	2

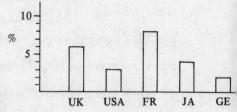

Fig 22.7

8) A survey of 60 people gave the following results about where they had gone for their holidays.

Holiday	Britain	Spain	Italy	Greece	Other
Numbers	23	12	6	5	14

Draw a bar chart on Fig 22.8 to show the same information.

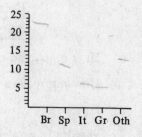

Fig 22.8

9) The bar-chart in Fig 22.9 shows the numbers of different types of ice-cream sold on a certain afternoon.

   a) How many vanilla flavoured ones were sold?

   b) What was the most popular flavour?

   c) How many ice-creams were sold in all?

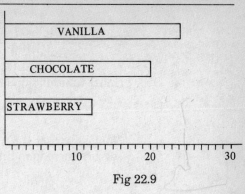

Fig 22.9

10) Out of 120 cars which were taken to a garage for repair, 30 had defective brakes, 45 were put in for service, 25 had to have body work done, and the remaining 20 were put in for other reasons. Complete the bar-chart of Fig 22.10 to illustrate this.

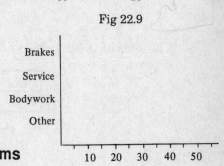

Fig 22.10

22.2 Construction of Pie-Charts, Frequency Tables. Histograms

A *frequency table* for different groups shows the numbers in each group.

A *histogram* is a special sort of bar-chart. The successive groups must be linked in a definite numerical order.

☐ **22.2.1 Examples.** *Foundation, levels 4, 5*

1) A girl did a statistical survey of the traffic passing along the road outside her house. She made up a tally sheet as shown. Convert it to a frequency table.

	Cars	Lorries	Buses	Motorcycles
Tally	̶HH̶ ̶HH̶	̶HH̶	̶HH̶	̶HH̶ ///
	̶HH̶ ̶HH̶	̶HH̶	/	̶HH̶
	̶HH̶ ̶HH̶	//		̶HH̶
	̶HH̶ /			̶HH̶

Solution   Each little bundle of figures ̶HH̶ corresponds to 5 units. Add up all the figures in each group, to obtain:

Cars	Lorries	Buses	Motorcycles
36	12	6	23

2) A survey into the tobacco habits of 400 people found how many people consumed different sorts of tobacco. The results are given by the following frequency table:

Cigarettes	Pipes	Cigars	Snuff	Non-user
100	60	40	10	190

Construct a pie-chart to illustrate these figures.

Solution   For each group we construct a sector, whose angle is proportional to the number of people in that group. The cigarette smokers are a quarter of the whole, so their sector has angle $\frac{360°}{4} = 90°$. The other figures are:

Pipe-smokers: $\frac{60}{400} \times 360 = 54°$ Cigar-smokers: $\frac{40}{400} \times 360 = 36°$

Snuff-takers: $\frac{10}{400} \times 360 = 9°$ Non-Users: $\frac{190}{400} \times 360 = 171°$

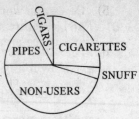

Fig 22.11

Now construct the pie-chart, dividing the circle into sectors corresponding to each group.

3) The ages of 100 members of a football club were given by the following frequency table:

Ages	15-20	20-25	25-30	30-35
Frequency	26	37	23	14

Construct a histogram to show these figures.

Solution   The construction of a histogram is the same as for a bar-chart.

There is a numerical relationship between the groups. Along the *x*-axis measure the ages of the players. The result is shown in Fig 22.12.

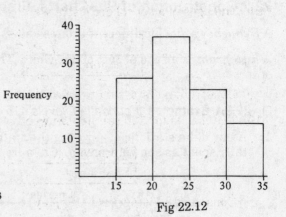

Fig 22.12

◻ **22.2.2 Exercises.** *Foundation, levels 4, 5*

1) A paragraph of text was analysed for the length of words occuring in it. The following tally chart was made:

Number of letters	1	2	3	4	5	6	7	8	9+
Tally	-IHT -IHT //	-IHT -IHT ///	-IHT -IHT /	-IHT -IHT ///	-IHT -IHT ////	-IHT -IHT	//// //	///	-IHT ///

Draw up a frequency table for the numbers of letters in a word.

2) A die is rolled 60 times. The following tally chart shows how often each of the numbers came up:

Score	1	2	3	4	5	6
Tally	-IHT -IHT ///	////	-IHT -IHT	-IHT -IHT //	-IHT /	-IHT -IHT -IHT

Draw up a frequency table for this information.

3) Throw a die 30 times, and draw up a table showing the frequency of each number.

4) Ask the members of your class how they travelled to school this morning. Record the information in a frequency table.

5) 60 people were asked what they drank first in the morning. 23 drank tea, 15 coffee, 12 chocolate and 10 other drinks. Construct a pie-chart to show this information.

6) An education authority spends its money as shown in the following table:

Type of Education	Expenditure in £millions
Primary	120
Secondary	135
Further and Higher	120
Other	165

Construct a pie-chart to show this information.

7) A woman calculates that of her income of £200 per week, £40 goes in tax, £50 on rent, £30 on food, £30 on entertainment, and £50 for other items. Draw up a pie-chart to show this.

8) The survey of a county showed that 20% of it was built-up, 50% was farmland, and 30% formed part of a National Park. Display this information on a pie-chart.

9) The ages of a school orchestra were found to be as given in the following frequency table:

Age	11-12	12-13	13-14	14-15	15-16
Frequency	2	7	11	15	12

Construct a histogram to show this information.

10) The marks for 120 candidates were as follows:

Marks out of 40	0-9	10-19	20-29	30-40
Frequency	20	35	43	22

Construct a histogram to display these marks.

11) The manager of a restaurant did a survey on how much his customers spent per meal. His results were as follows:

Price in £s	3.00-4.99	5.00-6.99	7.00-8.99	9.00-10.99
Frequency	10	39	27	14

Draw a histogram to show these results.

12) In an effort to cut down on its telephone bill, a company did a survey of the length of time of calls from its office. The results were shown on the following table:

Length in Minutes	0-3	3-6	6-9	9-12	12-15
Frequency in 10's of calls	53	25	15	7	2

Construct a histogram to illustrate these figures.

Shape of Histogram and Frequency Polygon

A *frequency polygon* is obtained from a histogram by drawing a line through the tops of the bars.

The thinness of a histogram or frequency polygon is an indication of how close together the measurements are.

☐ **22.2.3 Example.** *Intermediate, level 7*

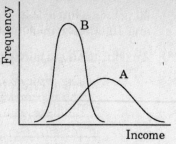

Fig 22.13

Frequency polygons showing the distribution of incomes in two countries are shown on the right. Comment on the difference.

> *Solution* Notice that the polygon for country A is to the right of that for country B, and is wider.

Incomes in A are higher, and more widely spread

☐ **22.2.4 Exercises.** *Intermediate, level 7*

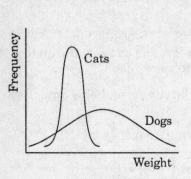

Fig 22.14

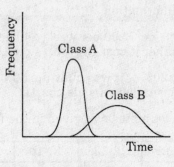

Fig 22.15

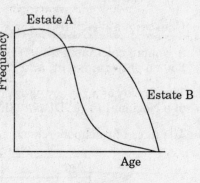

Fig 22.16

1) Frequency polygons shown the weights of groups of cats and dogs are shown. Comment on the difference. Fig 22.14.

2) Two classes each entered children in a race. The frequency polygons on the right show the times taken. Comment on the difference. Fig 22.15.

3) The age distribution of two council estates are shown on the right. Comment on the difference. Fig 22.16.

4) Class A and Class B were entered for an exam. Class A did better overall, and Class B showed a much wider spread of ability. Sketch possible frequency polygons for the distribution of marks for the two classes.

5) An opposition party claims that since the last election the rich have got richer and the poor have got poorer. Show how they might illustrate this by drawing frequency polygons of income distributions before and after the election.

22.3 Histograms with Unequal Intervals. *Higher, level 9*

The height of a bar on a histogram represents the frequency *per unit interval*. Hence if an interval is twice the normal width, then its height must be halved. If an interval is one third the normal width, then its height must be tripled.

☐ **22.3.1 Example.** *Higher, level 9*

The marks of 120 candidates in an exam were given by the following table:

% Marks	0-29	30-39	40-49	50-59	60-64	65-69	70-79	80-100
Frequency	12	10	16	24	17	14	19	8

a) Construct a histogram to show this information.

b) What is the modal class?

c) Estimate the proportion of candidates who got at least 55%.

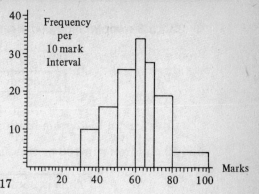

Fig 22.17

Solution   Notice that the intervals are not even. There must be a corresponding adjustment to the height of the bars.

The most common interval is of 10 marks.

The intervals at the end and the beginning are of 20 and 30 marks respectively. Divide each of their frequencies by 2 and 3 respectively.

The intervals in the 60's are of 5 marks each. Double the marks in those intervals.

a)   The histogram is in Fig 22.17.

b)   The modal class is the one in which the population is densest. This is equivalent to the highest bar on the histogram.

The modal group is 60-64

c)   Assume that half the candidates in the 50-59 range got at least 55%. This gives 12+17+14+19+8 = 70 candidates in all.

The proportion with at least 55% is $\frac{70}{120} = \frac{7}{12}$

❏ **22.3.2 Exercises.** *Higher, level 9*

1)   The ages of a group of 500 people at a cinema were as follows:

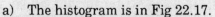

Age range	10-20	20-25	25-30	30-35	35-40	40-70
Frequency	74	73	104	116	67	66

a)   Construct a histogram, using a span of 5 years as the basic interval.

b)   What is the modal class?

c)   Estimate from your graph the proportion of the audience who were over $32\frac{1}{2}$.

2)   The averages of 80 cricketers were compared. The figures are shown in the following table:

Average	0-20	20-30	30-40	40-50	50-80
Frequency	22	18	16	12	12

a)   Draw a histogram to illustrate this data.

b)   Find the modal class.

c)   Estimate the percentage of cricketers whose average was less than 35.

3)   The salaries of 100 employees of a firm were analysed, and the following table shows the distribution:

Salary in £1,000's	7-9	9-10	10-11	11-12	12-13	13-16
Frequency	24	15	17	19	10	15

Draw a histogram to show the distribution.

4) 60 cats were weighed. The results were as follows:

Weight in kg.	1-3	3-3.5	3.5-4	4-4.5	4.5-6
Frequency	20	9	15	10	6

Construct a histogram to illustrate these figures.

5) 50 children ran 100 metres. Their times are given in the following table:

Time in secs.	12-14	14-15	15-15.5	15.5-16	16-17	17-20
Frequency	4	13	9	11	4	9

Construct a histogram to illustrate these times.

6) Out of 160 families, 33 had 0 children, 26 had one child, 45 had 2 children, 26 had 3 children, and 30 had 4, 5 or 6 children. Construct a histogram to show these figures.

7) A bridge player always counts the number of hearts in his hand. After 130 games, he finds that he has the following figures:

Number of hearts	0 or 1	2	3	4	5	6	7-13
Frequency	23	19	25	27	13	9	14

Show this information on a histogram.

22.4 Scatter Diagrams

A collection of points which are roughly but not exactly in a straight line is called a *scatter diagram*.

The straight line through the points which is closest to them is the *line of best fit*.

☐ **22.4.1 Example.** *Intermediate, level 7*

Different weights were put on the end of an elastic string, and the extension was measured. The weight in grams is given by x, and the extension in cm is given by y. The approximate results are given by the following table:

x	0	1	2	3	4	5
y	0.4	0.5	1.2	1.3	1.4	1.6

Plot these points on a scatter diagram. Draw a straight line through these points. Find the gradient of your line.

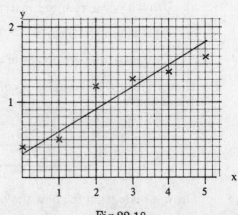

Solution The points are shown plotted in Fig 22.18. The line is shown.

The y change from end to end is $1.8 - 0.3 = 1.5$. The x change is 5. Divide the y change by the x change.

The gradient is $\dfrac{1.5}{5} = 0.3$

Fig 22.18

□ **22.4.2 Exercises.** *Intermediate, level 7*

1) For the following tables, plot the points as a scatter diagram on graph paper, and draw as close-fitting a straight line as you can. Find the gradient of your straight line.

a)

x	0	1	2	3	4
y	-0.2	1.0	3.0	3.9	5.4

b)

x	2	3	4	5	6
y	2.2	2.8	4.0	4.7	5.0

c)

x	1	2	3	4	5
y	27	20	13	14	3

2) The table below gives the heights and weights of 8 children. Plot them on a scatter diagram, draw a line of best fit and from the line estimate the weight of a 5 ft. child.

Height in inches	55	64	58	49	53	66	62	63
Weight in pounds	98	125	112	90	83	120	120	122

3) Ten pupils sat exams in Physics and Maths. The marks are given in the table below. Plot them on a scatter diagram, draw a line of best fit and from the line estimate the Physics mark of a pupils who scored 50 in Maths.

Maths mark	63	86	48	36	59	68	77	73	45	42
Physics mark	50	78	42	30	43	62	62	68	32	45

Common Errors

1) **Tables**

a) When measuring the frequencies of numerical data, be careful not to confuse a number with the frequency of that number. Suppose, for example, a die is thrown 100 times, and the number 5 comes up 20 times. Be sure that 20, not 5, goes into the frequency box.

b) When constructing a frequency table from a tally chart, be sure that each ////- symbol counts for 5, not 4. The cross stroke represents an entry as well as the vertical strokes.

2) **Bar-charts**

If you are drawing a bar-chart, and not a histogram, draw the bars separately and do not join up them up.

Suppose you are drawing a bar-chart representing car-production in different countries. Draw the bars for France and Japan apart from each other. It makes no sense to join up the bars, as there is no country 'half-way' between France and Japan.

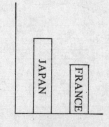

Fig 22.19

3) Histograms

a) If a histogram is measuring continuous data, such as time or height or weight, then the bars should be touching each other.

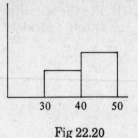

Suppose your histogram compares the weights of certain items. The bar corresponding to the range 30-40 grams should touch the bar corresponding to 40-50 grams.

b) If your histogram is measuring continuous data, then do not worry that the same figure is the right end of one interval and the left end of another, as in the example just above. It is almost impossible that an item weighs *exactly* 40 grams.

Fig 22.20

It would definitely be wrong to have intervals of 30-39 grams and of 40-50 grams. Where would you put an item of 39.5 grams?

4) Histograms with unequal intervals

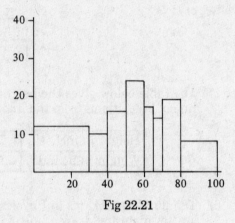

a) Be sure that you adjust the height of the bars to compensate for the different widths of the intervals. If you do not do so, the histogram will be misleading. The example of 22.3.1, without such compensation, would be as shown.

We can see that too much importance is given to the end intervals, and too little importance to the middle intervals.

b) Be sure that you label the vertical axis clearly. It is not enough to label it 'Frequency', you must also explain what the basic interval is. Label it 'Frequency per 5 marks', or 'Frequency per 10 cm.' etc.

Fig 22.21

c) The modal group is not necessarily the group with the greatest number in it. It is the group with the greatest density, i.e. it is the group with the tallest bar.

5) Scatter Diagrams

A line through a scatter diagram does not necessarily go through the first and last points. The gradient must be of the line you have drawn, not of the line through the first and last points.

Chapter 23

Averages and Cumulative Frequency

23.1 Means, Modes, Medians

An *average* summarizes a collection of numbers.

The *mean* of a set of numbers is what is normally meant by the word average. The mean age of ten children is the sum of all their ages divided by 10.

The *median* of a set of numbers is the middle number. Half of the numbers are less than it and half of the numbers are greater than it.

The *mode* of a set of numbers is the most commonly occuring number.

The *range* of a set of numbers is the difference between the highest and the lowest numbers.

❑ **23.1.1 Examples.** *Foundation, level 4*

1) 12 children took a test, and their marks out of 10 were as follows:

    8, 8, 7, 7, 6, 8, 9, 5, 6, 7, 7, 6.

    Find the mean, the mode and the range.

Solution For the mean, find the sum of all the marks.

    8 + 8 + 7 + 7 + 6 + 8 + 9 + 5 + 6 + 7 + 7 + 6 = 84.

    Divide by 12 to obtain:

$$\text{The mean mark is } \frac{84}{12} = 7$$

The most common mark is 7, which occurs 4 times.

The mode is 7

The highest score is 9, and the lowest is 5.

The range is 4

2) In 8 successive innings, a batsman scored the following:

    45, 65, 4, 0, 76, 12, 8, 30.

    Find his median score.

Solution Re-arrange the scores in increasing order:

    0, 4, 8, 12, 30, 45, 65, 76.

The median score is between his fourth best and his fifth best. Take halfway between 12 and 30:

The median is 21

❑ **23.1.2 Exercises.** *Foundation, level 4*

1) Find the mean, mode and range of the following numbers:

5, 6, 4, 5, 10, 3, 9, 4, 9, 5.

2) Find the mean, median and range of the following numbers:

23, 38, 20, 30, 31, 19, 27, 24, 32, 26.

3) 6 people were asked to estimate the distance of a church. They gave the answers:

100 m, 120 m, 90 m, 100 m, 150 m, 150 m.

Find the mean and the median of these distances.

4) 8 children were asked to measure the width of their right hands. The results were (in cm):

6.3, 7.1, 5.8, 6.0, 5.5, 5.8, 6.2, 5.3.

Find the mean and the median of these widths.

5) A golfer had the following scores in 9 rounds:

103, 96, 110, 99, 100, 92, 105, 109, 113.

Find the mean, median and range of these scores.

6) A football team scored the following number of goals in 12 matches:

0, 3, 1, 0, 2, 3, 6, 3, 4, 0, 2, 0.

Find the mean and mode of these scores.

7) Richard bought 10 matchboxes. He counted the number of matches in each. The results were as follows:

48, 50, 48, 51, 47, 45, 43, 47, 49, 52.

What is the mean number of matches?

8) The weights of 5 apples were as follows:

43 g, 50 g, 46 g, 50 g, 47 g.

Find the mean weight.

9) The salaries of 8 employees of a firm were:

£8,000, £7,500, £6,500, £10,000, £6,000, £7,400, £5,000, £22,000.

Find the mean salary and the median salary.

10) A boy surveyed the cars coming through the school gates, and noted down the number of people (including the driver) in each car. The results for the first 10 cars were:

1, 3, 1, 1, 2, 3, 4, 4, 1, 1.

Find the mean and mode of these figures.

11) A shopper compared the prices of a jar of honey in 8 different shops. The prices were:

80p, 75p, 88p, 83p, 81p, 77p, 90p, 86p.

Find the mean price and the median price.

12) 8 packets of butter were weighed and found to be:

252 g, 247 g, 253 g, 250 g, 250 g, 248 g, 247 g, 246 g.

Find the mean weight and the median weight.

13) For the seven days of his holiday George measured the hours of sunshine. His results were:

$$8.3, 7.3, 2.1, 9.0, 6.5, 7.3, 8.5.$$

Find the mean and median of these figures. Find the range of values.

14) On 5 successive days a television company estimated the number of people who had watched its news programme. The results, in millions of viewers, were as follows:

$$4.3, 5.3, 4.2, 4.6, 3.8.$$

Find the mean and the median number of viewers.

15) 9 batteries were used until they faded. The lifetimes in hours were:

$$53, 47, 38, 56, 22, 50, 45, 39, 35.$$

Find the mean lifetime and the median lifetime. Find the range.

16) For her first 7 litters, a cat has the following numbers of kittens:

$$1, 2, 3, 3, 5, 4, 3.$$

Find the mean number of kittens per litter and the modal number.

17) The 8 classes of a year at a school contain the following number of pupils:

$$27, 28, 24, 25, 26, 25, 26, 25.$$

Find the mean and median number of pupils per class.

18) The 24 children in a class were asked how they had come to school. Their answers were as follows:

Type of transport	Bus	Bike	Train	Walk	Car
Frequency	6	3	8	5	2

What is the modal type of transport? Why does it not make sense to ask for the mean or the median type of transport?

19) Over a period of 3 months Bill takes 15 books out of the library. 6 are Science-fiction, 5 are Crime books, and 4 are Thrillers. What is the modal group? Can one refer to the mean book or the median book?

23.2 Averages from Frequency Tables

If data is presented in the form of a frequency table the average can be found from it.

❒ **23.2.1 Examples.** *Intermediate, level 7*

1) A survey was made of 100 families, to investigate the number of children in each family. The results were given in the following frequency table:

Number of children	0	1	2	3	4	5	6	7	8+
Number of families	11	24	35	16	9	3	1	1	0

Find the mean number of children per family and the median number.

Solution The total number of children is found by multiplying the numbers of children by the frequencies.

$$0 \times 11 + 1 \times 24 + 2 \times 35 + 3 \times 16 + 4 \times 9 + 5 \times 3 + 6 \times 1 + 7 \times 1 + 8 \times 0 = 206$$

There are 100 families in all, which gives:

$$\text{The mean number is } \frac{206}{100} = 2.06$$

The median number is the number of children such that half the families have more than that number, and half the families have less. Only a very imperfect answer is possible here: 35 families have less than 2 children and 30 have more than 2. Hence:

The median number is 2

2) The Post Office conducted a survey of the weights of 500 letters in the lowest weight range. The results of the survey were as follows:

Weight in grams	0-10	10-20	20-30	30-40	40-50	50-60
Frequency	53	176	113	64	53	41

Find the modal class and estimate the mean weight of the letters. Estimate also the median.

Solution The modal class is the class with the largest number of letters in it.

The modal class is 10-20 grams

The table does not give the exact weight of all the letters. Without any more information the best that can be done is to assume that they are evenly spread along each interval.

Assume that the letters in the 0-10 range average at 5 grams each. Make a similar assumption in the other ranges. Work out the total weight:

$5 \times 53 + 15 \times 176 + 25 \times 113 + 35 \times 64 + 45 \times 53 + 55 \times 41 = 12{,}610$ grams.

There are 500 letters in all.

$$\text{Mean weight} = \frac{12\,610}{500} = 25.22 \text{ grams.}$$

The median will be the 250th weight. There are 229 letters less than 20 g., leaving 21 to go. Assuming that the 113 letters are evenly spread, the 21st of them will be $\frac{21}{113}$ of the way from 20 to 30.

$$\text{Median} = 20 + 10 \times \frac{21}{113} = 22 \text{ g}$$

☐ **23.2.2 Exercises.** *Intermediate, level 7*

1) The first 500 words of a book were examined to see how many letters they contained. The results are shown in the following table:

Number of letters	1	2	3	4	5	6	7	8	9	10	11+
Frequency	15	34	45	76	106	94	79	41	6	3	1

Find the mean, the mode and the median number of letters per word.

2) A class of 24 children were asked how many books they had read in the previous week. 6 had read none, 8 had read 1, 6 had read 2, 3 had read 3 and one had read 4. Find the mean number of books read.

3) The manager of a hotel investigated how long his guests stayed. The results were:

Number of nights	1	2	3	4	5	6	7	8	9	10+
Frequency	48	24	19	6	5	2	12	3	1	0

Find the mean and median number of nights stayed at the hotel.

4) The Ministry of Transport conducts a survey in a certain area to see how many driving tests people have before they pass. The results are as follows:

Number of tests	1	2	3	4	5	6	7	8+
Frequency	131	52	25	19	11	7	4	1

Find the mean number of tests needed to pass.

5) A traffic survey counted the number of people (including the driver) in cars during the rush hour period. The results were:

Number of people	1	2	3	4	5
Frequency	217	103	47	30	3

Find the modal number, and find the mean number.

6) A supermarket sells oranges in packs of 5. There are complaints about the number of over-ripe fruit. 150 packs are opened, and the number of rotten oranges in each pack are counted. The results are:

Number of bad oranges	0	1	2	3	4	5
Frequency	65	52	27	5	1	0

Find the mean and median number of bad oranges in a pack.

7) 20 children were asked how many foreign countries they had visited. The answers are given in the table:

Number of countries	0	1	2	3	4
Frequency	6	7	5	0	2

Find the mode and the mean number of countries visited.

8) Of 400 pupils entering a university, 50 had 2 A-levels, 297 had 3 A-levels, 42 had 4 and 11 had 5. Find the mean number of A-levels per pupil.

9) The 50 people at a party had ages given by the following frequency table:

Age range	15-20	20-25	25-30	30-35	35-40
Frequency	12	23	6	8	1

Find the modal class and estimate the mean age.

10) The heights of 100 boys were as follows:

Height in cm	50-60	60-70	70-80	80-90
Frequency	27	34	29	10

Find the modal class and estimate the mean height.

11) The salaries of 50 employees of a firm were as follows:

Salary in £1,000's	6-8	8-10	10-12	12-14	14-16	16-18
Frequency	8	13	17	7	3	2

Estimate the mean salary.

12) 30 children were asked how much television they watched each week. The answers were as follows:

Number of hours	0-2	2-4	4-6	6-8	8-10	10-12	12-14
Frequency	6	11	5	3	2	1	2

What is the modal class? Estimate the median number of hours watched.

13) British Telecom surveys the length of telephone calls in a particular area. The lengths of calls, measured to the nearest minute, are shown by the following table:

Length of call	0	1	2	3	4	5	6	7	8	9	10+
Frequency	52	197	110	85	42	36	23	19	15	8	13

Find the modal class and the mean length of a telephone call.

14) A gun was fired 80 times, and the range was measured to the nearest 100 m. The results were:

Range in 100 m's	23	24	25	26	27	28
Frequency	5	17	29	24	4	1

Estimate the mean range.

15) 120 candidates sat an exam, and the distribution of the marks is given by the following table:

% Marks	0-29	30-49	50-59	60-69	70-100
Frequency	16	33	31	26	14

Estimate the median mark.

23.3 Cumulative Frequency

The *cumulative frequency* at a value is the running total of all the frequencies up to that value.

A *cumulative frequency table* gives the cumulative frequencies at each value.

A *cumulative frequency chart* gives a graph of the cumulative frequency.

The cumulative frequency up to the median is $\frac{1}{2}$.

The cumulative frequency up to the *lower quartile* is $\frac{1}{4}$.

The cumulative frequency up to the *upper quartile* is $\frac{3}{4}$.

The *inter-quartile range* is the difference between the quartiles.

❏ **23.3.1 Example.** *Intermediate, level 8*

A man did a survey of the prices of second-hand cars advertised in his local newspaper. The results were as follows:

Price Range	0-499	500-999	1000-1999	2000-2999	3000-4000
Frequency	7	10	27	23	13

a) Fill up a cumulative frequency table.

b) Draw a cumulative frequency curve.

c) Find the median, the quartiles and the inter-quartile range.

d) What proportion of cars cost less than £1,500?

Solution a) Fill in the third row of the table, adding up the frequencies as you go along:

Price range	0-499	500-999	1000-1999	2000-2999	3000-4000
Frequency	7	10	27	23	13
Cum. Frequ.	7	17	44	67	80

b) The horizontal axis represents price. Label it from £0 to £4,000. The vertical axis represents cumulative frequency. Label it from 0 to 80. Now draw the graph.

The points on the graph correspond to the *endpoints* of the intervals. 44 cars cost less than £2,000, so the point (2000,44) must lie on the graph. The graph is shown in Fig 23.1

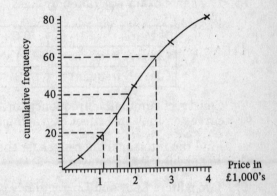

Fig 23.1

c) From the graph, 40 cars cost less than £1800.

The median is £1,800

20 cars cost less than £1,100, and 60 cars less than £2,600.

The lower quartile is £1,100. The upper quartile is £2,600

The interquartile range is £1,500

d) Take £1,500 on the horizontal axis. This corresponds to a cumulative frequency of 30.

The proportion of cars costing less than £1,500 is $\frac{30}{80} = \frac{3}{8}$

☐ **23.3.2 Exercises.** *Intermediate, level 8*

1) The heights of 60 boys were measured. The results are shown on the following table:

Height in cm	150-160	160-170	170-180	180-190
Frequency	10	23	18	9

a) Fill up a cumulative frequency table.

b) Draw a cumulative frequency curve.

c) Find from your curve the median, the quartiles and the inter-quartile range.

d) What proportion of boys were more than 165 cm in height?

2) The weights of 80 eggs were found to be as follows:

Weight in grams	50-60	60-70	70-80	80-100	100-120
Frequency	7	15	28	19	11

a) Construct a cumulative frequency table.

b) Draw a cumulative frequency graph.

c) From your graph find the median, the quartiles and the inter-quartile range.

d) What proportion of eggs lie between 60 and 90 grams?

3) to (7). For each of the situations in Questions (1) to (5) of Exercise 22.3.2 do the following:

a) Find the cumulative frequencies.

b) Draw a cumulative frequency chart.

c) Find the median, the quartiles and the inter-quartile range.

8) The same exam was taken by class A and class B. The results were as follows:

% Mark	0-29	30-39	40-49	50-59	60-69	70-100
Class A frequency	1	3	10	14	2	0
Class B frequency	2	5	7	9	4	3

Construct cumulative frequency tables for both the classes, and draw the cumulative frequency curves on the same sheet of graph paper. What is the difference between the curves?

Find the inter-quartile ranges for the two classes. What does the difference between the values tell you about the two classes?

9) Two golfers Lucy and Liz regularly play together over the same course. After they have each been round 40 times their scores are given by the following table:

Score	70-74	75-79	80-84	85-89	90-99	100-109
Lucy's frequency	2	3	23	7	5	0
Liz's frequency	5	8	14	5	4	4

Construct cumulative frequency tables for both players, and draw the cumulative frequency graphs on the same sheet of graph paper. Comment on the difference.

Evaluate the inter-quartile ranges for the two women. Explain the difference. Who do you think is the better player?

23.4 Standard Deviation

The interquartile range of a collection of data is a measure of how dispersed the figures are. Another measure is the *standard deviation*.

Suppose, for example, we have four numbers $a, b, c, d,$ whose mean is m.

The standard deviation is $\dfrac{\sqrt{(a-m)^2 + (b-m)^2 + (c-m)^2 + (d-m)^2}}{4}$

An alternative form is $\sqrt{\dfrac{a^2 + b^2 + c^2 + d^2}{4} - m^2}$

In many situations, the frequency polygon of a large amount of data has the bell-shape shown. This curve is the *normal* curve.

The centre of the curve is at the mean of the data. On both sides of the mean there are points, marked X, where the curve changes from convex to concave. The distance from the mean to these points is the standard deviation. (Fig 23.2.)

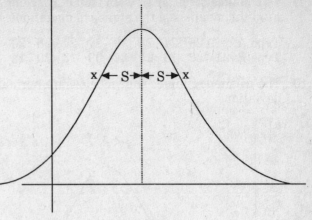

Fig 23.2

❏ 23.4.1 Example

The army is considering which make of field gun to order. Both were set for 1,000 m and were each fired 8 times. The results, in m, are below. Find the standard deviation for each gun. Which is the better gun?

A: 1008, 990, 989, 999, 1002, 1010, 1005, 1013
B: 1000, 1001, 999, 1004, 996, 999, 1003, 1002

Solution   Find the means for A and B by adding and dividing by 8.

Mean for A = 1002. Mean for B = 1000.5

Go through all the A figures, and find their differences from 1002. We obtain 1002 - 989 = 13, 1002 - 990 = 12 and so on. The *squared* differences for A are 13^2, 12^2 and so on. The sum of these squares is 552. The average of the squares is 69. Square root to obtain:

The standard deviation for A is $\sqrt{69} = 8.31$

Apply the same procedure for B, squaring the differences from 1000.5.

The standard deviation for B is $\sqrt{5.75} = 2.40$

The standard deviation for A is greater than for B. B is more accurate.

Gun B is better

❏ 23.4.2 Exercises

In Questions 1 to 4 find the standard deviation of the numbers.

1)   50, 70, 57, 63, 45, 48, 51, 64.

2)   90, 97, 97, 92, 94, 95, 92, 95.

3)   30, 26, 25, 36, 38, 32, 33, 27, 23.

4)   36.4, 35.2, 35.9, 36.1, 31.3, 34.2, 35.0, 35.6, 35.3.

5)   Eight cats and eight dogs were weighed. The results, in kilograms, are given below. Find the standard deviation for each set of figures, and comment on the result.

Cats: 3.3, 3.8, 2.9, 3.0, 3.4, 3.4, 3.1, 2.9.
Dogs: 5.6, 50, 20.1, 33, 2.0, 43, 10.2, 41.

6)   10 boys and 10 girls each took a test. The results are below. Find and compare the means and the standard deviations.

| Boys: | 12 | 13 | 9 | 14 | 17 | 20 | 5 | 14 | 8 | 18 |
| Girls: | 11 | 12 | 13 | 13 | 15 | 14 | 14 | 16 | 17 | 15 |

7) Two makes of batteries were tested by seeing how many hours they lasted. The results are below. Find the means and the standard deviations. Comment on your results.

Type A: 21 26 24 23 19 27 24 25 20 24 19 24
Type B: 21 33 15 25 32 39 13 20 12 14 31 27

8) The frequency curves shown below are normal. In each case estimate the mean and standard deviation.

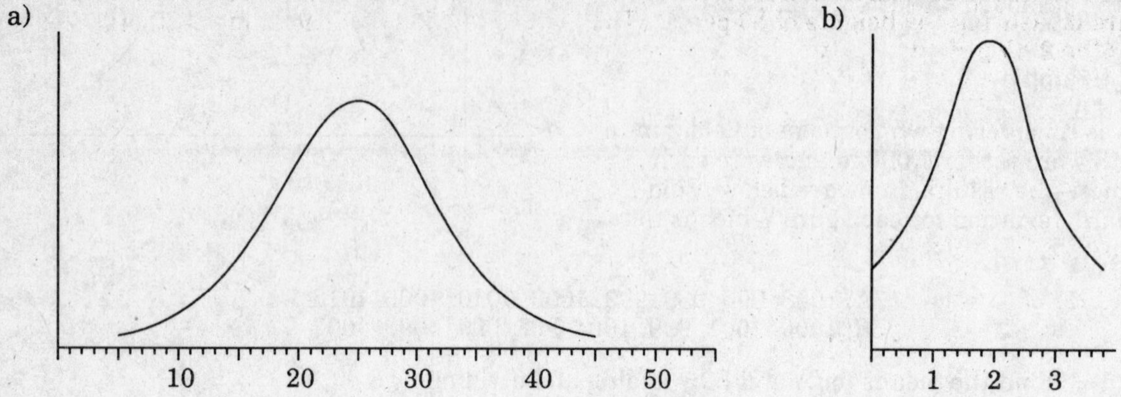

a)

b)

Fig 23.3

Common Errors

1) Be sure that you know the difference between the three averages, mean, median, mode. They are often confused.

2) **Mean**

   a) If the mean of a set of values is not an integer, it is wrong to round it to the nearest whole number. Though a family must have a whole number of children, it still makes sense to say that:

   'The mean family size is 2.4 children.'

   b) When finding a mean from a frequency table, be sure that you are finding the mean of the *figures*, not of the frequencies.

   c) When finding a mean from a frequency table, be sure that you divide by the number of figures, not by the number of classes. If your frequency table is:

Number of children in family	0	1	2	3	4	5
Frequency	10	27	34	22	15	12

   Be sure that you divide by 120 (the number of families) not by 6 (the number of classes). If you do the wrong thing you will get an absurdly large answer.

   d) When the figures in a frequency table are given by a range of values, make sure that you take the middle of the range for working out the mean. It may not be perfect, but it is the best that can be done in the circumstances.

3) **Median**

   When working out the median, be sure that you take the middle *value*. The median of 19 numbers is not 9, it is the *9th number*.

4) **Cumulative Frequency**

a) The points on a cumulative frequency curve must be at the *right endpoints* of the intervals, not at the middles.

   If your graph is wrong then all your subsequent calculations of the median and quartiles will be wrong also.

b) The median and quartiles must refer to the measurements, not to the frequencies. Suppose you are measuring the heights of 80 people. The lower quartile is not the number 20, it is the height of the *20th person*.

Chapter 24

Probability

Probability is measured on a scale between 0 and 1.

If an event is certain, then it has a probability of 1. If it is impossible, then it has a probability of 0.

If an event may or may not happen, then its probability is the proportion of times in which it does happen.

The probability of an event A is written P(A).

If the probability of an event is p, then the probability that the event does *not* happen is $1 - p$.

24.1 Probability of a Single Event

An experiment or game may have several different outcomes. If a card is drawn from a pack, then it may be any one of the 52 cards. When a six-sided die is thrown, any one of the six faces may be uppermost.

If there are n equally likely outcomes of an experiment or game, then the probability of each outcome is $\frac{1}{n}$.

☐ **24.1.1 Examples.** *Foundation, level 5*

1) A fair six-sided die is rolled. What is the probability that it shows a 5?

   Solution   Because the die is fair, each of the 6 numbers is equally likely to come up. The probability that the number is 5 is:

   $$\frac{1}{6}$$

2) In her cupboard a woman has 4 cans of tomato soup and 6 cans of celery soup. She shuts her eyes and picks a can at random: what is the probability that it is a can of tomato soup?

   Solution   She is equally likely to draw out any of the ten cans. The probability that she selects a tomato can is:

   $$\frac{4}{10} = \frac{2}{5}$$

3) A card is drawn at random from a well-shuffled pack. What is the probability that it is a club?

   Solution   Because the pack is well-shuffled, each of the 52 cards is equally likely to be drawn. The probability that one of the 13 clubs is drawn is therefore:

   $$\frac{13}{52} = \frac{1}{4}$$

4) A card is drawn from a well-shuffled pack. What is the probability that it is not an ace?

   Solution   The probability that the card is an ace is $\frac{1}{13}$. The probability that it is not an ace is therefore:

   $$1 - \frac{1}{13} = \frac{12}{13}$$

☐ **24.1.2 Exercises.** *Foundation, levels 4, 5, 6*

1) Make a copy of the probability line below. On it place the probabilities of the events:

    a) That the sun will rise tomorrow

    b) That it will rain tomorrow

    c) That there will be 32 days in this month.

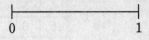

Fig 24.1

2) Estimate the probabilities of the following:

    a) That it will snow on next Christmas Day.

    b) That England will win the next World Cup for soccer.

    c) That you will be late for school or college tomorrow.

3) A fair six-sided die is rolled. What is the probability that a 3 is uppermost?

4) A fair coin is tossed. What is the probability that it comes up heads?

5) A solid in the shape of a regular tetrahedron (four-sides) has the numbers 1 to 4 on its faces. It is thrown. What is the probability that it lands on the face labelled 2?

6) A bag contains 8 red and 17 blue marbles. A child draws one at random. What is the probability that the marble is blue?

7) In an election 4,000 people voted Labour, 4,500 voted Conservative, and 2,500 voted Liberal Democrat. If a voter is picked at random, what is the probability that she voted Labour?

8) A multiple choice question has 5 possible answers. If a candidate picks the answer at random, what is the probability that it is correct?

9) Out of 100 batteries made by a certain firm, 7 are faulty. If I buy one battery, what is the probability that it works?

10) A roulette wheel has slots numbered 1 to 36, and a zero slot. Assuming that it is fair, what is the probability that the ball lands in the slot numbered 19?

11) A box of sweets contains 10 toffees, 8 liquorices and 7 chocolates. If a sweet is drawn at random what is the probability that it is a toffee?

12) A bingo caller has balls labelled 1 to 99. If he draws one at random, what is the probability that it is 11?

13) A card is drawn from a well-shuffled pack. What is the probability that it is an Ace?

14) The letters of the word MATHEMATICS are each written on squares of cardboard, which are then shuffled. If one is drawn, what are the probabilities that the letter is

    a) S   b) T?

15) In the fairground game shown in Fig 24.2 the pointer is spun until it comes to rest over one of the sectors. Find the probabilities that it points at:

    a) 7   b) An odd number.

Fig 24.2

16) A 4 sided die and a 6 sided die are thrown together. The following table gives the possible total score:

Score on 6 sided die

1 2 3 4 5 6

	2	3	4	5	6	7
1	2	3	4	5	6	7
2	3	4	5	6	7	8
3	4	5	6	7	8	9
4	5	6	7	8	9	10

Score on 4 sided die (rows 1, 2, 3, 4)

Find the probabilities that:

a) The total is 10

b) The total is 7

c) The total is less than 4.

17) Two fair six sided dice are thrown. Complete the following table of the total score:

Score on first die

	1	2	3	4	5	6
1	2	3	4	5	6	7
2	3					8
3				7		
4					9	10
5						
6					11	

Score on second die (rows 1–6)

Find the probabilities that the total is   a) 2   b) 7.

18) A fair six-sided die is rolled. What is the probability that the result is not a 6?

19) A roulette wheel has the numbers 1 to 36 and a zero. Find the probability that the zero does not come up.

20) A card is drawn from a well-shuffled pack of cards. What is the probability that it is not a heart?

24.2 Combinations of Probabilities

If two unconnected events have probabilities p and q, then the probability of them *both* happening is $p \times q$.

□ **24.2.1 Examples.** *Intermediate, levels 7, 8*

1) Two fair six-sided dice are thrown. What is the probability that the result is a double six?

Solution   For each single die, the probability of a six is $\frac{1}{6}$. The probability that both dice give sixes is therefore:

$$\frac{1}{6} \times \frac{1}{6} = \frac{1}{36}$$

2) A class contains 6 boys and 10 girls. Two are chosen at random. What is the probability that they are both boys?

Solution   The probability that the first is a boy is $\frac{6}{16}$.

There are now 5 boys out of 15 children.

The probability that the second is a boy is $\frac{5}{15}$.

These probabilities are shown on a tree diagram. The probability that both are boys is found by multiplying the probabilities along the top branch.

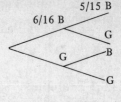

Fig 24.3

$$P(\text{Both are boys}) = \frac{6}{16} \times \frac{5}{15} = \frac{30}{240} = \frac{1}{8}$$

☐ **24.2.2 Exercises.** *Intermediate, levels 7, 8*

1) In a multiple choice exam, each question has 5 possible answers. A candidate answers two questions at random: find the probabilities that

   a) both questions are right   b) both questions are wrong.

2) A fair coin is tossed twice. Find the probabilities of   a) two heads   b) two tails   c) a head and a tail, in any order.

3) A coin is biased so that it is twice as likely to come up heads as tails. What is the probability that it comes up heads?

4) A roulette wheel has the numbers 1 to 36. For the first spin I bet that an even number will come up, and for the second spin that a number divisible by 3 will come up. What is the probability that I win both my bets?

5) A football team has probabilities $\frac{1}{2}, \frac{1}{3}, \frac{1}{6}$ of winning, losing, drawing respectively. It the team plays two matches, find the probabilities that:

   a) Both matches are drawn.

   b) The first match is won and the second lost.

   c) The team did not lose both matches.

6) Two fair dice are thrown and the sum is recorded. Find the probabilities that:

   a) Both are sixes

   b) Both show the same number

   c) The first number is less than the second.

7) Two letters are chosen from the word SCROOGE. What is the probability that both O's are chosen?

8) Two people are chosen from 5 women and 4 men. What is the probability that they are both men?

9) A box contains 7 red and 12 yellow counters. Two are chosen. What is the probability that they are both yellow?

24.3 Exclusive and Independent Events

Two events are *exclusive* if they cannot happen simultaneously.

If two event A and B are exclusive, then the probability of one or other of them occurring is the sum of their probabilities.

$$P(A \text{ or } B) = P(A) + P(B)$$

Two events are *independent* if one of them does not make the other either more or less likely.

If two events A and B are independent, then the probability of them both happening is the product of their probabilities.

$$P(A \text{ and } B) = P(A) \times P(B)$$

❒ **24.3.1 Examples.** *Higher, levels 9, 10*

1) A bag contains 5 red and 7 blue marbles. Two are drawn in succession. Find the probabilities that:

a) The first is red.

b) Both are red.

c) They are both the same colour.

d) The second is red.

Solution Draw a tree-diagram as shown. Mark in the probabilities at each fork.

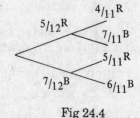

Fig 24.4

a) The probability that the first is red is $\dfrac{5}{12}$

b) The probability that both are red is the probability of the top branch.

$$P(\text{Both are red}) = \frac{5}{12} \times \frac{4}{11} = \frac{20}{132} = \frac{5}{33}$$

c) The probability that they are of the same colour is the sum of the probabilities of the first and last branches.

$$P(\text{Both same colour}) = \frac{5}{12} \times \frac{4}{11} + \frac{7}{12} \times \frac{6}{11} = \frac{62}{132} = \frac{31}{66}$$

d) The second marble is red in the first and third branches.

$$P(\text{Second red}) = \frac{5}{12} \times \frac{4}{11} + \frac{7}{12} \times \frac{5}{11} = \frac{55}{132} = \frac{5}{12}$$

2) In a class of 20 boys, 10 learn French, 12 learn German, and 7 learn neither. If a boy is selected at random, find the probabilities that:

a) He studies both languages.

b) He studies French but not German.

Solution The two events F (= he studies French) and G (= he studies German) may not be independent. Hence the two probabilities cannot be multiplied together.

12 + 10 + 7 = 29. 9 boys must have been counted twice. These boys must be the ones who study both languages.

$$P(F \text{ and } G) = \frac{9}{20}$$

Of the 10 who study French, 9 also study German. Hence only 1 studies French and not German.

$$P(\text{French but not German}) = \frac{1}{20}$$

□ **24.3.2 Exercises.** *Higher, levels 9, 10*

1) One in ten flash cubes is faulty. I buy 2. Complete the tree diagram.

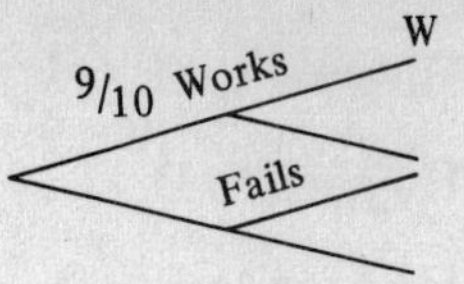

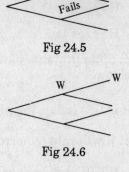

Fig 24.5

Find the probabilities that

a) Both work   b) exactly one works.

2) 3 white and 4 black balls are in a bag. 2 are drawn at random. Complete the tree diagram:

Find the probabilities that

a) the second is black   b) both are the same colour.

Fig 24.6

3) In his drawer a man has 8 left shoes and 11 right shoes. He picks out 2 at random. Find the probabilities that:

a) Both are left shoes.

b) Both are left or both are right.

c) He draws a left and a right shoe.

4) Three quarters of a batch of tulip bulbs give flowers. If I plant 2, find the probabilities that:

a) Both flower. b) At least one flowers.

5) Two fair six-sided dice are rolled. Find the probabilities that:

a) Both are sixes.

b) Neither is a six.

c) At least one is a six.

6) A box of chocolates contains 12 soft-centered and 14 hard-centered chocolates. Two are selected; find the probabilities that:

a) Both are soft-centred.

b) One is soft-centred and the other hard-centred.

7) A woman goes to work by bus, by car or on foot with probabilities $\frac{1}{2}, \frac{1}{3}, \frac{1}{6}$ respectively. For each type of transport, the probabilities that she will be late are $\frac{1}{10}, \frac{1}{5}, \frac{1}{50}$ respectively. Find the probabilities that:

a) She will go by car and be late.

b) She will go by bus and be on time.

c) She will go by bus or by car and be late.

d) She will be late.

8) The diagram shows a street map of a village. A motorist enters the village at A, and at each crossroads he drives straight ahead with probability $\frac{1}{2}$, and turns left or right with equal probability $\frac{1}{4}$. Find the probabilities that:

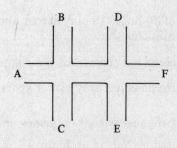

Fig 24.7

a) He goes to C.

b) He goes to D.

c) He goes to F.

9) Two different letters are chosen from the word COMPUTER. Find the probability that at least one is a vowel.

10) If it is fine today, the probability that it will rain tomorrow is $\frac{1}{5}$. If it is rainy today, the probability that it will rain tomorrow is $\frac{2}{3}$. The probability that today is fine is $\frac{1}{2}$. Find the probabilities that:

a) It will be fine on both days.

b) There will be different weather today and tomorrow.

11) Two cards are drawn from a well-shuffled pack. Find the probabilities that:

a) Both are hearts.

b) They are of the same suit.

c) The first is a heart and the second is a King.

12) In order to start at a certain game each player must throw a six with a fair die. Find the probabilities that a player starts after:

a) One throw   b) Two throws c) Three throws.

13) A number is chosen at random from 1 to 20 inclusive. Find the probabilities that:

a) It is even.

b) It is greater than 17.

c) It is odd and greater than seventeen.

d) It is a prime greater that 10.

14) A poker hand consists of 5 cards. The first four cards I receive are the 6, 7, 8, 9 of hearts. If the fifth card is another heart, then I have a *flush*. If the fifth card is a 5 or a 10, then I have a *straight*. Find the probabilities that I get:

a) A flush.

b) A straight.

c) A straight flush.

d) Either a flush or a straight.

15) Out of 30 boys, 18 play soccer and 12 play rugger. 8 play both. If a boy is picked at random, find the probabilities that:

a) He plays soccer but not rugger

b) He does not play either sport.

16) The probability that chips are on the menu for dinner is $\frac{2}{3}$, and the probability of beans is $\frac{1}{4}$. The probability of both is $\frac{1}{5}$. Find the probabilities of:

a) Chips but not beans.

b) Neither chips nor beans.

Common Errors

1) Simple probability

If there are n outcomes to an experiment, then the probability of each individual outcome is $\frac{1}{n}$ if all the outcomes are *equally likely*. Do not forget the "equally likely" condition.

If two coins are tossed, then there are either 0, 1 or 2 heads. But it does not follows that P(1 head) = $\frac{1}{3}$. This outcome is more likely than the others.

2) Addition of probabilities

Probabilities are only added together if the events cannot happen together. If events A and B can happen together, then it is not true that P(A or B) = P(A) + P(B).

If your answer gives a probability of more than 1, then you must have made a mistake of this sort.

3) Multiplication of probabilities

Probabilities are only multiplied together if the events do not affect each other.

If event A makes event B either more likely or less likely, then it is not true that P(A & B) = P(A) × P(B).

4) Exclusive and independent events

Do not confuse these terms. They refer to *pairs* of events, not to single events.

They are not opposites – if two events are not exclusive then it does not follow that they are independent.

5) Independence of successive events

Read a question carefully to make sure that you do not confuse the following sorts of probability:

Rolling a die twice. If the first roll is a six then the second roll is just as likely to be a six.

Drawing two cards. If the first is an Ace then the second is less likely to be an Ace.

Chapter 25

Diagrams, Networks, Flow Charts

25.1 Two-way tables and Networks

The choices made by different people can be represented in a *two-way table*.

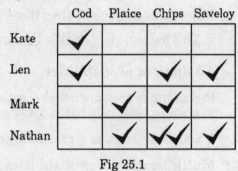

☐ **25.1.1 Examples.** *Foundation, level 6*

1) Four people enter a fish and chip shop. Kate and Len have cod, Mark and Nathan order Plaice. Nathan has double chips, Kate has no chips, the rest have single chips. Nathan and Len have a saveloy. Draw up a two way table to show their orders.

   Solution   Draw a table, with the names down one side and the items of food down the other. Put a tick to show who has what.

Fig 25.1

2) The diagram on the right shows the roads connecting four villages. The distances are in miles. What is the shortest route from A to C?

   A shop at A must deliver items to B, C and D, and then return to A. How long is the shortest route?

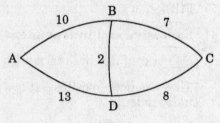

   Solution   Going from A to C through B takes 17 miles. Through D it takes 21 miles.

   The shortest route is A – B – C

   After going to B, C and D, it is quicker to return by B.

   The shortest route is 37 miles

Fig 25.2

☐ **25.1.2 Exercises.** *Foundation, level 6*

1) Anne, Brian and Cecile can speak French. Donna and Brian can speak German. Spanish is spoken by Cecile and Donna. Draw up a two-way table to show who speaks which language.

2) In a school, Maths is taught by Mr Robinson, Miss Smith and Mrs Taylor. Miss Smith also teaches Physics, and Computing is taught by Mrs Taylor and Mr Urquart. Draw up a two-way table to show who teaches what subject.

3) In each of the following diagrams, distances are in miles. Find the shortest routes between A and C.

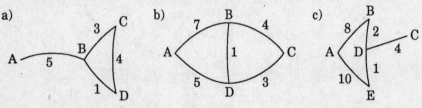

Fig 25.3

4) In the diagrams of Question 3, find the lengths of the shortest routes which start at A, visit all the other places, and return to A.

5) In the diagrams of Question 3, find the lengths of the shortest routes which start at A, cover all the roads at least once, and return to A.

25.2 Flow Charts

A sequence of operations can be described by a *flow-chart*. If the same operation is repeated many times the chart will contain a loop.

❐ **25.2.1 Examples.** *Intermediate, level 7*

1) The flow chart shown converts temperature in Fahrenheit to Centigrade. Use the flow chart to find the Centigrade equivalent of 32°, 50°, 131°.

Write a new flow chart which will convert Centigrade to Fahrenheit.

$$\boxed{\text{Read F}°} \rightarrow \boxed{\text{Subtract 32}} \rightarrow \boxed{\text{Divide by 9}} \rightarrow \boxed{\text{Multiply by 5}} \rightarrow \boxed{\text{Write C}°}$$

Solution Put in the values and apply the operations.

32°F	0	0	0	0°C
50°F	18	2	10	10°C
131°F	99	11	55	55°C

To convert Centigrade to Fahrenheit reverse the flow chart.

$$\boxed{\text{Read C}°} \rightarrow \boxed{\text{Divide by 5}} \rightarrow \boxed{\text{Multiply by 9}} \rightarrow \boxed{\text{Add 32}} \rightarrow \boxed{\text{Write F}°}$$

2) The following flow-chart evaluates $\sqrt{2}$. Use it to fill in the table of values. How would you find $\sqrt{2}$ to a greater degree of accuracy?

x	y	z
1	2	1.5
1.5	$\frac{4}{3}$	

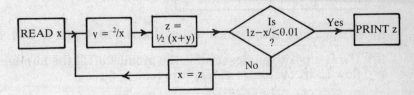

Fig 25.4

Solution Fill in the table as shown:

x	y	z
1	2	1.5
1.5	$\frac{4}{3}$	1.417
1.417	1.412	1.414

The answer 1.414 will be printed.

To get a greater degree of accuracy change the entry in the diamond shaped box. Changing it to $|z - x| < 0.00005$ will make the answer accurate to 4 decimal places.

❐ **25.2.2 Exercises.** *Intermediate, level 7*

1) The following flow-chart works out values of the expression $3(x + 5)$. Use it to evaluate the expression when $x = 1, 3, 10$.

$$\boxed{\text{Read } x} \rightarrow \boxed{\text{Add 5}} \rightarrow \boxed{\text{Multiply by 3}} \rightarrow \boxed{\text{Write } y}$$

2) The standing charge for a telephone is £12.50, and the cost per unit used is 2 p. The following flow-chart finds the total bill £C in terms of the number x of units used.

$$\boxed{\text{Read } x} \rightarrow \boxed{\text{Multiply by 0.02}} \rightarrow \boxed{\text{Add £12.50}} \rightarrow \boxed{\text{Write £}C}$$

Find the charge for bills which have used

a) 300 units   b) 500 units.

Write the flow chart which will convert the charge £C to the number x of units used.

3) Write down the flow-chart which will work out values of the expression 2(x - 7). Use your flow chart to evaluate the expression when x = 12, 20, 100.

4) Tracy writes a flow-chart to describe the making of toast. She puts the instructions in the wrong order. Re-arrange them in the right order.

| Put bread in | Cut slice | Turn toast | Take out toast | Light grill |

5) Re-order the following instructions in a flow-chart to describe the making of a telephone call.

| Lift receiver | Wait for answer | Dial number | Find Number |

6) Re-arrange the following instructions so that they form a flow-chart for washing hands.

| Pull plug out | Wait | Turn tap on | Turn tap off | Wash hands | Put plug in |

7) The following flow chart evaluates √3. Use it to fill in the table, and hence to find √3 to 3 decimal places.

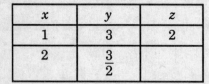

x	y	z
1	3	2
2	$\frac{3}{2}$	

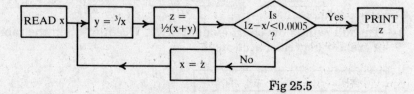

Fig 25.5

8) Write a flow-chart which will evaluate √10. Use it to find √10 to 3 decimal places.

9) If n is a positive integer, n! is the product of all the numbers up to and including n. The following flow-chart evaluates n! Use it to find 4!, 6!.

n	P	i

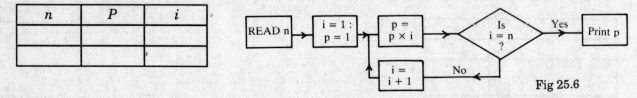

Fig 25.6

10) The following flow-chart evaluates the sum of all the numbers up to and including n. Fill in the empty boxes.

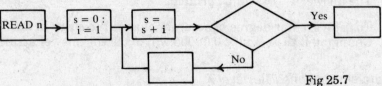

Fig 25.7

11) Apply the following flow chart to the numbers 5, 8, 15, 17. Which sort of numbers end in List A and which in List B?

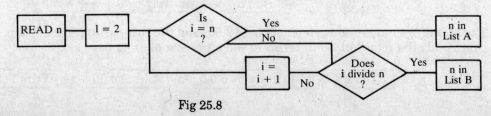

Fig 25.8

12) When the score at tennis is "Deuce", the winner is the first player to be 2 clear points ahead. Fill in the boxes of the following flow-chart to describe the scoring at Deuce.

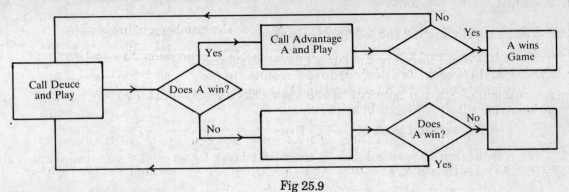

Fig 25.9

25.3 Critical Path Diagrams

Networks can be used to show the structure of complicated jobs.

Each individual task of a job takes a certain time to complete. This can be shown on the diagram.

Suppose that Task B depends on Task A. Then the earliest beginning time for B is the finishing time for A. The whole diagram can then be filled in with beginning times and finishing times, and the total time for completion can be found.

The tasks which must be completed on time are *critical* tasks. There is a path through the whole network consisting of critical tasks, which is called a *critical path*.

If a task is not critical, then it can be delayed, without extending the time for the whole job. The time by which it can be delayed is the *slack* time.

A					
B					
C					
D	A	B			
E	C				
F	A	B	C	D	E

Fig 25.10

❐ **25.3.1 Examples.** *Higher, level 10*

1) The first column of the diagram lists the tasks A, B, C, D, E and F of a job. The second column shows the tasks which must be done first. Draw the network diagram for the job.

Solution Notice that Tasks A, B and C do not depend on any other. So they can be put at the beginning of the network. D depends on A and B, so it must follow from them. E follows from C. Finally F depends on all the tasks, and so is shown at the end of the network. The diagram is Fig 25.11.

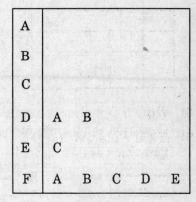

Fig 25.11

2) The diagram on the right shows the network for a job. The time in minutes for each task is indicated. Find the completion time for the whole job, and find the critical path. Find the slack time for the first non-critical task.

Solution A, B and C can all begin after 0 minutes. A takes 5 minutes, so the finishing time for A is 5. Similarly the finishing times for B and C are 7 and 8 minutes.

D can begin when A is finished, after 5 minutes. E has to wait until both B and C are done, after 8 minutes. The ending times for D and E are found by adding on the times for the tasks.

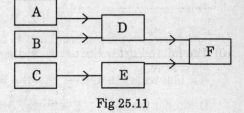

Fig 25.12

Finally, the starting time for F is the larger of the finishing times for D and E. The whole diagram is Fig 25.13.

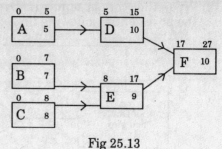

Fig 25.13

The total time for the job is 27 minutes

Notice that F began immediately after E was finished. But there could be a delay of up to 2 minutes in the completion of D. So E is critical and D is not. Similarly C is critical, while B is not.

The critical path is C – E – F

A is the first non-critical task. It could be delayed for an extra 2 minutes, without altering the start time of F.

The slack time of A is 2 minutes

❑ **25.3.2 Exercises.** *Higher, level 10*

1) The listings for the tasks of three jobs are shown below. Draw the network diagrams.

a)
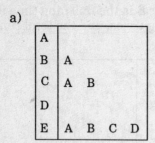

b)
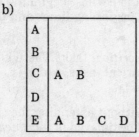

c)
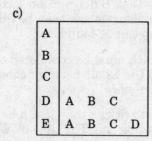

Fig 25.14

2) Construct the listings of the tasks in the networks below, showing which tasks must be done first.

a)
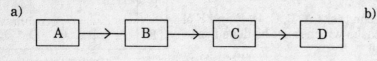

b)
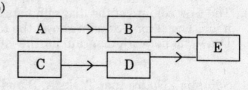

Fig 25.15

3) The job of preparing breakfast is broken down into tasks as below. Show which tasks depend on others, and draw the network diagram.

A = boil water in kettle   B = boil water in saucepan   C = cut bread

D = toast bread   E = butter toast   F = put food on table

G = warm and put tea in pot   H = infuse tea   I = boil egg

4) The job of changing a car wheel is broken up into tasks as below. List the tasks showing their precedence, and draw the network.

A = loosen nuts   B = get spanner and jack   C = get spare wheel

D = jack up car   E = remove old wheel   F = put on new wheel

G = lower car   H = put away tools and old wheel   I = tighten nuts

5) A room has to be cleared up after being used for a sports exhibition. The tasks are:

A: re-lay carpets   B: return furniture   C: clear away cups and plates

D: wash up cups and plates   E: remove rows of chairs   F: sweep up

G: put cups and plates in cupboard   H: lock up   I: switch off lights.

List the tasks showing their precedence, and draw a network.

6) Before going on holiday the following must be done.

    A: apply for passport   B: collect passport C: apply for foreign money   D: collect foreign money

    E: book flight   F: find dates of flights   G: book hotel   H: go to airport.

    List the tasks showing their precedence, and draw a network.

7) In each of the following diagrams, find the completion times and the critical path.

    a)

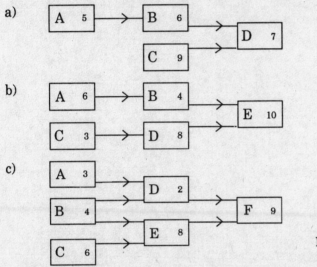

                                   Fig 25.16

8) List the non-critical tasks of the jobs of Question 7. Find the slack times of these tasks.

9) Suppose the tasks of Question 4 took the following times (in minutes). Find the completion time and the critical path.

    A: 2   B: 5   C: 2   D: 15   E: 1   F: 3   G: 4   H: 6   I:3

10) Suppose the tasks of Question 5 took the following times (in minutes). Find the completion time and the critical path.

    A: 10   B: 6   C: 8   D: 15   E: 12   F: 10   G: 5   H: 2   I: 1

11) Estimate the times for the tasks of Question 3. Hence find the completion time and the critical path.

Common Errors

1) Networks

    Read the question carefully. Don't confuse the following:

        A route which visits each town once.

        A route which covers each road once.

2) Flow-charts

    a) When a number is going through a flow-chart, its value may be changing. Be careful that you do not use the wrong value of the number.

b) The convention for flow-charts is that *orders* are in rectangular boxes, and *questions* in diamond boxes. Make sure that you use the correct shapes.

c) Do not be confused by the statement "S = S + 3". It can be read: "Change S to S + 3" or "Add 3 to S".

3) **Critical Path**

If a task depends on two or more previous tasks, make sure you take their latest finishing time, not their earliest.

Test papers

Exam 1 *Foundation*

Answer all the questions.. Time: one hour thirty minutes

1. A train leaves Camford station at 11 48. It reaches Deeton 4 minutes later.
    When does it reach Deeton? (1)
    The train reaches Effville at 12 10. How long did the train take between Deeton and Effville? (2) [3]

2. Make a copy of the shape below and shade a third of it. (2) [2]

3. Alexandra's recipe for vinaigrette uses three parts of oil to one of vinegar. How much oil is there in 12 ounces of vinaigrette? (2) [2]

4. From the list of numbers 4, 5, 6, 8 find:
    a) a prime number   b) a square number. (2)
    Show how three numbers from the list can be added to obtain a total of 18. (2) [4]

5. Fill in the boxes below to make the equations true.

    $\Box + 3 = 9$ $\Box \times 4 = 12$ $18 \div \Box = 9$ (3) [3]

6. a) Evaluate 2^3. (1)
    b) Evaluate 1.37^3, giving your answer to 2 decimal places. (2) [3]

Moussaka	£5.25
Klefliko	£6.85
Dolmades	£4.50
Afelia	£5.55

7. 'Dial a Meal' is a meal delivery service. Part of its menu is shown on the right. What is the cost of the ringed items? (2)
    The delivery charge is 10% of the cost of the food. What is the total cost of the items chosen? (2) [4]

8. Continue the tessellation shown with six more triangles. (3)
    What is each angle of an equilateral triangle? (2) [5]

    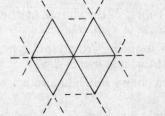

9. If $x^2 + x = 3$, then x lies between 1 and 2. Use a calculator to find the value of x to one decimal place. (5) [5]

    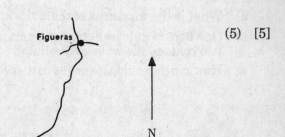

10. The map shown is in the scale of 1 cm per 10 km.
    a) What is the distance from Gerona to Figueras? (2)

243

b) What is the bearing of Figueras from Gerona? (2) [4]

11. A bag contains 12 marbles, of which 5 are red, 3 brown and the rest blue. A marble is drawn out at random.

   a) What is the probability the marble is blue? (2)

   b) What is the probability the marble is white? (1) [3]

12. A 'Magic Square' is a square of numbers in which every line, horizontal, vertical or diagonal, has the same total.

   A 3 by 3 magic square contains the numbers 1 up to 9. Find the total of these numbers, and hence find how much each line must add up to. (4)

   Complete the 3 by 3 square below. (3)

4		
	5	7

   How much does each line on a 4 by 4 magic square add up to? (4) [11]

13. Write down the next two terms of the sequence:

   $$1, 3, 6, 10, 15, \ldots$$ (3)

   Write down the cubes of the first five whole numbers. (2)

   Write down the first cube, the sum of the first two cubes, the sum of the first three cubes, the sum of the first four cubes, the sum of the first five cubes. (2)

   What is the connection between these numbers and the original sequence? (2) [9]

14. Two fair four-sided dice are rolled. Complete the table below for the total score.

First die

		1	2	3	4
	1	2			
Second die	2			5	
	3				
	4			7	

(4)

   What is the most likely total, and what is its probability? (3) [7]

15. Make copies of the shapes on the right, and draw their lines of symmetry. (4)

   Which shape has rotational symmetry, and what is the order of its rotational symmetry? (3)

   Draw a shape which has rotational symmetry of order 3. (2) [9]

16. A room is 2 m high, $3\frac{1}{2}$ m long and 3 m wide. There is a window which is 1 m by $1\frac{1}{2}$ m.

   a) What is the total area of the walls and ceiling? (3)

   b) One litre of paint will conver a surface of 20 m². How much paint is needed to give two coats to the walls and ceiling? (2)

   c) How much would this paint cost at £5.50 per litre? (2) [7]

17. The dots shown are 1 cm apart. Draw a net for a cuboid which is 4 cm by 1 cm by 1 cm. (4)
What is the volume of this cuboid? (1) [5]

18. The maximum temperatures in 60 days of the summer were as follows.

83	85	77	66	65	81	74	90	84	72
63	69	80	72	72	91	78	89	82	88
82	68	75	69	77	84	72	90	92	85
67	69	64	77	84	88	84	86	83	65
62	64	74	66	75	72	84	89	89	75
78	73	84	81	80	79	77	84	82	86

Fill up the frequency table below. (4)

Temperature	60 – 64	65 – 69	70 – 74	75 – 79	80 – 84	85 – 89	90 – 94
Frequency							

Draw a histogram to show this information. (4) [8]

19. A taxi firm charges 40p per mile for the first 10 miles, then 35p per mile for the next 30 miles, then 30p per mile thereafter. Complete the following table: (4)

Miles	5	10	20	30	40	50	60	100
Charge	£2		£7.50			£17.50		

Draw a graph of Charge against Miles, using 1 cm per 10 miles along the x-axis, and 1 cm per £5 up the y-axis. (4)

How far can I get for £11? (1)

What is the cost per mile for journeys of a) 10 miles b) 100 miles? (4) [13]

20. A pizza parlour sells four types of pizza, A, B, C, D. During an evening the sales were:

B B A C C A D D D A A B A C D B A D A A B B D C D B A A D C A A B C A D.

Complete the following table:

Type of pizza	A	B	C	D
Charge				

(2)

Which is the most popular pizza? (1)

Complete the bar-chart shown. (3)

Make an accurate pie-chart showing the sales, indicating the angles in each sector. (4)

A prize is to be awarded to one customer. What is the probability it is awarded to someone who ordered an A pizza? (2) [12]

245

Exam 2 *Foundation*

Answer all the questions. Time 1 hour

1. a) Find a prime number between 21 and 28. (1)
    b) Find a square number greater than 30. (1)
    c) Write down an odd cube number greater than 1. (1) [3]

| LESSONS £5.50 each |
| OR |
| £41 for a course of 8 |

2. Belinda wants 16 driving lessons. How much will she save if she books two sets of 8, instead of buying them singly? (2) [2]

3. a) A class contains 20 of children, of which 12 are girls. What fraction of the class are boys? Express your answer in its simplest form. (2)
    b) What fraction of the shape below is shaded? (2) [4]

4. Sid buys a jacket at £23 and two shirts at £9 each. He is given a discount of 10p in the £. How much does he pay? (2) [2]

5. A film on television began at 7.35, and ended at 9.03.
    a) How long does the film last? (2)
    b) Rachel saw only the last 27 minutes. When did she start watching? (2) [4]

6. Anne is 5 feet and 6 inches high. What is her height in centimetres? (1 inch is approximately 2.54 cm.) (3) [3]

7. a) What is the cost of 12 cakes at 15 p each? (1)
    b) Joan, Jackie and Liz share a bill of £15.60. How much does each pay? (1) [2]

8. Write down the coordinates of the points A, B and C shown below. (3)

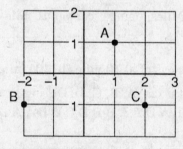

Point D (not shown) lies on the *x*-axis. If D has coordinates (x,y) what is the value of *y*? (1)

Point E (not shown) lies in the second quadrant, i.e. the top left quarter of the graph. If E has coordinates (x,y) are *x* and *y* positive or negative? (2) [6]

9. The figure shows a pattern of triangles. Complete the table below.

Length of side	1	2	3
Number of triangles			

Describe the sequence of the second row. (2)
Find the next two terms of the second row. (2) [6]

10. a) Evaluate the expression $23 - 4x$ when $x = 3$. (1)

    Solve the equation $23 - 4x = 9$. (2)

b)   Evaluate the expression $(23 - 4) \times x$, when $x = 3$. (1)

Solve the equation $(23 - 4) \times x = 38$. (2)

c)   Why is the equation $2x + 3 = 2x + 5$ not soluble? (2) [8]

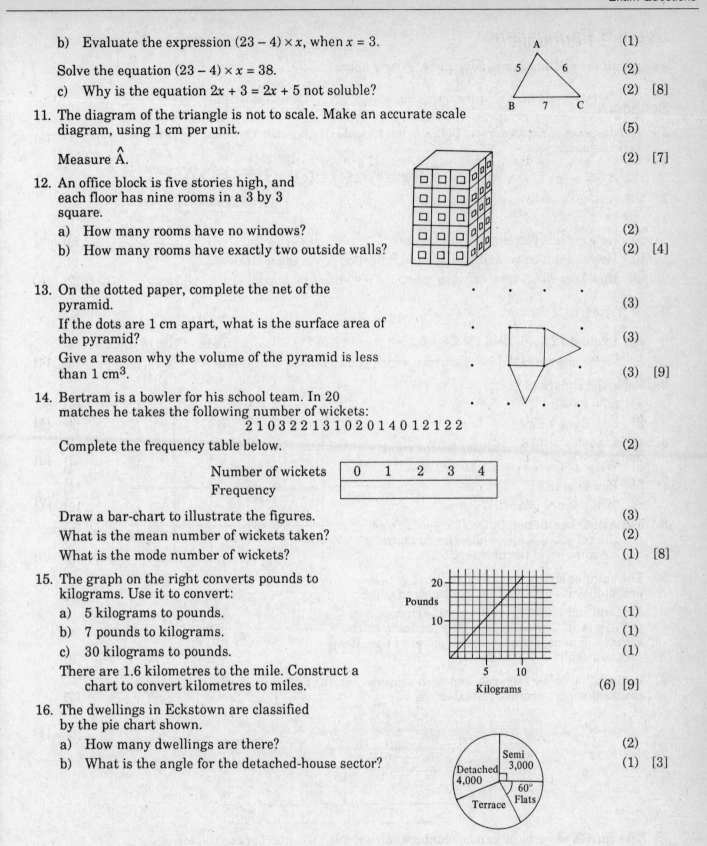

11. The diagram of the triangle is not to scale. Make an accurate scale diagram, using 1 cm per unit. (5)

Measure $\hat{A}$. (2) [7]

12. An office block is five stories high, and each floor has nine rooms in a 3 by 3 square.
    a)   How many rooms have no windows? (2)
    b)   How many rooms have exactly two outside walls? (2) [4]

13. On the dotted paper, complete the net of the pyramid. (3)

If the dots are 1 cm apart, what is the surface area of the pyramid? (3)

Give a reason why the volume of the pyramid is less than 1 cm^3. (3) [9]

14. Bertram is a bowler for his school team. In 20 matches he takes the following number of wickets:

2 1 0 3 2 2 1 3 1 0 2 0 1 4 0 1 2 1 2 2

Complete the frequency table below. (2)

Number of wickets	0	1	2	3	4
Frequency					

Draw a bar-chart to illustrate the figures. (3)
What is the mean number of wickets taken? (2)
What is the mode number of wickets? (1) [8]

15. The graph on the right converts pounds to kilograms. Use it to convert:
    a)   5 kilograms to pounds. (1)
    b)   7 pounds to kilograms. (1)
    c)   30 kilograms to pounds. (1)

There are 1.6 kilometres to the mile. Construct a chart to convert kilometres to miles. (6) [9]

16. The dwellings in Eckstown are classified by the pie chart shown.
    a)   How many dwellings are there? (2)
    b)   What is the angle for the detached-house sector? (1) [3]

Exam 3 *Foundation*

Answer all the questions in both sections. Time 2 hours

Section A

1. The diagram of the motorway below is not to scale. It gives the distances in miles between various junctions.

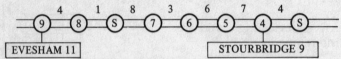

   a) How far is it between service areas? (2)
   b) How far is it between Evesham and Stourbridge via the motorway? (2)
   c) How long will it take between junctions 6 and 8 at 60 m.p.h.? (2) [6]

2 a) Write $\frac{1}{8}$ as a decimal. (2)

   b) Evaluate $2.3 \times 1.7 + 4.1 \times 2.8 + 5.2 \times 1.3$. (2)

   c) Write down a calculator sequence which you could use to evaluate $8.1 - (4.2 - 5.3)$ (3) [7]

3. Solve the equations:
   a) $35 - 2x = 3$. (2)
   b) $3x - 6 = 4 - 2x$. (3) [5]

4. Make a copy of the grid shown on the right.
   a) Write down the coordinates of A. (1)
   b) Plot D at (3,1). (1)
   c) What figure is ABCD? (2)
   d) ABCD is reflected in the line $y = 2$. Draw the reflected shape, and write down the coordinates of the new vertices. (4) [8]

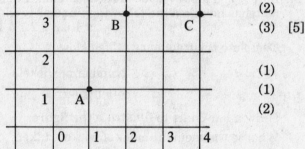

5. The spinner shown was twirled. What is the probability it lands on an odd-numbered edge? (2)

   Ron and Raj play a game with this spinner. Ron pays 4p to Raj, and Raj pays Ron the score on the spinner in pence. What is the probability that Ron makes a profit? (3) [5]

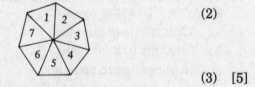

6. In the table below each row and each column contains numbers which increase at a constant rate. Complete the table. (7)

	1	2	3	4	5
1	3	5	7		
2	5	8		14	
3	7	11	15		
4					
5		17			35

The entries in each cell can be found as follows. Take the numbers at the top of the column and at the left of the row. Add their sum and their product. So the entry in the 2nd column and 5th row is $2 + 5 + 2 \times 5 = 17$.

If the table were extended, what would be the entry in the 7th column and 8th row? (3)

Obtain a formula for the entry in the xth column and yth row. (3) [13]

7. A bill at an Indian restaurant is shown.

1 Meat Dhansak	£4.45
1 Bombay Chicken Curry	£4.65
2 Rice	50p each
2 Popadoms	20p each

a) What is the total cost of the meal? (2)

b) If the bill is paid with pound coins, how much change will there be? (2)

c) The change is left as a tip. What is the tip as a fraction of the bill? (3)

d) If the diners had given a 10% tip, what would have been the total bill? (2) [9]

8. Samantha keeps a record over 10 weeks of how much she spends per week. She prepares a frequency table and a bar chart. Complete them both.

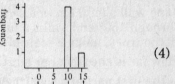

Amount	£0-£5	£5-£10	£10-£15	£15-£20
Frequency	3	2		

(4)

Suppose that in each week her expenditure was at the bottom of the range. What is the least amount she could have spent in the 10 weeks? (4)

Over the next 5 weeks she spends £4.32, £16.60, £12.41, £6.37, £13.66. What has been her average expenditure over these 5 weeks? Give your answer to the nearest penny. (3) [11]

9 A wall is made from bricks which are 3 by 3 by 6 inches. A brick which is laid along the wall is a *stretcher*, and a brick which is laid across the wall is a *header*. A layer of bricks is called a *course*.

a) If the wall is 6 inches thick and 12 feet long, how many bricks are there in each course? (2)

b) If the wall is 4 feet high, how many bricks does it contain? (2)

c) If the courses are alternately of headers and stretchers, how many bricks have a face on the front of the wall? (4) [8]

10. Ben starts off to school walking, then realizes he might be late and starts to run. The graph of his distance against time is shown.

a) How far did he walk? (1)

b) For how long was he walking? (1)

c) What were his speeds when walking and running? (4)

d) What was his average speed over the whole journey? (2)

e) If he had kept on walking, how long would the whole journey have taken him? (3) [11]

11. a) What is the bearing of Ashdon from Ashqelon? (3)

b) Measure the distance between these towns on the map. (2)

c) If the actual distance between these towns is 17.5 km, what is the scale of the map in km per cm? (2) [7]

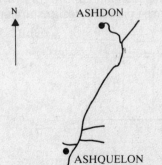

12. The dots shown below are 1 cm apart. Make a copy of the dots, and use them to draw a net for a cuboid which is 2 cm by 2 cm by 1 cm. (4)

What is the volume of the cuboid? (1)

Estimate, showing your calculations, how many of these cuboids would fit inside an empty milk carton. (4) [9]

13. Jane walks for 3 km in a straight line. She then turns through 150° and walks a further 4 km. Make a diagram of her journey, using a scale of 1 cm per km. (3)

How far is she from her starting point? (3)

If the first part of her journey was due North, what is her bearing from her starting point? (3) [9]

14. Jeff thinks a die might be unfair. He throws it 60 times, and records the frequency of each score.

Score	1	2	3	4	5	6
	ⅢⅢ ⅢⅢ ⅢⅢ	ⅢⅢ /	ⅢⅢ ⅢⅢ ⅢⅢ	ⅢⅢ ⅢⅢ ⅢⅢ	ⅢⅢ /	ⅢⅢ ⅢⅢ ///

a) Make a frequency table for the scores. (2)

b) Draw a bar-chart for the scores. (2)

c) Find the average score. (3)

d) How convinced are you that the die is biased? Give *brief* reasons. (4)

e) Assuming that the die is biased, what is the probability of obtaining a 5 with it? (3) [14]

Section B

Using and Applying Mathematics

15. The national colours of Ruritania are purple, orange and scarlet. They must all appear on the national flag. Regions of the flag which are next to each other must be coloured differently. Three possible designs are suggested:

a)

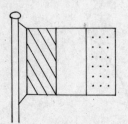

b)

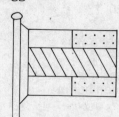

c)

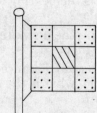

a) A tricolour. One possible colouring is Purple Orange Scarlet. (abbreviated to POS). List all the possible colourings. (3)

b) In the second design, the flag must be symmetrical about the horizontal line throught the middle. Describe the possible colourings. (4)

c) In the third design, which is square, the flag must have rotational symmetry of order 4 about the centre. Describe the possible colourings. (4)

Which of the results above would be altered if we dropped the rule that all three colours must be included? Give reasons. (4) [15]

16. A garden has been allowed to run wild. You are going to re-turf the lawn, which is rectangular. Explain the measurements you would make and the calculations you would do to find out the cost. Show how you would estimate the time it would take. [15]

Exam 4 *Foundation*

Using and Applying Mathematics

Attempt as many of the questions as you can. Time: 1 hour

1. Below is a network showing the roads connecting four places. The distances are in miles.

   a) Leroy is about to enter a 20 mile race. How could he run between the places shown and cover a distance of exactly 20 miles? (3)

   b) Janice is going to enter a race of 15 miles. Why is it not possible for her to run between the places shown and cover a distance of exactly 15 miles? (3) [6]

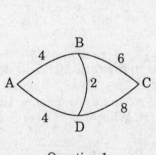

Question 1

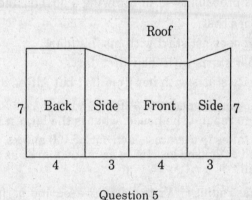

Question 5

2. You are in charge of producing the printed programme for a concert. What measurements and calculations would you make to ensure that it is sold at a modest profit? [6]

3. You have several identical squares of wood, which can be joined up edge to edge.

   a) With two squares, show that only one shape can be made. (2)

   b) With three squares, show that there are two different shapes. (3)

   c) How many shapes can be made with 4 shapes? (5) [10]

4. Prepare an observation sheet which will find out how long the pupils in your school or college take going home in the evening. (6) [6]

5. Above is the net of a shed. Draw the plan and the front and side elevations once it is assembled, labelling the lengths. (5)

   A person whose eyes are 6 feet above the ground is standing 20 feet in front of the shed. Draw a sketch of what he might see. (5) [10]

6. You have a large supply of 1p and 2p coins. You can use them to make a total of 3p in two different ways, either three 1p coins or one 2p coin and one 1p coin.

How many ways can you make totals of 2p, 4p, 5p, 6p, 7p? (4)

How many ways can you make totals of 100p, 101p? (4)

Describe how can find out the number of ways of making any total. (4) [12]

Exam 5 *Intermediate*

Answer all the questions. Time: two hours

1. a) A number is divisible by 9 if the sum of its digits is divisible by 9. What is the first number after 2135 which is divisible by 9? (2)

    b) How can you tell from its digits whether a number is divisible by 5? (3) [5]

2. Solve the equation: $3(x + 1) - 2(x - 3) = 8$ (2)

    Expand and simplify $(x - 5)(x + 3)$ (2) [4]

3. A paving slab is of length 2 ft. It is propped up by a vertical stick as shown.

    a) If the stick is of length 1 ft, what is the angle does the slab make with the horizontal? (3)

    b) What length should the stick be to keep the slab up at 45° to the horizontal? (4) [7]

4. Alfred and Beatrice play darts: each must throw a double before they can start scoring. Alfred has probability $\frac{1}{10}$ of throwing a double, and Beatrice has a chance of $\frac{1}{8}$. Find the probabilities that:

    a) Alfred does not start with his first dart. (1)

    b) Neither starts with their first dart. (2)

    c) Beatrice starts with her first dart but Alfred does not. (3) [6]

5. a) A house is divided into 3 flats, and expenses are shared in the ratio 10:9:7. If it costs £520 to repaint the house, what is the largest share? (3)

    b) Jane takes two exams, each out of 100 marks. She got 72% in her first exam. In the second exam she lost twice as many marks as in the first. What was her percentage overall? (3) [6]

6. A circle has radius r. Write down its area and perimeter. (1)

    The area is 6π greater than the perimeter. Write down an equation in r, and show that it reduces to:
    $$r^2 - 2r - 6 = 0$$
    (3)

    Plot a graph of $y = r^2 - 2r - 6$, taking values of r from 0 to 5. (4)

    From your graph solve the equation $r^2 - 2r - 6 = 0$. (2) [10]

7. Make a copy of the diagram on the right. The shape S is transformed to T by reflection in the line $x = 1$. Draw the shape T on your diagram. (3)

    The shape T is transformed to U by reflection in the line $x = 0$. Draw U. (2)

    Describe the single transformation which will take S to U. (3) [8]

8. The diagram shows a pattern of triangles.

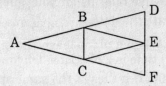

a) Write down a triangle congruent to ABC. (1)

b) Write down a triangle similar but not congruent to ABC. (2)

c) List 5 different routes which go from A to E. (4)

d) A rat starts at A and travels round the diagram, never going along a route which it has covered before. At each junction it takes each of the available routes with equal probability. What is the probability it follows the route A-B-E-F? (3) [10]

9. Miss Shirasu took her car to be serviced. Fill in the blanks (a, b, c, d) in the bill.

Oil filter	6.50
Air filter	5.45
Spark plugs	5.80
Contacts	a
Material Total	25.62
Labour	b
Total	41.12
VAT @ 17½%	c
TOTAL	d

a) . . . . (2)   b) . . . . (1)   c) . . . . (2)   d) . . . . (1) [6]

10. a) Mrs Layton drives down the motorway, stopping at a service station during the journey. The graph below shows her journey.

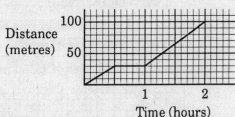

i) How long did she stop at the service station? (1)

ii) What were her speeds during the two parts of the journey? (2)

b) For the return journey she drives at a steady speed of 50 m.p.h. without a break. Draw a distance-time graph to show the return journey. (5) [8]

11. In 10 innings, Geoffrey scores 0, 45, 55, 23, 0, 43, 6, 34, 49, 10.

What is his mean score? (1)

What is his median score? (2) [3]

12. The lengths of 100 novels were found. The results are in the frequency table below.

Number of pages	0-99	100-199	200-299	300-399	400-499	500-600
Frequency	5	15	31	28	13	8

Estimate the mean number of pages. (2)

Guess the median number of pages. Give reasons for your guess. (3)

Draw a frequency polygon to illustrate the data. (4) [9]

13. Athens is 2 hours ahead of London. When a plane leaves London the time is 1600, and when it arrives in Athens the (local) time is 2200. How long has the flight been? (1)

The flight back leaves Athens at 0900. If it flies at the same speed, what is the time in London when it arrives? (2)

A flight leaves London at 1200, and arrives 8 hours later in New York, where the local time is 1500. What is the time difference between London and New York? (1)

A flight leaves London at 0800, and arrives in Moscow where the time is 1500. The return flight leaves Moscow at 2000 and arrives in London at 2100. What is the time difference between Moscow and London? (3) [7]

14. Describe the transformation which takes S to S' in the picture shown. (2)

Make a copy of the diagram. On it draw S', where S' has been obtained from S by enlargement from the origin with a scale factor of 2. (2) [4]

15. The diagram shows the cross-section of a tree. It is a circle of diameter 20 cm. A plank of width 19 cm is to be cut from it. Find the thickness of the plank. (4) [4]

16. Express the region shown shaded on the right in terms of three inequalities in x and y. (3) [3]

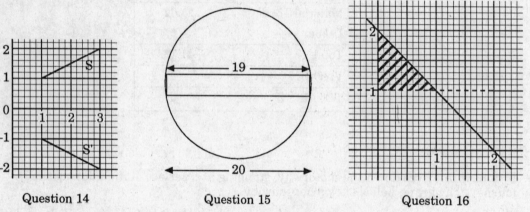

Question 14   Question 15   Question 16

17. Show that there is a solution to the equation $x^3 = x + 1$ between $x = 1$ and $x = 2$. (2)

By trial and improvement find this solution to one decimal place. (3) [5]

18. The cumulative frequency table below shows the salaries of 1,000 people.

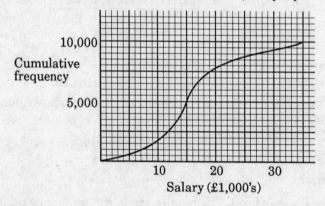

Salary (£1,000's)

a) Find the median salary. (1)

b) Find the interquartile range of the salaries. (3)

c) If two people are picked at random, what is the probability that both their salaries are less than the median salary? (2) [6]

19. Which names best describe the quadrilaterals with the following properties?
    a) With all angles equal and diagonals equal. (1)
    b) With all angles equal and diagonals perpendicular. (1)
    c) With all sides equal and diagonals perpendicular. (1) [3]

20. You don't have a calculator with you. How do you find the approximate value of:

$$\pi \sqrt{\frac{3.78}{1.02}}$$

(3) [3]

21. a girls and b boys each bought a copy of the same book. The total cost was X. What was the cost of each book? (3) [3]

Exam 6 *Intermediate*

Answer all the questions. Time: 1 hour 30 minutes

1. a) A length is divided in the ratio 3:5. What fraction of the length is the larger part? (2)
   b) Three sevenths of a school are boys, of whom a quarter are over 16. What fraction of the school are boys over 16? (2)
   c) A bag contains marbles which are either red, white or blue. A quarter are red and two fifths are white. What fraction are blue? (2) [6]

2. In a racing sport the winner, 2nd and 3rd of each race gain 5 points, 3 points and 1 point respectively. At the end of the season all the points are added up and an overall champion is declared. When there is only one race to be run A and B have 23 and 20 points respectively, and their nearest rival has 10 points.
   If B wins the last race and A comes second, what are the final totals? (1)
   If A comes third in the last race, what must B do to win the championship? (2)
   Write down all the possible results of the last race which will ensure that A ends the championship with more points than B. (4) [7]

3. a) What is the whole number before 100,000? (1)
   b) What is the least whole number greater than -1,000,000? (1)
   c) Multiply your answers together and give the answer in standard form, to 5 significant figures. (2) [4]

4. Write the following as integers:

   a) 3^2 b) 2^{10} c) $\left(\frac{1}{4}\right)^{-1}$ (3)

   Write 81 as a power of 3. (2) [5]

5. Solve the equations:
   a) $3x + 2 = 23$ (1)
   b) $2x + 3 = x + 5$ (2) [3]

6. A man is four times as old as his son. In 16 years time he will be twice as old. Letting the son's age be x, form an equation in x and solve it. (6) [6]

7. Simplify the following:
   a) $x + 5y + x - 2y$ (1)
   b) $3x^2y^4 \times 5 \times y^4$ (1)
   c) $6p^3s^2 \div 2ps^4$ (1) [3]

8. Fill up the following table for the function $y = x^2 - 2x + 2$.

x	-1	0	$\frac{1}{2}$	1	$1\frac{1}{2}$	2	3
y							

(3)

Draw the graph of the function, indicating the axis of symmetry. (5)

Solve the equations

$$x^2 - 2x + 2 = 2 \qquad (1)$$

$$x^2 - 2x + 2 = 1.5 \qquad (2) \ [11]$$

9. Write down the coordinates of the vectors **a** and **b** shown on the diagram. (2)

   Find **a + b**. Make a copy of the diagram, and on it illustrate **a + b**. (3)

   Explain the geometrical connection between **a**, **b** and **a + b**. (3) [8]

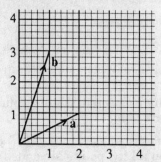

10. Let x and y represent lengths, A represent area, and V represent volume. Which of the following represent lengths?

   a) $x + \dfrac{A}{y}$ b) $\dfrac{V}{x}$ c) $\dfrac{A^2}{V}$ (3) [3]

11. In $\triangle ABC$, BC = 8 cm, $\hat{B}$ = 73°, $\hat{C}$ = 64°. Make an accurate scale diagram of the triangle. (3)

   What is the length of AC? (2)

   If the diagram is a map in the scale of 1:50,000, which represents three towns A, B, C, what is the distance of B from A? (3) [8]

12. The figure shows a river of width 20 metres. Angie stands directly opposite Bert. If Bert now walks 30 metres along the river, how far is he from Angie? (2)

   What angle does the line between them make with the bank? (2)

   Angie now walks away from Bert until the line between them makes 20° with the bank. How far has she walked? (3) [7]

13. A cube is placed so that the vertices of its lower face are at (1,0,0), (1,3,0), (4,3,0), (4,0,0). Find the coordinates of the vertices of the top face. (2)

   Find the coordinates of the centre of the cube. (2) [4]

14. A coin is tossed 3 times. Complete the following list of the possible results.

   HHH   HHT   ——   ——

   ——   ——   ——   TTT. (2)

   What are the probabilities of:

   a) Exactly one head. (2)

   b) At least one head. (2) [6]

15. The mileages of 60 second-hand cars were found to be as follows. (In 1,000's of miles.)

53	43	48	51	67	21	43	19	9	71
21	28	42	31	80	37	43	27	19	64
5	41	84	32	49	30	51	47	32	64
16	19	63	34	27	53	27	82	39	70
19	36	72	57	93	74	48	74	34	60
29	37	42	75	45	63	32	54	23	7

Group these data into a frequency table, using intervals of width 10,000 miles. (3)

From the frequency table find the average mileage. (3)

Plot a frequency polygon to show your results. (3) [9]

16. The population of a new town is given by the following figures.

Year	1985	1986	1987	1988	1989	1990	1991
Population in 1,000's	16	21	23	29	33	37	42

Plot these figures on a graph. Draw a straight line through the points. (3)

Use your line to answer the following:

At what rate is the population growing? (4)

When was the town founded? (3) [10]

Exam 7 *Intermediate*

Answer all the questions in both sections. Time 2 hours 30 minutes.

Section A

1. Light travels at 186,000 miles per second. It takes light 2 years to reach us from the star Sirius. How far away is Sirius? (5) [5]

2. a) Solve the inequality $2x + 3 > 13 - 3x$ (2)

   b) Illustrate the inequality $-1 < x \leq 2$ on the number line. (2)

   (c) List the integers x for which $x^2 < 8$. (2) [6]

3. a) With the money I have, I could buy 10 cakes. If the cakes were 3p cheaper I could buy 13 of them. How much does each cake cost? (3)

   b) Solve the simultaneous equations:

   $$2x + 3y = 7, \quad 3x + 4y = 9$$ (3) [6]

4. Azaeez and Neville play a game by each rolling a die. If the dice show the same number, Azaeez pays Neville 40 p. If they show different numbers Neville pays Azaeez 10 p.

   By considering the table below, or otherwise, find the probability that Neville wins a game.

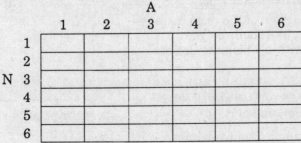

(3)

257

They play this game 60 times. How many times would you expect Neville to win? (3)

Who is likely to gain overall when they play 60 times, and by how much? (4) [10]

5. a) Evaluate $5.35 \times 1.72 + 5.28 \times 1.44 - 8.12 \times 0.23$. (2)

b) Evaluate to 3 decimal places: $\dfrac{6.46 - 2.85}{6.65 + 4.23}$ (2)

c) If $y = x^2 - 9x + 3$, find y when x is (i) -3 (ii) $\dfrac{2}{3}$ (3)

d) Write down possible sequences of calculator buttons to evaluate: (i) $\sqrt{3.2 + 4.1}$

(ii) $\dfrac{3}{4.2 + 5.7}$ (4) [11]

6. There is 120 m of fencing with which to make a rectangular field. If the breadth of the field is x m, find the length of the field in terms of x, and show that the area A of the field is given by:

$$A = 60x - x^2 \qquad (5)$$

Draw a graph of A against x, taking for your values of x 10, 20, 30, 40, 50, 60. (7)

For what values of x is the area 730 m²? (3)

What is the greatest area of the rectangle? (3) [18]

7. Mark is standing 20 metres due West of Tom. Mark walks North at 1 m/sec, and Tom walks North at 1.2 m/sec. After 1 minute they both stop. Make a diagram of their journeys, using a scale of 1 cm for 10 metres. (7)

How far apart are they after 1 minute? (4) [11]

8. Over a period of 8 years the average price in p of a pint of milk was calculated. The results are below.

Year	1	2	3	4	5	6	7	8
Price	23	25	26	29	30	33	36	40

Plot these values on a graph, taking the year along the x-axis and the price up the y-axis. Draw a straight line through the points. (4)

From your line, what do you expect the average price to be in year 9? When will the price reach 45 p? (4) [8]

9. The figure shows a right-angled triangle. Write down the relationship between the sides a, b, c. (2)

If a = 7 and b = 24 find c. (1)

If a = 4 and c = 7 find b. Give your answer to 3 significant figures. (2)

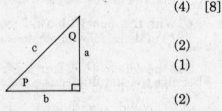

If a = 39 find whole numbers b and c such that $a^2 + b^2 = c^2$. (3) [8]

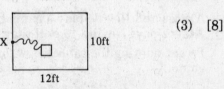

10. a) A room is 10 ft by 12 ft. The only power point is at X, midway along one of the shorter walls. The flex of an electric fire is 6 ft long.

Make an accurate diagram of the room, using a scale of 1 cm per foot, and shade on your diagram the region within which the fire could be used. (6)

b) A and B are connected cog wheels, with 30 and 60 teeth respectively. A rotates anticlockwise at 20 r.p.m. Predict the motion of B. (2) [8]

11. The diagram shows a triangle S and two lines $y = x$ and $y = 2$.

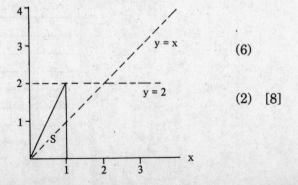

a) Make a copy of the diagram. S is reflected in $y = 2$ to S'. Draw S' on the graph and write down its coordinates. (4)

b) S is reflected in $x = y$ to S''. Draw S'' on the graph and write down its coordinates. (3)

c) S is enlarged to S''' by a factor of 2 from the point (1,0). Draw S''' on your diagram and write down its coordinates. (4) [11]

12. The ages of 160 members of a squash club are given in the frequency table below.

Age	15-19	20-24	25-29	30-34	35-39	40-44	45-49	50-54
Frequency	24	42	28	24	18	10	8	6

Find the cumulative frequencies. Plot a cumulative frequency graph. (5)

Find the median and the interquartile range. (4)

Two members are picked at random. What is the probability that they will both be under 25? (3) [12]

13. Certain gold coins are $\frac{1}{8}$ in thick and $\frac{3}{4}$ in in diameter. What is the volume of each coin? (2)

One cubic inch of gold weighs 11.6 ounces. Gold costs £300 per ounce. A forger buys gold to make 50 coins, which he then sells for £240 each. How much profit does he make? (4) [6]

Section B

Using and Applying Mathematics

14. Make a tracing of the dots below.

Use the dots to draw four pairs of lines, a), b), c), d), which are:

a) Parallel, but not horizontal or vertical. (2)

b) Perpendicular, but not horizontal or vertical. (2)

c) At 45° to each other. (2)

d) At an angle of $\tan^{-1} 2$ to each other. (2)

If the dots can be extended as far as we like, what other angles could be constructed by lines joining pairs of dots? (7) [15]

15. The game of 'Battle of Numbers' is played as follows. The first player A picks any digit 1, 2, 3, 4, 5, 6, 7, 8 or 9. The second player B also picks a digit (which may be the same as A's) and adds the two numbers together. And so on – each player picks a digit and adds it on to the sum of the previous digits. A player wins when he brings the total to 40.

Show that there is a strategy by which B can always win. (5)

If the winning total was 43, who could win, and by what strategy? (5)

How could the game be varied? (5) [15]

Exam 8 *Intermediate*

Using and Applying Mathematics

Attempt as many of the questions as you can

1. How many grains of sand would fill a matchbox? (3) [3]

2. 'A survey shows that 66.7% of doctors think that spicy food is dangerous for health.'
   Is there anything about this announcement which makes you suspicious? (3) [3]

3. You are drawing lines on a large sheet of paper. None of the lines are parallel (so each pair of lines crosses.) No three lines go throught the same point.

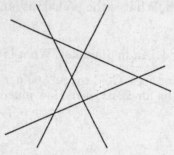

   How many crossing points are there? How many separate regions are there? (8) [8]

4. You have a collection of cubes, which can be joined together face to face.
   a) Show that with two cubes only one shape can be formed. (1)
   b) Show that with three cubes two different shapes can be formed. (2)
   c) See how many shapes you can form out of four cubes. (7) [10]

5. You have a robot which can be given the following instructions, for any n:
   Move n.   (After which the robot moves forward n cm.)
   Turn n.   (After which the robot turns through $n°$ clockwise.)
   Write down the instructions to make the robot go round a square of side 120 cm. (3)
   Show how to write instructions to make the robot go round a path which is approximately a circle of radius 120 cm. (5) [8]

6. Marcia and Nancy take turns to roll a die, Marcia having first go. The first to obtain a six is the winner.
   Is the game fair? (2)
   What would you guess is the probability that Marcia wins? (3)
   How would you find the probability that Marcia wins? (3) [8]

7. You are asked to investigate the importance of computers in people's lives. Devise a questionnaire for this, and explain how you would obtain responses. Explain how you would present the results. (10) [10]

Exam 9 *Higher*

Answer all the questions. Time: two hours

1. Find a, b and c in the identity below.
   $$2x^2 + 8x - 3 \equiv a((x + b)^2 + c)$$
   (2) [2]

2. a) After a pay rise of 12%, Robert is earning £17,248. How much was he earning before the rise? (3)

    b) Mr King buys a car for £8,000 and reckons that every year its value decreases by 10%. What is its value after 5 years? (3) [6]

3. In the diagram shown AD = DB = BC. $\hat{EBC}$ = 117°. Letting $\hat{BCD}$ = x express $\hat{ABD}$ and $\hat{DBC}$ in terms of x. Hence find $\hat{BCD}$. (4) [4]

4. a) Find an obtuse angle x for which sin x = 0.4. (2)

    b) y is an angle between 0° and 360°, for which sin y = $-\frac{3}{5}$ and cos y = $\frac{4}{5}$. Find y. (2) [4]

5. a) If interest is chargeable at 10%, how much each year will a man have to pay on a loan of £28,000? (1)

    b) If the man pays income tax at 30%, how much does he have to earn to pay the interest? (3)

    c) The interest rate rises to 12%, and the tax rate falls to 25%. What is the percentage change in the amount he has to earn to pay the interest payments? (3) [7]

6. In triangle ABC, $\underline{AB}$ = **b** and $\underline{AC}$ = **c**. X, Y Z are the midpoints of BC, AC, AB respectively.

    a) Express in terms of **b** and **c**, $\underline{AZ}$, $\underline{BC}$, $\underline{BX}$, $\underline{AX}$. (4)

    b) What is the relationship between the triangles ABC and XYZ? (2)

    c) Triangle ABC has area 24. What is the area of XYZ? (2) [8]

7. The proportion of Tombola tickets that will win a prize is x. Corinne claims that by buying two tickets she will be twice as likely to win at least one prize as with one ticket. Show that in general she is incorrect. (5)

    In fact, with two tickets she is $1\frac{3}{4}$ times as likely to win at least one prize as with one ticket. Form an equation in x and solve it. (7) [12]

8. For the digits 1, 2, 3, 4, 5, 6, 7, 8, 9, find their mean and median. (2)

    Find, in terms of n, the mean of the whole numbers from 1 to $2n + 1$. (2)

    Find, in terms of n, the mean of the whole numbers from 1 to $2n$. (2) [6]

9. The Hamburger Harriers ran a cross-country race of 5 miles. Their times are given below.

Time (minutes)	30-32	32-34	34-35	35-36	36-38	38-40	40-50
Frequency	5	8	6	5	7	4	15

    Draw a histogram to illustrate these figures. What was the modal time? (6)

    The Ravioli Runners also had 50 entrants for the race. Their average time was 33 minutes, and they finished much more closely together. On your histogram *sketch* a possible histogram for the Ravioli Runners. (4) [10]

10. Richard walks 10 miles North. He wants to return home, but he turns through 160° instead of 180° and walks for 15 miles. The diagram on the right shows his journey.

    a) How far is he from his starting point? (4)

    b) What is his bearing from his starting point? (4)

    c) During the 15 mile stretch, what was the closest he came to his starting point? (3) [11]

11. An isosceles triangle is 8 cm high and 4 cm wide. A rectangle which is $2x$ wide by y high is drawn inside the triangle as shown.

a) By considering similar triangles, show that $y = 4(2 - x)$. (6)

b) Show that the area A of the rectangle is $A = 16x - 8x^2$. (2)

c) Plot A against x, taking values of x from 0 to 2, and using a scale of 2 cm per unit along the x-axis and 1 cm per unit up the y-axis. (4)

d) What is the greatest possible area of the rectangle? (2)

e) What is the value of x when the area is 3 cm²? (2) [16]

12. a) The temperature is given as 16.4°, and then falls by 2.8°. Both figures are given to one decimal place. What is the least possible value of the new temperature? (3)

b) There were 5,400 people at a funfair, to the nearest hundred, and on average they spend £8 each, to the nearest £1. What were the total receipts? Give your answer to an appropriate degree of accuracy. (3) [6]

13. F varies inversely as the square of d. $F = 4$ when $d = 5$.

a) Find an equation giving F in terms of d. (3)

b) Find the value of d when $F = 400$. (2) [5]

14. a) Define what is meant by an *irrational* number. (2)

b) Which of the following are true? If true, give a justification. If false, give a counterexample.

(i) The product of two rational numbers is rational.

(ii) The product of two irrational numbers is irrational.

(iii) A number with a terminating decimal is rational. (6) [8]

15. a) The heights of 10 plants were as follows:

23 34 35 29 31 26 33 41 30 28

Find the mean and standard deviation of these heights. (3)

b) The frequency table below gives the heights of 100 plants. Find the mean of these figures, explaining why your answer is only an approximation. (3)

Height	10-20	20-30	30-40	40-50	50-60
Frequency	5	19	37	22	17

[6]

16. a) Express the following as single fractions in their simplest forms.

i) $\dfrac{2}{p} + \dfrac{3}{p^2}$ (2)　　ii) $\dfrac{5q}{r} \times \dfrac{q^2}{10r^2}$ (2)

b) Factorize the expression $x^2 + 11x - 60$. (2) [6]

17. x and y are digits which could be 1, 2 or 3. Evaluate the following, leaving your answer in standard form.

$x00{,}000 + y00{,}000$　　　　$x00{,}000 \times y00{,}000$ (3) [3]

Exam 10 *Higher*

Answer all the questions. Time 2 hours.

1. a) By rounding each term to 1 significant figure, obtain a rough estimate of the following expression. Show all your working. (3)

   $$\left(\frac{4.75 + 1.31}{3.13}\right)^2$$

   b) Evaluate the expression of part a), giving your answer to three significant figures. (1) [4]

2. Five schools play each other in a league. 2 points are awarded for a win, 1 for a draw, and 0 for a loss. The results of the first 8 matches are:

   A beat B : C drew with D : E drew with A : A beat D : B beat C : E drew with D : E beat C : A beat C

   After 10 games B had 2 points. What were the results of the last two games? (2)

   Fill in the table below:

Teams	A	B	C	D	E
Points					

   (3)

   If six schools play each other, how many games will there be? (2) [7]

3. Solve the equation $3.2x + 7.1 = 23.5$, giving your answer to 1 decimal place. (2)

   The three numbers in the equation above are accurate to 1 decimal place. Find the range of values within which the solution x must lie. (4) [6]

4. Find the square roots of the following, giving your answers to three significant figures where appropriate.

   a) 4   b) 40   c) 4×10^{18} d) 0.0000004 (2)

   Which of the square roots above are rational? (2)

   For what values of n is 4×10^n a rational number? (2)

   Find a number, bigger than 1, whose square root and cube root are both rational. (2) [8]

5. a) Lorna tells Martha to think of a number, then to add 5, then to square the result. Martha's answer is 1. What are the two possible values of Martha's original number? (3)

   b) Solve the equation $x^2 + 3x - 2 = 0$, giving your answers to three decimal places. (3) [6]

6. The graph shows the level of water in a rain barrel.

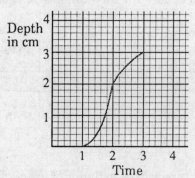

   a) When was it raining most heavily? (2)

   b) Draw a tangent to the curve at the point where it was raining most heavily. What was the flow in cm hr⁻¹? (3)

c) At 3 o'clock a plug at the bottom of the barrel was pulled out, and the barrel drained in 30 minutes. Extend the graph to show this. (2) [7]

7. a) Find the equation of the straight line which goes through (1,1) and (5,2). (3)

b) Find where the line $2y + 3x = 6$ crosses the axes. Find the gradient of this line. Find the area enclosed between the line and the axes. (3) [6]

8. (a) below is the graph of $y = f(x)$. On (b), (c) and (d) sketch the graphs of $y = f(x + 2)$, $y = f(2x)$, $y = f(x) + 2$ respectively. (6) [6]

a)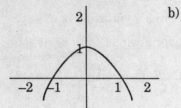
b)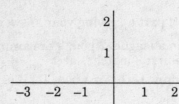

c)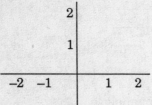
d)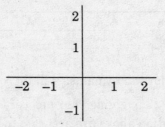

9. A tree is 30 m high. P stands 100 m south of the tree, Q stands 200 m east of P.

a) Find the distance of Q from the top of the tree. (2)

b) In a 3-dimensional coordinate system, the x, y and z axes point east, north and upwards respectively. If the origin of coordinates is at Q, find the coordinates of the top of the tree. (2) [4]

10. The triangle T is transformed to the triangle T' by an enlargement of scale factor 3 from the origin.

a) Make a copy of the diagram, and draw T'. (2)

b) Write down the matrix which will perform the enlargement. (2) [4]

11. a) $n!$ is defined as $n \times (n - 1) \times (n - 2) \times \ldots \times 3 \times 2 \times 1$. Find 3! and 5! (2)

b) Follow through the flow chart below for the numbers 5, 4, 6, 4, 6, 3, –2. What number is printed at the end? (4)

What is the purpose of the flow chart? (3) [9]

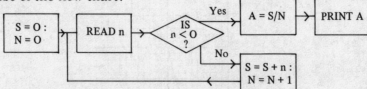

12. An ice-cream consists of a wafer cone with a cylinder of ice-cream on top. The radii of both cone and cylinder are 3 cm, the height of the cone is 7 cm and the height of the cylinder is 4 cm. Find the total volume. (2)

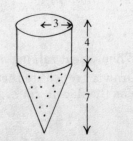

The ice-cream melts, and sinks into the wafer. Assuming there is no change in volume, how much ice-cream spills over the top of the wafer? (2) [4]

13. Let M be the matrix $\begin{pmatrix} -1 & 0 \\ 0 & 1 \end{pmatrix}$.

Let T be the triangle with vertices A(1,1), B(3,1), C(2,5). Plot T on graph paper. (1)

T' is obtained by applying M to T. Find the vertices of T', and plot them on the same paper. (3)

Describe the action of the matrix M. (2)

Write down a matrix which will perform a reflection in the x-axis. (2) [8]

14. The heights of 100 girls are given by the table below.

Height in cm	120-130	130-140	140-150	150-160
Frequency	12	33	38	17

a) Find the cumulative frequencies. (1)

b) Draw a cumulative frequency graph, using a scale for 1 cm per 10 cm along the x-axis, and 1 cm per 10 girls up the y-axis. (4)

c) Find the interquartile range. (2)

d) If a girl is selected at random, what is the probability she is taller than 145 cm? (2) [9]

15. An electric power station can use either oil or coal. It has a contract with a nearby mine to use at least 100 tons of coal per week. Each ton of coal provides 6 units of energy, and each ton of oil produces 10 units. The station must produce at least 1500 units per week.

The pollution produced by coal and oil is 5 and 2 units per ton respectively. Local regulations restrict pollution to 1,000 units per week.

Let x and y be the numbers of hundreds of tons of coal and oil respectively used per week. Write down inequalities in x and y which describe the above restrictions. (4)

Illustrate the inequalities on graph paper, leaving unshaded the region of possible options. (4)

The cost of coal is £250 per ton, and of oil £350 per ton. What is the cheapest permissible way to supply the power? (4) [12]

Exam 11 *Higher*

Answer all the questions in both sections. Time: 2 hours 30 minutes

Section A

1. a) Expand $(5x + 3y)(2x - 7y)$, simplifying your answer as far as possible. (2)

   b) Make x the subject of $a = \sqrt{2 + x}$. (2) [4]

2. ABCD is a rectangle 6 cm by 4 cm. P consists of those points nearer BC than AD, and Q consists of those points within 3 cm of D.

   Make an accurate scale diagram of ABCD, and shade the region of those points which are in both P and Q. (4) [4]

3. Two shopping areas were investigated, by visiting 10 stores in each area and finding the prices of a standard basket of goods. The results rounded to the nearest £ were:

A:	32	28	25	36	34	30	31	24	27	33
B:	28	27	29	30	28	26	27	29	27	29

Find the mean and standard deviation for each set of figures. (4)

What do your results tell you about the areas? (3) [7]

4. One third of the tickets in a tombola are blank, and two thirds win prizes. Ann, Bert and Charles each buy a ticket. Find the probabilities that:
   a) Ann wins a prize. (1)
   b) Not all of them win. (3)
   c) Charles is the only one to lose. (3) [7]

5. The coordinates of two points on a graph are (1.3,1.7) and (3.8,4.9), where each number is given correct to 1 decimal place. Find the least possible gradient of the line joining the two points. (7) [7]

6. The current in the sea is flowing at 2 km hr^{-1} due North. A man can row at 5 km hr^{-1}. If he points the boat North East, find the direction in which he travels. (4)

   Suppose the man wishes to travel due east. Find the direction in which he should point the boat and the net speed with which he will travel. (4) [8]

7. Below are 4 graphs and 4 equations. Which graph fits which equation? In each case give brief reasons for your choice.

   a) $y = 3 - x^2$ b) $y = x + 1$ c) $y = x^2 + 1$ d) $y = 1 - x$ [8]

   i)   ii)   iii)   iv)

   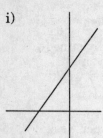 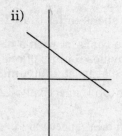 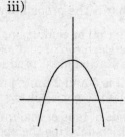 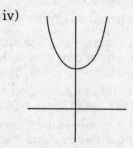

8. In his pocket Jo has many 5p and 10p pieces. To pay a bill of 15p he pulls out coins at random until he has 15p or more. List the possible ways he could pay the bill. (4)

   If he is equally likely to pull out a 5p as a 10p piece, find the probability he overpays. (3) [7]

9. Stephanie bought 1160 shares at 253 p each. The commission was £48.42 and the transfer stamp was £15. What was the total cost? (2)

   Later the price reached 308 p per share, and she sold them at this price, paying a commission of £66.10. What was (a) her actual profit (b) her percentage profit? (4)

   The proceeds of the sale were invested in a building society at 8% compound interest. How much will it have amounted to after 5 years? (4) [10]

10. A train accelerates at 2 m s^{-2} for 15 seconds, then it travels at a constant speed for 60 seconds, then it brakes to a halt at 3 m s^{-2}.
    a) What was the greatest velocity? (1)
    b) Draw a graph of the speed against time. (4)
    c) Find the time taken and the distance covered. (3) [8]

11. a) Solve the equation $2^{3x} = 4^{x+1}$. (3)

    b) Express $5^3 \times 25^2 \times 125^2$ as a power of 5. (3)

    c) It is thought that T is proportional to a power of S. The following values were obtained.

S	2	3	4
T	6	13.5	24

    Find an equation giving T in terms of S, and hence find S when $T = \frac{1}{6}$. (8) [14]

12. A ship is slowing down. Values of its speed are given below.

Time in minutes	0	1	2	3	4	5
Speed in m s^{-1}	8.0	7.1	6.2	5.0	4.9	4.1

   a) Plot these points on a graph, and draw a straight line throught them. (3)

   b) Find the equation of your straight line. (4)

   c) When will the ship come to a halt? (2) [9]

13. A cone C is of height 12 cm, and its base radius is 6 cm. A horizontal cut is made 4 cm from the top. The upper cone is A, and the lower frustum is B.

   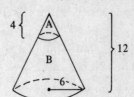

   Where relevant leave your answers as multiples of π.

   a) Find the ratio of the heights of A and C. (1)

   b) Find the base radius of A. (2)

   c) Find the volumes of A and B. (3)

   d) What is the ratio of the volumes of A and C? How is this connected to your answer to a)? (3)

   e) What is the ratio of the curved surface areas of A and B? (3) [12]

14. Complete the table below for the function:

$$y = \frac{1}{8}x^3 - \frac{1}{4}x^2 + 2$$

x	-3	-2	-1	0	1	2	3
y							

(3)

   Plot the graph of this function. (3)

   Use your graph to solve the equation:

$$\frac{1}{8}x^3 - \frac{1}{4}x^2 + 2 = 1 - \frac{1}{2}x$$

(3)

   Draw a tangent to the curve at $x = 2$, and hence find the gradient of the curve at this point. (4) [13]

Section B

Using and Applying Mathematics

15. In Ancient Greece scientists knew that the Earth is round. They were able to find the radius of the Earth by measuring the angle of elevation of the Sun at different places at the same time.

   Give an outline of how this experiment could be done and how the radius of the Earth could be found from it. [15]

16. The number 4 can be written as the sum of 2's and 1's in five different ways.

   $1 + 1 + 1 + 1$

   $2 + 1 + 1$

   $1 + 2 + 1$

   $1 + 1 + 2$

   $2 + 2$.

   Show that 5 can be written as the sum of 2's and 1's in 8 different ways.

   Investigate this situation. [15]

Exam 12 *Higher*

Using and Applying Mathematics

Attempt as many of the questions as you can. Time: 1 hour 30 minutes

1. The diagram below shows a 'House of Cards'. Write down the number of cards necessary to build a house with a) 2 storeys b) 3 storeys.

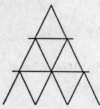

   The number of cards necessary to build a house of n storeys is $an^2 + bn$, where a and b are constant. Use your results so far to find a and b, and check the formula for a house with 4 storeys. [6]

2. Suppose you want to conduct a survey in which you cannot trust people to answer honestly. For example, you might want to investigate the extent of drug-taking among athletes.

   A way to get round the difficulty is as follows. Give each respondent a die, and ask them to roll it without showing you the result. They are to respond as follows:

   Say 'A'   The die shows 1, 2 and they *have* taken drugs.
   Say 'A'   The die shows 3, 4, 5, 6 and they *haven't*.

   Say 'B'   The die shows 3, 4, 5, 6 and they *have* taken drugs.
   Say 'B'   The die shows 1, 2 and they *haven't*

   How would you analyse the results to find the overall extent of drug-taking? [6]

3. A sequence of fractions obeys the rule $\dfrac{x}{y} \to \dfrac{y}{x+y}$, i.e. at each

   stage the new numerator is the old denominator, and the new denominator is the sum of the old numerator and denominator.

   Investigate this sequence. (8) [8]

4. You are given a square piece of paper, 8 cm by 8 cm. On it you are to draw the net of a cylinder as shown. How can you draw the net so that the volume of the cylinder is as large as possible? [8]

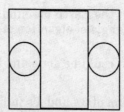

5. In a wind-tunnel experiment, the speed of the wind and the force on an object in the tunnel were measured. The result are below.

Speed v	2	4	6	8
Force F	3	24	64	124

   Dr Noakes thinks that F is proportional to a power of v. Using the first two values only, i.e. for $v = 2$ and $v = 4$, find the power and the formula giving F in terms of v. (4)

Professor Moakes thinks that the relationship is of the form $F = av + bv^2$. Use the first
two values to find two equations in a and b, and solve them. (4)

Which formula gives a better fit to the actual values? (4) [12]

6. You and some friends have been asked to paint a room. List the tasks that are necessary
before you can start painting. They might include:

> Clear room of furniture.

> Buy paint.

> Remove or cover carpets.

> Choose colour.

> Clean surface to be painted.

Estimate the time needed for all the tasks. Arrange the tasks in a critical path diagram to
find the shortest time for completion. [10]

Mental Test 1. *Foundation level*

1. Write thirty three over seventy seven as a fraction.

2. A rectangle has length 7 m and width 5 m. What is its area?

3. A square has area 64 cm². What is its side?

4. A sweet is taken from a box containing three toffees and five acid-drops. What is the probability it is a toffee?

5. How many vertices does a cube have?

6. Two angles of a triangle are 85° and 35°. What is the third angle?

7. Find two fifths of £120.

8. What is the cost of 12 m of flex at 30p per metre?

9. Simplify seven x plus eight x.

10. If x plus 3 is eight, what is x?

Questions 11 to 14 refer to the calendar on the opposite page.

11. How many days are there between April 3rd and June 17th?

12. How many Saturdays are there between 1st January and 7th July?

13. What is the date 27 days after 10th August?

14. What is the date 21 days before 11th January?

Questions 15 and 16 refer to the meter dials on the opposite page.

15. Read the top row of dials.

16. Mark the second row of dials to show a reading of 7392.

Questions 17 and 18 refer to the telephone bill on the opposite page.

17. What is the total exclusive of VAT?

18. What is the total including VAT at $17\frac{1}{2}$%?

Questions 19 and 20 refer to the pie chart on the opposite page.

19. How many children were there in all?

20. What is the angle in the tennis sector of the chart?

1989		January					February					March			
Monday		2	9	16	23	30		6	13	20	27		6	13	20	27
Tuesday		3	10	17	24	31		7	14	21	28		7	14	21	28
Wednesday		4	11	18	25		1	8	15	22		1	8	15	22	29
Thursday		5	12	19	26		2	9	16	23		2	9	16	23	30
Friday		6	13	20	27		3	10	17	24		3	10	17	24	31
Saturday		7	14	21	28		4	11	18	25		4	11	18	25	
Sunday	1	8	15	22	29		5	12	19	26		5	12	19	26	

		April				May					June			
Monday		3	10	17	24	1	8	15	22	29		5	12	19	26
Tuesday		4	11	18	25	2	9	16	23	30		6	13	20	27
Wednesday		5	12	19	26	3	10	17	24	31		7	14	21	28
Thursday		6	13	20	27	4	11	18	25		1	8	15	22	29
Friday		7	14	21	28	5	12	19	26		2	9	16	23	30
Saturday	1	8	15	22	29	6	13	20	27		3	10	17	24	
Sunday	2	9	16	23	30	7	14	21	28		4	11	18	25	

		July					August					September			
Monday		3	10	17	24	31		7	14	21	28		4	11	18	25
Tuesday		4	11	18	25		1	8	15	22	29		5	12	19	26
Wednesday		5	12	19	26		2	9	16	23	30		6	13	20	27
Thursday		6	13	20	27		3	10	17	24	31		7	14	21	28
Friday		7	14	21	28		4	11	18	25		1	8	15	22	29
Saturday	1	8	15	22	29		5	12	19	26		2	9	16	23	30
Sunday	2	9	16	23	30		6	13	20	27		3	10	17	24	

		October					November					December			
Monday		2	9	16	23	30		6	13	20	27		4	11	18	25
Tuesday		3	10	17	24	31		7	14	21	28		5	12	19	26
Wednesday		4	11	18	25		1	8	15	22	29		6	13	20	27
Thursday		5	12	19	26		2	9	16	23	30		7	14	21	28
Friday		6	13	20	27		3	10	17	24		1	8	15	22	29
Saturday		7	14	21	28		4	11	18	25		2	9	16	23	30
Sunday	1	8	15	22	29		5	12	19	26		3	10	17	24	31

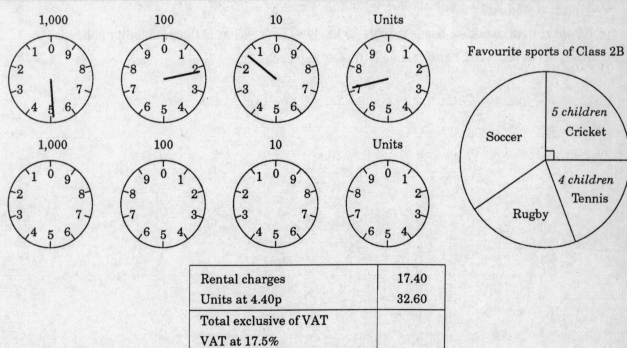

1,000	100	10	Units

1,000	100	10	Units

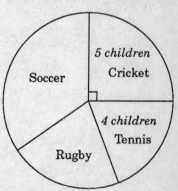

Favourite sports of Class 2B

Soccer

5 children Cricket

4 children Tennis

Rugby

Rental charges	17.40
Units at 4.40p	32.60
Total exclusive of VAT VAT at 17.5%	
Total payable	

Mental Test 2. *Intermediate level*

1. Express ten thousand two hundred and forty three in figures.

2. A rectangle has length 12 cm and width 4 cm. What is the perimeter?

3. £200 is put in the bank at 9% interest. How much is there after one year?

4. How many edges does a tetrahedron have?

5. A triangle has 3 axes of symmetry. What can you say about the triangle?

6. What is the square root of 0.04?

7. Write the ratio $\frac{1}{2} : \frac{1}{3}$ as the ratio of integers.

8. Convert 192 cm to m.

9. Sweet boxes contain 20 sweets each. How many can be made up from 177 sweets, and how many are left over?

10. The instructions for roasting meat tell one to roast it for 30 minutes plus 20 minutes for each pound. For how long should a 5 pound joint be roasted?

11. Three consecutive numbers add to 63. What is the greatest number?

12. What is the price of a £60,000 house after a 20% price rise?

13. To win at darts one must score 301. How much is left after double 10, 19 and 3?

14. Take 3, square it and add 7. What is the result?

15. Convert x feet to inches.

16. The temperature rose from -12° to 7°. What was the rise?

17. A cubit is 18 inches. How many cubits are there in 9 feet?

18. The seller of a car wants £1700 for it. The buyer offers £1400. If they split the difference what is it sold for?

19. A bent coin is such that heads is twice as likely as tails. What is the probability of heads?

20. What is the cost of 12 aerogramme letters at 28p each?

Mental Test 3. *Higher level*

1. The perimeter of a rectangle is 36 cm, and its length is 10 cm. What is the width?

2. Find the product of the square root of 2 with the square root of 8.

3. Convert 30 gallons to pints.

4. The average of 12 weights is 11 kg. What is the total weight?

5. The product of three consecutive numbers is 60. What are they?

6. Time in California is 6 hours behind London time. Find the time in London when it is 10 p.m. in California.

7. The fuel consumption of a car is 30 m.p.g. How far will it go on 5 gallons?

8. A model of a car is in the ratio 2:35. If the car is 7 ft wide how wide is the model?

9. Write a hundred cubed in standard form.

10. Write down the prime factors of 30.

11. Take $-2\frac{1}{2}$, add 1 and then double. What number do you have now?

12. Simplify: a half of x plus a third of x plus a sixth of x.

13. Find x, given that twice x is the same as x plus three.

14. A biased coin is such that the probability of heads is x. What is the probability of tails?

15. At a pay rate of £4.50 per hour, how long must I work to earn £90?

16. A typist can work at 20 words per minute. How long will a document of 10,000 words take?

17. Find the total cost of £200 worth of goods after VAT at $17\frac{1}{2}\%$ has been added.

18. To convert Centigrade temperature to Absolute temperature add 273°. Convert 200° Absolute to Centigrade.

19. Accelerating at 2 m sec^{-2}, how long does it take to increase speed from 1 m/sec to 10 m/sec?

20. A man earns £156 for his work during weekdays, and he works on Saturday for four hours at £6.50 per hour. What is his total pay?

Solutions to exercises

Chapter 1

Exercise 1.1.2 page 1

1. a) $14 + 19$ b) 3×14
   c) $2 \times 14 + 19$ d) $3 \times 14 + 2 \times 19$

2. $3 \times 10 + 5 + 2 \times 2$ or $2 \times 10 + 3 \times 5 + 2 \times 2$

3. a) $3B$ b) $A + 2B$ c) $A + 4B$
4. £7   5. 1975   6. £125
7. 58   8. 119, 120
9. a) 13, 1   b) 6, 6   c) 14, 2
10. 7, 2   11. 6, 11p   12. 36, 5   13. 14, 6
14. a) 4,363   b) 227,412   c) 500,062
15. a) 2,902 b) 3,981 c) 4,850 d) 2,000
16. a) 212   b) 4,209 c) 1,999 d) 8,999,999
17. a) 9   b) 7   c) 4   d) 12
18. a) $23 + 97 = 120$ b) $48 - 35 = 13$
    c) $241 + 162 = 403$ d) $746 - 324 = 422$

19. a,c,e,g. $6 = 2 \times 3 : 25 = 5 \times 5 : 35 = 5 \times 7$.

Exercise 1.1.4 page 3

1. a) 2   b) 2   c) 3   d) 6   e) 1
2. a) 30   b) 28   c) 45   d) 60   e) 210
4. 10p   5. 15 pints   6. 24 secs
7. 30 grams   8. 30×30 cm.

Exercise 1.2.2 page 4

1. a) 11, 13   b) 26,31   c) 22,26
   d) 32, 64   e) 81,243   f) 48,96
   g) 8, 4   h) 6,2   i) 36, 49
   j) 125, 216   k) 50,72   l) 37, 50
2. 1, 11   3. 10, 5   4. $1 + 6 + 21$
5. $5 + 10 + 26$
6. a) $5 \times 6 + 1 = 31, 6 \times 7 + 1 = 43$
   b) $4 \times 6 - 1 = 23, 5 \times 7 - 1 = 34$
   c) $2 \times 2 \times 2 \times 2 \times 2 = 32, 2 \times 2 \times 2 \times 2 \times 2 \times 2 = 64$
   d) $1 \times 2 \times 3 \times 4 \times 5 = 120, 1 \times 2 \times 3 \times 4 \times 5 \times 6 = 720$
7. a) 4,9,16,25,36   b) 2,6,12,20,30
   c) 3,5,7,9,11
8. 3,6,9,12,15
9. a) 4,8,12,16,20   b) 3,7,12,18,25
10. 13,21,34.

Exercise 1.3.2 page 6

2. a) -5 b) Friday
3. 150 miles 4. £205   5. 10 hours 6. 53 m
7. a) Dead Sea b) 1,770 m
   c) 398 m   d) 1,802 m
8. 2512 miles
9. a) 1 in, -15 in b) 12 in   c) 16 in
10. 520, 184 BC   11. 35 AD
12. 74 BC   13. 32 BC
14. a) 1797, 1774   b) 11, -43, -2002.

Exercise 1.4.2 page 8

1. $\frac{1}{2}, \frac{1}{3}, \frac{1}{4}, \frac{3}{4}, \frac{5}{8}$
2. fifth, ninth, five sixths, seven eighths
3. 0.5, 0.25, 0.375, 0.7
4. $\frac{1}{2}, \frac{1}{4}, \frac{7}{10}, \frac{7}{20}$ 5. $\frac{1}{2}, \frac{1}{3}, \frac{1}{3}, \frac{1}{2}, \frac{1}{4}$
6. a) $\frac{2}{4}$ b) $\frac{4}{6}$ c) $\frac{3}{4}$ d) $\frac{1}{7}$
7. $\frac{1}{5}$ 8. $\frac{3}{8}$, 0.3
9. a) $\frac{1}{2}$ b) $\frac{1}{4}$ c) $\frac{3}{8}$ d) $\frac{4}{9}$
10. $\frac{1}{10}$ p, $\frac{1}{8}$ p   11. 5 gallons. At the $\frac{3}{4}$ mark.
12. a) 30, 45   b) $\frac{1}{6}, \frac{1}{12}$
13. $\frac{9}{20}$ 14. 1,200

Exercise 1.4.4 page 9

1. a) $1\frac{1}{4}$ b) $2\frac{1}{3}$ c) $6\frac{2}{3}$ d) $1\frac{1}{2}$ e) $4\frac{1}{2}$
2. a) $\frac{3}{2}$ b) $\frac{11}{4}$ c) $\frac{27}{8}$
   d) $\frac{19}{5}$ e) $\frac{16}{7}$ f) $\frac{53}{10}$
3. a) 0.02 b) 0.08   c) 0.7   d) 0.75
4. a) $\frac{3}{2}, 1\frac{1}{2}$ b) $\frac{23}{10}, 2\frac{3}{10}$
   c) $\frac{13}{4}, 3\frac{1}{4}$ d) $\frac{9}{8}, 1\frac{1}{8}$
5. 24   6. 55   7. $\frac{3}{8}$, £2,000.

Exercise 1.4.6 page 10

1. a) $\frac{1}{9}$ b) $\frac{6}{11}$ c) $3\frac{4}{9}$
   d) $\frac{7}{30}$ e) $\frac{5}{37}$ f) $2\frac{19}{55}$
2. b) $\frac{2}{1}$ d) $\frac{11}{1}$ f) $\frac{1}{2}$
   h) 4   j) $\frac{4}{9}$ k) $\frac{2}{3}$
   l) $\frac{1}{3}$ rest irrational
3. e.g. $\sqrt{2}, \sqrt{8}$ 4. e.g. $\sqrt{8}, \sqrt{2}$ 5. e.g. $\sqrt{2}, \sqrt{-2}$
7. e.g. 4.5   8. e.g. $\sqrt{17}$

Chapter 2

Exercise 2.1.2 page 12

1. a) 10   b) 15   c) 12
   d) 10   e) 10   f) 3
   g) 5   h) 2   i) 0
2. a) 35   b) 48   c) 36
   d) 24   e) 27   f) 56
3. a) 7   b) 6   c) 5
   d) 6   e) 7   f) 7
4. a) 800   b) 6,000   c) 800,000
   d) 1,200   e) 360,000   f) 5,600,000
5. a) 40   b) 300   c) 20

d) 4   e) 40   f) 70

Exercises 2.1.4 page 13

1. a) 80   b) 1,500   c) 3,500
   d) 1,200   e) 4   f) 420
2. a) 20,000   b) 20,000   c) 50,000
   d) 9,000   e) 4,000,000   f) 2,000,000

Exercises 2.2.2 page 14

1. a) 894   b) 986   c) 939
   d) 1,237   e) 1,047   f) 1,705
2. a) 350   b) 427   c) 18
   d) 183   e) 263   f) 254
3. a) 936   b) 1,611   c) 4,473
   d) 2,310   e) 5,751   f) 3,000
4. a) 102   b) 208   c) 246
   d) 163   e) 56   f) 65
5. a) 4,354   b) 8,400   c) 17,472
   d) 22,140   e) 65,491   f) 98,901
6. a) 58   b) 46   c) 24
   d) 15   e) 9   f) 88

Exercises 2.3.2 page 15

1. a) 9.4   b) 22.7   c) 79.92
   d) 3.9   e) 3.78   f) 1.5
   g) 12.3225   h) 1932.832   i) 110.7
   j) 4.7   k) 3.5   l) 11.5   m) 20
2. £8.81   3   479 g   4. 20.1 km   5. £2.11
6. 6.35 m   7. 96 g   8. £7.62   9. £11.64
10. £12.90   11. £1.71   12. £14.80   13. 37 p
14. £1.43   15. £667   16. 7
17. a) £7   b) £9.62   c) £10.38
18. a) £18.60   b) £45.60
19. a) £27.55   b) £37.28
20. 169,344,294

Exercises 2.3.4 page 17

1. a) 13   b) 25   c) 16   d) 6   e) 7
   f) $3\frac{1}{4}$ g) 12   h) 2
2. a) $3 \times (4 + 6) = 30$ b) $(12 - 5) \times 2 = 14$
   c) $(7 - 1) \div 2 = 3$ d) $6 \div (17 - 14) = 2$
3. a) 49.1   b) 99.3   c) 68.1549   d) 238.76
4. a) 24.42   b) 2.9502   c) 512.7432   d) 0.224
5. £87.95   6. £24.19   7. 16

Exercise 2.4.2 page 18

1. a) $\frac{11}{12}$ b) $\frac{23}{30}$ c) $\frac{17}{21}$ d) $\frac{7}{20}$
   e) 3,4   f) $\frac{7}{10}$ g) $\frac{8}{21}$ h) $\frac{27}{40}$
2. a) $\frac{8}{21}$ b) $\frac{1}{9}$ c) $\frac{9}{10}$ d) $\frac{10}{21}$ e) 3
   f) $3\frac{7}{12}$ g) $\frac{8}{15}$ h) $\frac{-3}{8}$ i) $-4\frac{3}{20}$ j) $1\frac{1}{6}$
   k) $\frac{5}{6}$ l) $\frac{5}{27}$ m) 22
3. a) 12.8   b) 2   c) 9   d) 0.45
   e) 24   f) −12   g) 9   h) −32

4. $\frac{5}{6}$ 5. $1\frac{5}{12}$ yards   6. $12\frac{3}{4}$ pints   7. $\frac{1}{20}$
8. 40   9. 24   10. 6 hours   11. $\frac{1}{2}$ mile
12. £210   13. 125 m   14. 6   15. $\frac{7}{30}$
16. $\frac{11}{240}$ lb

Chapter 3

Exercise 3.1.2 page 21

1. a) 4   b) 4   c) 130   d) 987   e) 6
   f) −8   g) −77   h) 1   i) 0   j) −1
   k) 0   l) 9   m) 76   n) 1
2. a) 1,097,000   b) 1,097,400   c) 1,097,380
3. a) 460   b) 500   c) 0
4. a) − 67   b) − 70   c) − 100
5. a) 90   b) 1   c) 13   d) 5
6. a) 17   b) 86   c) 13   d) 7   e) 5   f) 1
7. £8.13   8 £107   9 £13.74   10. 2p
11. 190 miles   12. 3 hours   13. £590

Exercise 3.2.2 page 23

1. a) 2,300   b) 660,000   c) 0.0047
   d) 0.025   e) −3.5   f) 1.0
   g) 2.1   h) 500   i) 600
   j) 5.1   k) 4.0
2. a) 4,740   b) 4,700   c) 5,000
3. a) 3.576   b) 5.235   c) 23.038
   d) 65.476   e) 0.346   f) 0.123
   g) 0.005   h) 0.000   i) 0.120
   j) 0.270
4. a) 0.1358   b) 0.136   c) 0.14
5. a) 6.94   b) 1.51   c) 7.97
   d) 0.71   e) 2.72   f) 1.82
   g) 0.00
6. 1.47 g/c.c.   7. 660 g   8. £13
9. a) 3.571   b) 3.57

Exercises 3.2.4 page 24

1. 34.5°, 33.5°   2. 915 kg   3. 85,000, 75,000
4. 0.545, 0.535   5. 325 miles, 315 miles
6. 3.45, 3.55   7 9.735, 9.745

Exercises 3.2.6 page 25

1. 60.4, 60.8   2. 691 kg   3. 6 miles   4. £620
5. 4312.5 miles   6. 3.4 hours
7. a) 21   b) 7   c) 30   d) 2,860   e) 30   f) 2.8

Exercises 3.3.2 page 26

1. a) 8.8   b) 11.1   c) 15.47
   d) 5.07   e) 76.15   f) 133
   g) 79.54   h) 13.86   i) 6014
   j) 4.56   k) 3.11   l) 32.9
2. £4   3. 100 grams   4. 100 seconds
5. 2 hours   6. 1,000 grams

Exercises 3.3.3 page 26

1. a) 6.90   b) 73.0   c) 196   d) 0.926

2. £200   3. 5,000,000 grams

Exercise 3.4.2 page 27

1. 1.3 in   2. 360 g   3. 70 g
4. 600,000,000 miles

Chapter 4

Exercise 4.1.2 page 29

1. a) 9   b) 64   c) 4   d) 81
    e) 100   f) 144   g) 10,000
2. a) 2   b) 3   c) 4   d) 8   e) 10   f) 11
3. a) 8   b) 64   c) 1,000   d) $2\frac{1}{4}$
    e) $\frac{1}{4}$ f) $\frac{1}{9}$ g) 0.125
4. a) 100   b) 9   c) 12   d) 1   e) $\frac{1}{2}$ f) $\frac{3}{2}$
5. a) 36, 85 b) 16,9   c) 9   d) 2
6. 144 sq. ft.   7. 1,600 cm^2 8. 2 m
9. 5 cm   10. 27 c.c.   11. 4 cm

Exercise 4.2.2 page 31

1. a) 81   b) 256   c) 1   d) 1,024
    e) $\frac{1}{8}$ f) 0.027   g) 3.375   h) −8
2. a) $\frac{1}{2}$ b) $\frac{1}{27}$ c) $\frac{1}{125}$ d) $\frac{1}{64}$
    e) 1   f) 2   g) 9
3. a) 3^6 b) 2^7 c) 4   d) 3   e) 7
4. a) 4^5 b) 3^6 c) 1   d) $(3/2)^9$
5. 21   6.   3–3
7. a) 1.414   b) 1.732   c) 8.062
    d) 0.447   e) 0.007
8. a) 30   b) 32   c) 31.623
9. a) 3.73   b) 32.35   c) 1.22   d) 6.97
12. a) 2^3 b) $2^2 \times 3$ c) 3×5^2
    d) $2^2 \times 3^2$ e) 2^{10} f) $2^2 \times 3^2 \times 5$
13. $2^3 = 8$ 14. 45

Exercise 4.3.2 page 33

1. a) 3   b) 10   c) 2
    d) $\frac{1}{7}$ e) $\frac{1}{4}$ f) $\frac{1}{10}$
2. a) 4   b) 27   c) 100
    d) 16   e) 2   f) 10
3. a) 2^3 b) 3^3 c) 10
4. a) x b) y c) 1
5. a) 2   b) 4   c) $1\frac{2}{3}$

Exercise 4.4.2 page 35

1. a) 2.3×10^4 b) 8.76×10^8 c) 1.23×10^{-4}
    d) 1×10^{-3}
2. a) 1×10^5 b) 2.1×10^9 c) 9×10^{-2}
    d) 8×10^7 e) 1.04×10^{-12} f) 3×10^2
    g) 2×10^3 h) 2.5×10^7 i) 1.6×10^{17}
    j) 3.1×10^{-10} k) 6.4×10^7

3. a) 8.4×10^6 b) 3.57×10^6 c) 1.03×10^6
    d) 6.4×10^7 e) -2.6×10^9 f) 9.8×10^{-3}
4. 6.696×10^8 miles, 5.9×10^{12} miles
5. 5×10^{27}

Chapter 5

Exercise 5.1.2 page 37

1. 4:3   2. 5:6   3. 4:1   4. 2:1
5. 2:1   6. 3:1   7. 10 litres 8. 40 kg
9. 30 kg   10. £32   11. 3 km   12. 5 cm
13. $\frac{1}{2}$ m   14. 25 m   15. 1:100   16. 9 cm

Exercise 5.1.4 page 38

1. £9,000   2. 4 kg   3. 26 cm   4. 150
5. a) 2:1,   b) £4,000
6. £10,000   7. 40
8. £6,000, £8,000, £10,000 9. 240 c.c.
10. 8 hours   11. 10 min, 4 min, 6 min

Exercise 5.1.6 page 39

1. $\frac{1}{4}$ km^2 2. 20 cm^2 3. $\frac{3}{25}$ km^2
4. 5 cm^2 5. 0.6 m^3
6. a) 1 km   b) $\frac{1}{2}$ km   c) $\frac{1}{4}$ km
7. 1 cm to $\frac{1}{4}$ km. or 1:25,000
8. 1 cm to 2 km. 1:200,000   9. 4,000 m^3
10. a) 480 cm b) $27\frac{7}{9}$ cm c) 15.12 m^3 d) 2
11. a) $16\frac{2}{3}$ cm   b) 16.2 m   c) 110 g
12. a) 1:20   b) 20 cm^2 c) 10
13. 22.5 cm2   14 1 cm to $\frac{1}{2}$ km. or 1:50,000

Exercise 5.2.2 page 41

1. 300 ml   2. £12   3. 272 miles
4. 90 g   5. £48   8. $1\frac{1}{2}$ hours
9. £6   10. 2 days   11. 2 hours

Exercise 5.2.4 page 43

1. $y = \frac{1}{2}x, 2\frac{1}{2}$
2. a) 72   b) 30
3. a) $3\frac{1}{2}$ amps   b) 120 volts
4. 100 km   5. $p = 70/q, 17\frac{1}{2}$
6. a) 4   b) 16
7. $V = \frac{200}{P}, 5$ m^3
8. a) 50,000 cps   b) 250 m
9. a) $y = 4x^2$ b) 36   c) 1
10. a) $T = 2s^2$ b) 72   c) 9
11. $E = 2s^2$, 10 m s^{-1} 12. $P = 8I^2$, 200 Watts
13. $E \propto e^2$, E $= \frac{7}{9}e^2, \frac{7}{9}$

14. a) $T = 20/R$ b) 1.25 c) $\frac{1}{2}$

15. $B = 2/c^2$, 8, $\frac{1}{2}$ 16. $1\frac{7}{9}$

17. $R = \frac{3}{4}/r^2$, 12 ohms, 1 cm 18. $2\frac{2}{3}$ cm

19. 689 days 20 2, 108 21 3, 27

Chapter 6

Exercise 6.1.2 page 46

1. £1.50 2. £140 3. £1,000
4. £11 5. £40 6. 8 litres
7. 12 kg 8. 90 9. 12 ft
10. £16,100 11. 12 stone 12. £214,000
13. £9.60 14. £1,600 15. £600
16. £1.47 17. £81.78 18. £32.43
23. a) 2.5 m b) 127.5 m c) 122.5 m

Exercise 6.2.2 page 48

1. a) $\frac{1}{10}$ b) $\frac{1}{5}$ c) $\frac{3}{5}$
   d) $\frac{1}{20}$ e) $\frac{1}{8}$ f) $\frac{3}{2}$ g) 1
2. a) 50% b) 25% c) 75% d) 60%
   e) 5% f) 85% g) 175%
3. a) 12% b) 74% c) 3%
   d) 60% e) 154% f) 270%
4. a) 0.23 b) 0.54 c) 0.05
   d) 0.2 e) 1.5 f) 1.04
5. 65% 6. 25% 7. 85% 8. $66\frac{2}{3}$%
9. 55% 10. 40% 11. $66\frac{2}{3}$% 12. 10%
13. 25% 14. 10% 15. 20% 16. 4%

Exercise 6.3.2 page 49

1. £266.2 2. £1,157.63 3. £524.32
4. £3,070.63 5. 21,900,000
6. a) £726 b) £1,556.25
7. a) £472.06 b) £2,306.99
8. a) £1,425.72 b) £3,375.20
9. 22.3 million 10. £3,596
11. a) 2.43 kg b) 1.046 kg
12. £300 13. £1,500 14. £1,600
15. 20 million 16. £11,000 17. £500
18. £25 19. £2.80

Chapter 7

Exercise 7.1.2 page 54

1. a) 300 b) 360 c) 3.6
   d) 6,500 e) 12.7 f) 103.632
2. a) 0.4 b) 0.213 c) 1,390
   d) 0.6858 e) 4.572
3. a) 0.3 b) 15,340 c) 5,584.2
   d) 104 e) 482
4. a) 0.867 b) 0.065 c) 1.362
   d) 1.94312 e) 0.96475
5. a) 4 b) 0.277 c) 13.62 d) 2.8375

6. 72,846 m^2 = 7.3 hectares
7. a) 22 km b) 16.5 m
8. a) 55 b) £7.50
9. a) 1.95, 2.1 b) 25
10. a) 60 b) 576 c) 3.94 d) 9.65 e) 13.4
11. a) 42 b) 2 c) 10.1 d) 50.7
12. a) 10.872 b) 1,235.5
13. 40, £4 14. 9.1 kg, 11 lb
15. a) 74$ b) £37, £13

Exercise 7.2.2 page 56

1. a) £2 b) £1.05 c) £7.125 d) £3.60
2. a) 16 galls b) 10 lb c) 8 m d) 40 m
3. a) 86p b) £6 c) 3 p d) 19p
4. a) 5 hours b) 8 cm c) $2\frac{1}{2}$ hours
5. a) 20 b) $\frac{1}{3}$ c) 200 cm^3.
6. 22 min 7. 40 m.p.h.
8. 13 km, $4\frac{1}{3}$ km/hr 9. 13 p
10. 520 c.c. 11. 135 miles, 4 hours
12. a) 54 m.p.h. b) 1.30
13. 13 l 14. 32 m.p.g., 256 miles
15. £18.75 16. £57
17. $2\frac{1}{2}$ hours 18. £40

Exercise 7.3.2 page 58

1. a) £4,420 b) £9,620 c) £10,800
   d) £8,528 e) £6.916
2. a) £112 b) £230
3. Yes, by 10p per hour 4. £172.05
5. £128.7, 4 hours 6. £215.8
7. a) £1,950 b) £1,950
8. £10,500
9. a) £7 b) £3.5 c) £73.5
10. a) £388 b) £2,710
11. £5

Exercise 7.4.2 page 60

1. a) 1100 b) 0730 c) 1600
   d) 1200 e) 1945 f) 2305
2. a) 8 a.m. b) 10.15 a.m.
   c) 12.15 p.m. d) 4.45 p.m. e) 8.30 p.m.
3. a) 0300 b) 1700 c) 0730 d) 23.45
4. a) $4\frac{1}{2}$ hours b) $3\frac{1}{4}$ hours c) 11 hrs 38 mins
5. a) 7 b) 31 July, 28 Aug, 4 Sept
6. 13
7. a) 5217 b) 2936
9. a) 24 miles b) 120 miles
   c) 138 miles, 159 miles
10. a) 1 hr 44 min, 5 min
    b) 18.41 c) 2042
11. a) £1,577.6 b) £1,427
12. £14

13. a)   1 hr 40 min b)   1 hr 5 min

Chapter 8

Exercise 8.1.2 page 64

1. 16 　　　　2. $p=4, q=28$ 　　3. $7+x$
4. $£(x+2)$ 　　5. $10-x$ m 　　6. $180-d$
7. $8x$ 　　　　8. $10b$ 　　　　9. $20-x$
10. $100h$ 　　　11. $10p$ 　　　　12. $£p/8$
13. $x+y=10$ 　　14. $N=M+5$
15. a) 4 　　b) 14 　　c) 3 　　d) 6

Exercise 8.1.4 page 65

1. $LB, 2(L+B)$ 　2. $x=68t$ 　　3. $£(n-m)$
4. $180n-360$ 　5. $x+120$ 　　6. $301-s$
7. $£(30+20m)$ 　8. $2N+5$ 　　9. $sr-t^2$
10. $£(4.50x+6y)$ 　11. $£(10x+8y), \dfrac{£(10x+8y)}{18}$
12. $£54x$ 　　13. $x=180-y-z$ 　14. $6x+4y+z$

Exercise 8.2.2 page 66

1. 17 　　2. 5 　　3. 42 　　4. 18
5. 14 　　6. 2 　　7. 2 　　8. 154
9. 32 　　10. 6 　　11. 28 　　12. 26

Exercise 8.2.4 page 67

1. 4 　　2. a) 60,000 b) 2,250
3. 18 　　4. 120 　　5. 132 　　6. $-61, 7$
7. 3 　　8. -5 　　9. 30 　　10. 1
11. 35, -2 　12. 4 　　13. 4 　　14. 132

Exercise 8.3.2 page 69

1. $5x + 35$ 　　2. $4a - 4b$ 　　3. $14x + 7y$
4. $6x - 10y$ 　　5. $8x + 18$ 　　6. $5y - 27$
7. $-2z + 22$ 　　8. $37w - 21$ 　　9. $7x - y$
10. $8a + 11b$ 　　11. $2x + 2y$ 　　12. $c - 14d$
13. $11r$ 　　14. $2p + 14q$ 　　15. $x^2 + 6x +$
16. $y^2 + 10y + 21$ 　17. $p^2 - p - 12$ 　18. $z^2 + 2z - 24$
19. $w^2 - 18w + 77$ 　20. $q^2 - 7q + 10$ 　21. $x^2 - y^2$
22. $4x^2 - y^2$ 　　23. $9p^2 - 4q^2$ 　　24. $18r^2 - 8s^2$
25. $3x^2 + 19x - 40$ 　26. $6y^2 + 13y - 28$ 　27. $9a^2 - 4b^2$
28. $6w^2 + 11wz - 10z^2$ 　　29. $x^2 + 6x + 9$
30. $4y^2 - 12y + 9$ 　　31. $x^2 + 4xy + 4y^2$
32. $9z^2 - 12zw + 4w^2$ 　　33. $ac + ad + bc + bd$
34. $xz + xw - yz - yw$ 　　35. $6tr + 2tp + 3sr + sp$
36. $fj - 5fk - 3gj + 15gk$
37. $10yz - 14yw + 15xz - 21xw$
38. $15sr - 10st + 3r - 2t$

Exercise 8.4.2 page 70

1. $y = \dfrac{x - 2}{3}$ 　　2. $d = a + c - b$ 　　3. $C = \dfrac{5(F - 32)}{9}$
4. $s = \dfrac{rt}{3p}$ 　　5. $a = \dfrac{y}{3x^2}$ 　　6. $m = \dfrac{(59 - j)}{n}$
7. $a = 2b - c$ 　　8. $b = 2a(d + 3)$ 　　9. $r = \dfrac{(3V + 2)}{4}$
10. $X = 20Y - 2Z$ 　　11. $x = 2z - y$
12. $p = \dfrac{7}{(5(q - r))}$

Exercise 8.4.4 page 71

1. $x = \sqrt{(y/4a)}$ 　　　2. $r = \sqrt[3]{(3V/4)}$
3. $S = \sqrt{(3 + T + R)}$ 　　4. $h = \sqrt{(l^2 - 4r^2)}$
5. $n = ((h + 3)/3m)^2$ 　　6. $l = g(t/2\pi)^2$
7. $b = \dfrac{(c + 1)}{a} + 3$ 　　8. $u = \dfrac{1}{(1/f + 1/v)}$
9. $m = \dfrac{T}{(g - a)}$ 　　10. $x = \dfrac{(-5a - 2b)}{2}$
11. $q = \dfrac{p(c - a)}{(b - d)}$ 　　12. $x = \dfrac{(7b - 3a)}{(a + b)}$
13. $R = \dfrac{rm}{(1 - m)}$ 　　14. $x = \dfrac{(3y + 1)}{(y - 1)}$

Exercise 8.5.2 page 72

1. $x(y + z)$ 　　2. $q(p - 3)$ 　　3. $s(3 + r)$
4. $t(q + r)$ 　　5. $r(t - 5s)$ 　　6. $x(3y - 4z)$
7. $2(x + 2y)$ 　　8. $3(t - 3r)$ 　　9. $2r(s + 2q)$
10. $3y(3x - z)$ 　11. $3f(2g + 3h)$ 　12. $7t(1 + 3q)$
13. $a(a + b)$ 　　14. $z(1 - z)$ 　　15. $5x(x - 3)$
16. $7y(3y - 2)$ 　17. $4z(2z + c)$ 　18. $3d(1 + 3cd)$

Exercise 8.5.4 page 73

1. $(p - q)(p + q)$ 　　　2. $(t - 1)(t + 1)$
3. $(2 - n)(2 + n)$ 　　　4. $(q - 3)(q + 3)$
5. $(3 - s)(3 + s)$ 　　　6. $7(1 - s)(1 + s)$
7. $2(1 - 2x)(1 + 2x)$ 　　8. $9(1 - 3a)(1 + 3a)$
9. $(7t - 3s)(7t + 3s)$ 　　10. $(4m - 5n)(4m + 5n)$
11. $(y - \frac{1}{2})(y + \frac{1}{2})$ 　　12. $(\frac{1}{3} - z)(\frac{1}{3} + z)$
13. $(x + 2)(x + 1)$ 　　　14. $(x + 2)^2$
15. $(x + 3)(x + 4)$ 　　　16. $(x - 3)(x - 8)$
17. $(x + 3)(x - 2)$ 　　　18. $(y - 4)(y + 3)$
19. $(z + 6)(z - 1)$ 　　　20. $(w - 6)(w + 2)$
21. $(x-8)(x+3)$ 　　　22. $(p+10)(p-3)$
23. $(2x-1)(x+3)$ 　　　24. $(3x+2)(x-4)$
25. $(5y+4)(y+1)$ 　　　26. $(6x-1)(x+6)$
27. $(6a+5b)(a+b)$ 　　　28. $(2p-3q)(3p+2q)$
29. $(t-q)(s+r)$ 　　　30. $(x-y)(a-b)$
31. $(3t-2)(2s+1)$ 　　　32. $(z-3)(2x+y)$
33. a) $(x-1)^2+4.$ 4 　　b) $(x+2)^2-3.$ -3
    c) $(x-5)^2-14.$ -14 　d) $(x+3)^2+4.$ 4
    e) $(x+1.5)^2-1.25.$ -1.25 　f) $(x-2.5)^2-9.25.$ -9.25

Exercise 8.6.2 page 75

1. $\dfrac{ac}{bd}$ 　　2. $\dfrac{2x^2}{5}$ 　　3. $\dfrac{4x^2}{3y^2}$
4. $\dfrac{wx}{yz}$ 　　5. $\dfrac{7}{2}$ 　　6. $\dfrac{q^2}{2}$
7. $\dfrac{11x}{15}$ 　　8. $\dfrac{5a}{4}$ 　　9. $\dfrac{7}{r - 4}$
10. $\dfrac{(-11x-18)}{12}$ 　11. $\dfrac{(27b-1)}{6}$ 　12. $\dfrac{(7c-25)}{12}$
13. $\dfrac{(2y+3)}{xy}$ 　14. $\dfrac{(3bc+4a)}{abc}$ 　15. $\dfrac{1}{(3x(x-1))}$
16. $\dfrac{6z^2}{(z^2-1)}$ 　17. $\dfrac{2(x+2)}{(3(x-1))}$ 　18. $\dfrac{2(2-y)}{(3(1-y))}$

19. $\dfrac{(7x+3)}{(x(x+1))}$ 20. $\dfrac{(-7z+5)}{(z(1-z))}$ 21. $\dfrac{(z^2+z+6)}{(3(z+1))}$

22. $\dfrac{(17y+8)}{(y(3y+2))}$ 23. $\dfrac{(3y+5)}{((y-1)(y+3))}$

24. $\dfrac{(-2w+1)}{((3w+1)(w+2))}$ 25. $\dfrac{(16p+4)}{((2p-1)(2p+3))}$

26. $\dfrac{(-3q+1)}{((3-2q)(1-q))}$

Chapter 9

Exercise 9.1.2 page 78

1. 2	2. 17	3. 7	4. −2
5. 6	6. 4	7. 3	8. 2
9. 1	10. 1	11. 3	12. 4
13. 5	14. 1	15. 147	16. 3
17. 28	18. 7	19. 2	20. 5
21. −6	22. 3		

Exercise 9.2.2 page 79

1. a) 5.4   b) 4.1   c) 4.8
   d) 4.1   e) 0.9   f) 3.9
   g) 3.1   h) 2.9   i) 3.0
2. a) 0.37   b) 0.53   c) 1.37

Exercise 9.3.2 page 80

1. 5, −2	2. 6, −9	3. 2, −1
4. 1, 4	5. 4, 3	6. 27, −37
7. 6, 2	8. 5, 5	9. 4, 1
10. 3, −2	11. 2, 7	12. 4, −3
13. 2, 3	14. 8, 7	15. 12, 18
16. 8, 7		

Exercise 9.4.2 page 82

1. 1, −6	2. −3, −4	3. 9, 1
4. −4, −15	5. 30, −2	6. 1.5, −0.5
7. 5, −2	8. 5, 3	9. 5, −4
10. 6, 2	11. 1, −6	12. 4, 3

13. −0.349, −2.151   14. 4.562, 0.438
15. 1.193, −4.193   16. 2.303, −1.303
17. 2.618, 0.382   18. 3.815, −6.815

Exercise 9.5.2 page 82

1. £$(20+\dfrac{x}{10})$, 170 miles   2. 16
3. £$(2x+1,500)$, £13,100, £14,600   4. 50°
5. $2x-7$, 9   6. 40

Exercise 9.5.4 page 83

1. $(6y-3)=11(y-3)$, 6
2. $n+1$, $n+2$, $n+n+1+n+2=441$, $n=146$
3. 8
4. $3x$, $6x$, £180, £540, £1,080
5. £$(9x+50)$, 63   6. £500   7. 320
8. 92p   9. $x=96$, $y=24$ 10. 21

Exercise 9.5.6 page 84

1. 260   2. 90   3. 120 miles   4. 13
5. 3   6. 8   7. 8, 9   8. 8, 12
9. 98   10. 3 m.p.h.

Chapter 10

Exercise 10.1.2 page 87

1. $6\dfrac{3}{4}$ 2. $4\dfrac{1}{3}$
3. $1\dfrac{1}{4}$, $\dfrac{9}{7}$, 1.3, 1.42
4. $\dfrac{1}{100}$, 0.0011, thousandth, millionth, 0
5. 1 kg jar   6. £3.20 per roll
7. method 2, £400   8. second
9. yes, by 60p   10. second
11. £17 rise   12. England
13. $\dfrac{1}{8}\dfrac{1}{4}\dfrac{3}{8}\dfrac{1}{2}\dfrac{5}{8}\dfrac{3}{4}\dfrac{7}{8}1$

Exercise 10.2.1 page 88

1. a,d,e   2. 4.45<4.6<14/3<19/4
3. 0.2>1/8>0>−1/8>−3/17>−1/4>−1/2
4. 4 or 5   5. −4&−3
6. a) 1,2,3,4,5   b) 2,3,4,5,6,7
   c) 14,15,16,17   d) 7,8,9
7. a) −1,0,1   b) −7,−6,−5,−4,−3
   c) −1,−2,−3,−4   d) 3,2,1,0,−1,−2
8. a) $x\geq1$ b) $x\leq1$ c) $x>-1$ d) $x<-6$

Exercise 10.2.3 page 90

1. a) $\{x:x<7\}$ b) $\{x:x>1\}$ c) $\{x:x\geq3\}$
   d) $\{x:x<-4\}$ e) $\{x:x>7\}$ f) $\{x:x\leq3\}$
2. a) $x>12$ b) $x<3$ c) $x<6$
   d) $x\leq13$ e) $x\geq-\dfrac{1}{5}$ f) $x>33$
3. a) $\{x:2<x<6\}$ b) $\{x:4<x<6\}$
   c) $\{x:-2\leq x\leq6\}$ d) $\{x:2\leq x\leq11\}$
   e) $\{x:-5<x\leq-2\}$ f) $\{x:-2<x\leq4\}$
4. $\dfrac{x}{10}\leq1$, $x\leq10$ 5. $25y<150$, $y<6$
6. $30z\geq100$, $z\geq\dfrac{10}{3}$ 7. $\dfrac{m}{50}\leq5$, $m\leq250$
8. $17x+24x<500$, $x<13$ 9. $15+0.05x<50$, $x<700$
10. $4x+4<60$, $x<14$

Exercise 10.3.2 page 92

3. a) $y\geq1$ b) $x+y\geq1$
   c) $x+2y<2$ d) $x>2y$
4. $10x+15y\leq100$ 5. $x+y\leq8$ 6. $3x+4y\geq15$

Exercise 10.4.2 page 93

1. 10 type 1, 5 type 2   2. 300 car, 50 lorries
3. $x + y \leq 5$. $200x + 300y \leq 1,200$. £280
4. $4X + 6Y \geq 16$. $5X + 3Y \geq 11$. $X = 1$, $Y = 2$
5. $x \leq y$. $5x + 4y \leq 44$. $150x + 80y \geq 900$. $x = y = 4$
6. $3x + 4y \leq 600$. $20x + 50y \leq 5,000$. $x \geq 50$. 10,000

Chapter 11

Exercise 11.1.2 page 96

1. a) A3   b) C3, C4, C5, B5, A5
2. (1,3), (2,2), (3,1), (4,2)   4. (3,2)
5. (4,3), $(2\dfrac{1}{2}, 2\dfrac{1}{2})$

7. A(1,–2) B(–1,–1) C(–2,2) D(0,–2)
8. P(1.6,0.4) Q(0.5,0.5) R(1.4,1.9) S(0.7,1.2)

Exercise 11.1.4 page 99

1. (0,0,20), (0,30,20), (40,30,20), (40,0,20)
2. (5,4,3). (2.5,2,3)
3. (0,0,±1), (1,0,±1), (1,1,±1), (0,1,±1).
4. cuboid. 6
5. prism. 30
6. (1,1,0), (–1,1,0), (–1,–1,0), (1,–1,0), (0,0,3). (Many others)
7. (0,0,5), (1,0,5), (0,1,5) or (0,0,–1), (1,0,–1), (0,1,–1)

Exercise 11.2.2 page 100

1. a) 1 mile, 20 min
   b) 11 miles, 30 min c) 10 min
2. a) $\frac{1}{2}$ mile   b) 35 min
3. a) 1 hr, $\frac{1}{2}$ hr, $1\frac{1}{2}$ hr
   b) 30 m, 40 m, 30 m
   c) 30 m.p.h., 80 m.p.h., 20 m.p.h.
   d) $\frac{100}{3}$ m.p.h.
4. a) 10 min b) 4 miles c) 24 m.p.h.
5. a) 4 min b) 30 m.p.h. c) $2\frac{1}{2}$ min
6. a) 4 years b) 1984 c) 1989
7. a) 21°, 2° b) 3–4 a.m. c) 2 hours
8. a) 1980–81 b) 1982–83 c) Late 1982
9. a) 111 m, 4.8 secs
   b) 3.5 secs, 6.1 secs c) 9.3 secs
10. 70 mins, 90 mins, 110 mins
11.

Units	100	400	800	1200	1500
A in £	25	40	60	80	95
B in £	13	37	69	101	125

By B. Over 500 units.

Exercise 11.2.4 page 103

1. a) 3 m.p.h b) 6 m.p.h. c) $1\frac{1}{2}$ hours
2. 3 m.p.h., 22 m.p.h.
3. 2 m.p.h., 6 m.p.h., 6 m.p.h.
4. 20 m/sec, 0 m/sec
5. 0.75 ins/month, 0.45 ins/month
6. a) 2,000 galls b) 10–11, 1 hour c) 9 a.m.
   d) 160 galls/hr

Exercise 11.2.6 page 105

1. a) $\frac{1}{3}$ m s^{-2}, $\frac{2}{3}$ m s^{-2} b) 1,125 m
2. a) 2 m s^{-2} b) 45 m
3. a) 1 m s^{-2} b) 9 m
4. after $\frac{3}{4}$ hour, 2.6 miles
5. 2 hours, 100 miles 6. 1.10 & 1.00, at 1.30
9. c

Chapter 12

Exercise 12.1.2 page 109

1. 2, 3, 4, 5
2. a) 1, 2, 3, 4 b) 3, 2, 1, 0 c) 1, 3, 5, 7
3.

x	–2	–1	0	1	2
y	5	2	1	2	5

4. 8, 2, 0, 2, 8, 18

Exercise 12.1.4 page 109

1.

$x =$	–1	0	$\frac{1}{2}$	1	$1\frac{1}{2}$	2	3
$y =$	4	1	$\frac{1}{4}$	0	$\frac{1}{4}$	1	4

2.

$x =$	$\frac{1}{4}$	$\frac{1}{2}$	$\frac{3}{4}$	1	$1\frac{1}{2}$	2	3
$y =$	9	5	3.7	3	2.3	2	1.7

$x = 0.7$

3.

$x =$	–3	–2	$-1\frac{1}{2}$	–1	$-\frac{1}{2}$	0	1
$y =$	2	–1	$-1\frac{3}{4}$	–2	$-1\frac{3}{4}$	–1	2

$x = 0.4$ or –2.4

4.

$x =$	$\frac{1}{8}$	$\frac{1}{4}$	$\frac{3}{8}$	$\frac{1}{2}$	$\frac{3}{4}$	1	$1\frac{1}{2}$	2	3
$y =$	$8\frac{1}{8}$	$4\frac{1}{4}$	3	$2\frac{1}{2}$	2.1	2	2.2	$2\frac{1}{2}$	3.3

$x = 2.6$ or 0.4

5.

$x =$	1	2	3	4
$y =$	1	4	7	10

6.

$x =$	0	1	$1\frac{1}{2}$	2	3
$y =$	1	–1	$-1\frac{1}{2}$	–1	1

2.6 and 0.4

7.

$x =$	–2	–1	$-\frac{1}{2}$	0	$\frac{1}{2}$	1	2
$y =$	–1	2	$2\frac{3}{4}$	3	$2\frac{3}{4}$	2	–1

1.4 and –1.4

8. $x = 0.7$

Exercise 12.1.6 page 112

1.

$x =$	–2	–1	$-\frac{1}{2}$	0	$\frac{1}{2}$	1	$1\frac{1}{2}$	2
$f =$	–6	0	$\frac{3}{8}$	0	$-\frac{3}{8}$	0	$1\frac{7}{8}$	6

$x = –1.5.$ grad. = 2

2.

$x =$	–1	$-\frac{1}{2}$	0	$\frac{1}{2}$	1	$1\frac{1}{2}$	2
$f =$	–3	$-\frac{5}{8}$	0	$-\frac{3}{8}$	–1	$-1\frac{1}{8}$	0

$x = –1$ or 1 or 2.

grad. = –1

3.

$x =$	$\frac{1}{4}$	$\frac{1}{2}$	$\frac{3}{4}$	1	$1\frac{1}{2}$	2
$y =$	-4	$-1\frac{3}{4}$	$-.77$	0	1.58	$3\frac{1}{2}$

$x = 1.2$. grad. $= 5$

4.

$x =$	$\frac{1}{4}$	$\frac{1}{2}$	$\frac{3}{4}$	1	$1\frac{1}{2}$	2	3
$f =$	$16\frac{1}{4}$	$4\frac{1}{2}$	2.5	2	1.9	$2\frac{1}{4}$	3.1

$x = 0.6$. grad. $= 0.75$

6. $-\frac{1}{2}$

7. a) 60°, 300°   b) 120°, 240°
   c) 50°, 310°   d) 140°, 220°

Exercise 12.2.2 page 114

1. a) 1   b) 2   c) –1   d) $-\frac{1}{2}$
2. 3   3. $\frac{1}{2}, x = 1, y = 4$
4. a) 4   b) 1   c) $\frac{1}{2}$ d) $-\frac{3}{4}$
5. a) (3,4)   b) (3,2)   c) (2,3)   d) (4,1)
6. parallel

Exercise 12.2.4 page 115

1. a) $y = 3x$ b) $y = x+1$
   c) $y = 7-3x$ d) $y = \frac{1}{2}x + \frac{1}{2}$
   e) $y = 2x$ f) $y = -1-2x$
2. a) $y = 2x$ b) $y = \frac{1}{2}x+1$ c) $y = 2-2x$

Exercise 12.3.2 page 116

1. a) 22m   b) 3.5 sec and 0.5 sec   c) 4.1 sec
2. a) 205 ft   b) 54 m.p.h.
3. $x = 5$ cm   4. 3.3 cm   5. A = 12 sq. in.

Exercise 12.4.2 page 117

4. $p = 1, q = -1, r = -1, s = -1$

Chapter 13

Exercise 13.1.2 page 119

1. a) 110°   b) 120°   c) 10°   d) 130°
2. a) 60°   b) 80°   c) 30°
3. a) 180°   b) 90°   c) 1,080°
4. a) 2   b) 10   c) $1\frac{1}{2}$ d) $\frac{1}{3}$
5. 270°
6. a) 90°   b) 180°   c) 150°
7. 60°   8. 4 hours
9. $y = x = 145°, z = 35°$ 10. 20°, 160°

Exercise 13.1.4 page 121

1. $a = c = e = 130°$ $b = d = f = 50°$
2. $a = 140°$ $b = 70°$ $c = 60°$ $d = 50°$
3. 90°   4. $a+b = 180°$
5. 50°   6. c&d

Exercise 13.2.2 page 123

1. a) 80°   b) 75°   c) 70°   d) 65°

2. a) 30°   b) 110°   c) 85°   d) 70°
3. a) 140°   b) 60°
4. a) 7   b) 4
5. $x = 40°, y = 100°$ 6. $z = 100°$
7. 30°   8. Both 61°
9. Equilateral   12. Square
13. Rhombus

Exercise 13.2.4 page 124

1. a) 540°   b) 720°   c) 1080°   d) 3240°
2. a) 108°   b) 144°   c) 150°   d) 168°
3. a) 45°   b) 30°   c) 20°   d) 12°
4. 7   5. 9   6. 5   7. 12
8. 88°   9. 60°   10. 110°
11. a) 36   b) 18   d) 12
12. 9   13. 10
14. 108°, 36°, 72°
15. 120°, 90°, 60°, 30°, rectangle
16. 100° rhombus   17. 108°
18. 3, 4, 6   19. 150°, 12

Exercise 13.3.2 page 126

1. a   2. ALM & ABC
5. a) 3   b) 4   c) 6
   d) 1
6. 30 cm   7. 80 cm, 160 cm
8. 1.5 m, 1 m   11. 1:4, 2, 30
12. AEF & CDF, 1:9   13. AXY & ABC, 1:3, 64
14. ABC & AQP, $A\hat{P}Q = \hat{C}$, $A\hat{Q}P = \hat{B}$, 36, 45
15. ABC & DAB a) $\frac{20}{3}$ b) 4

Exercises 13.4.2 page 129

1. a&c
2. BAC & BED, $A\hat{B}C = E\hat{B}D$, $B\hat{A}C = B\hat{E}D$
3. BAC & BDE, $B\hat{A}C = B\hat{D}E$, $B\hat{C}A = B\hat{E}D$, parallel
4. XPQ & XRS, XP = XR, XS = XQ
5. ACD & CAB, $\hat{D} = \hat{B}$, $D\hat{A}C = A\hat{C}B$, $A\hat{C}D = C\hat{A}B$, parallel, parallelogram
7. AC, ADC & ABC, ADX & ABX, CDX & CBX, $\hat{D} = \hat{B}$

Chapter 14

Exercise 14.1.2 page 132

2. 100 m, 314 m   3. 25 cm
4. 7.85 cm   5. 9.4 cm
6. 9.4 cm   7. 127 m, 63.5 m
8. 3.18 cm   9. 7.85 cm
10. 10 cm   11. 1,244 in, 20.7 in
12. 52.5 cm, 26.5 times
13. a) 50°   b) 140°   c) 150°

Exercise 14.2.2 page 134

2. a) $x = 90°, y = 20°$
   b) $z = 45°$ c) $p = 30°, q = 120°$
4. 70°, 20°, 20°   5. 2.62   6. 82°
7. 5:4:3   8. ACO = 90°, congruent, 40°

9. OA, 40°

10. AC = BC, AD = DB, AM = MB, CMA≡CMB, kite

11. $\hat{C} = \hat{D} = 100°$, $\hat{B} = 80$

12. congruent, equal

Exercise 14.3.2 page 136

1. a) 50°   b) 61°

2. a) 43°   b) CDB

   c) AXD & BXC, DXC & AXB

3. a) 124°   b) 106°, 74°

4. a) 100°   b) 85°   c) 100°   d) 55°   e) 70°

6. 184°   7. 70°, 70°, 35°

8. both 44°   9. 70°, 110°, 70°, parallel

11. rectangle

Exercise 14.4.2 page 138

1. a) 67°   b) 135°   c) 10°

2. 35°   4. 108°   6. square   8. similar

11. 10 cm

12. a) 43°   b) 80°   c) 45°

13. 60°, 57°, 63°   14. 70°, 54°, 56°   15. 20°

16. 82.5°, 97.5°, 97.5°, 82.5°

17. 140°, 70°, 45°

18. a) 40°   b) 100°   c) 67°

Chapter 15

Exercise 15.1.2 page 141

1. 5, 8, 5

2. 5, 9, 6, ABC & DEF, AD & CF etc

4. 12,4   5. 27, 3, 12, 12, 4

6. a) 4   b) 12   c) 8

7. cube, F, H   9. b&d   10. cuboid

11. a) pyramid   b) cuboid

   c) prism   d) pyramid

Exercise 15.2.2 page 142

2. a) tetrahedron, D, C

   b) BED   c) 4, 6, 4

7. a) 3   b) 5   c) 9

   d) ∝   e) 4   f) ∝

   g) ∝   h) 4

8. a) 3   b) 5   c) 13

   d) ∝   e) 1   f) ∝

   g) 1   h) 4

10. a) pyramid, 5, 8, 5   b) tetrahedron, 4, 6, 4

11. a) cuboid, 6, 12, 8   b) prism, 5, 9, 6

   c) tetrahedron, 4, 6, 4

12. BH, DF, EC

13. a) 8, 12, 6

Chapter 16

Exercise 16.1.2 page 145

1. 1.5   3. sum is 180°   5. 5.4, 45°

6. a) 3.2 cm   b) 72°   c) 4

7. 81°, 55°, 44°

8. 70°   9. 90°

10. 7.6 cm, 6.4 cm   11. 15.8 cm, 15.2 cm

12. 3.0. 2.3   13. 5.1 cm, 83°

14. 7.1 cm, 52°   15. 9.3 cm

Exercise 16.1.4 page 147

5. 3.6 cm   6. 2.1 cm   7. 12.7 cm, 44°

9. 3 cm   10. 8.6 cm

11. hexagon, right–angled triangle, 8.7 cm

14. GC=GC'

21. a) 30 r.p.m., acw   b) 120 r.p.m., acw

   c) 120 r.p.m., cw

22. a) up, 1 m s⁻¹   b) up, $\frac{1}{4}$ m s⁻¹   c) up, $\frac{1}{3}$ m s⁻¹

Exercise 16.2.2 page 151

1. a) 090°   b) 180°   c) 000°   d) 045°   e) 225°

2. a) E   b) W   c) SE   d) NW

3. a) 20 km   b) 060°

4. 56 km, 216°

5. a) 70 km, 060°   b) Bristol c)   Norwich

6. 22 miles, 027°   7. 025°, 1.9 miles, 72°

8. 126 m   9. 35 m

10. 10.5 km   11. 60.5 miles, 100°

Chapter 17

Exercise 17.1.2 page 155

1. a) 12   b) 9   c) $15\frac{3}{4}$ d) 8

2. a) 14   b) 12   c) 16   d) 18

3. a) 1,890 sq ft, 186 ft b) £9.45   c)   £65.10

4. £16.10   5. 6 m², £96   6. 30 cm   7. 625 m²

8. a) 6   b) $3\frac{1}{2}$ c) 4   d) 6

9. a) 2   b) 3   c) 2   d) 2

10. 12 ft   11. 10 cm   12. 6,10

13. 6   14. 6   15. 314 m

16. 191 m   17. 40,000 km   18. 1,634 cm

19. a) 28 cm²   b) 12.6 sq ft   c) 452 m² d) 0.20 sq in

20. a) 3 cm   b) 1.9 in   c) 1.7 m   d) 1.3 ft

21. 100 sq in   22. 6 m²   23. 576 sq ft

24. 759 cm²   25. 25 cm

26. a) 16 cm²   b) 2 cm²   c) 8 cm²

27. a) 6 cm²

Exercise 17.1.4 page 157

1. a) 24   b) 10   c) 12

2. a) 15   b) 22.5   c) 12

3. a) 4 cm²   b) 3 cm²   c) 6 cm²   d) 4 cm²

4. 3   5. $3\frac{1}{2}$ 6. 14 cm²

Exercise 17.2.2 page 159

1. a) 112   b) 72   c) 108

2. 120   3. 150,000 kg   4. 3.12 m³

5. 4 cm   6. 0.4 m   7. 18

8. 512, 5.0625 g

9. a) 6 cm², 1 cm³   b) 10 cm², 2 cm³

10. a) 141   b) 154   c) 9.42

11. 77 cm^3

Exercise 17.2.4 page 160

1. a) 7   b) 162

2. 1.96 cm^3 3. 51.5 m^2, 5,150 cm^3

4. 54 cm^2 5. 3.28 m^2

6. a) 7.07 cm^2 b) 28,300 cm^3

7. a) 17,700 cm^3 b) 31,400 cm^3 c) 13,700 cm^3

8. 1.99 cm   9. 2,550 cm   10. 50 cm^2 11. 6.37 cm

12. a) 2.8125 m^2 b) 8.4375 m^3

13. 1,000 cm, 39.8 times

14. c), e) length a) area b), d), f) volume

Exercise 17.2.6 page 162

1. a) 524 cm^3 b) 33.5 m^3 c) 1,767 in^3

2. a) 314 cm^2 b) 50.3 m^2 c) 707 sq in

3. a) 151   b) 352   c) 39.3

4. 2.88 cm   5. 2.02 cm

6. a) 2.22 cm   b) 1.42 in   c) 1.10 mm

7. 11 cm^2 8. 50.4 mm^3

9. 2,610,000 m^3 10. 264 cm^3

11. a) 50.27   b) 50.29

12. a) 0.427 cm   b) 0.043   c) 0.54 cm

13. 9.15 cm   14. 35,200 cm

15. a) 61.3 cm^3 b) 3,063 cm

16. 283 cm sec^{-1} 17. 0.75 cm

18. a) 402 cm^3 b) 2 cm   c) 50.3 cm^3 d) 352 cm^3

19. 1:63

20. a) 20 cm   b) 293 cm^3

21. a) cone   b) rectangle

Exercise 17.3.2 page 165

1. a) 14   b) 14   c) 14   d) 10

2. a) 4 cm^2 b) 5 cm^2 c) 4 cm^2 d) 7 cm^2

3. 163 m, 1,314 m^2 4. 16 cm^2 9,600 cm^3

5. 6 m^2, 15 m^3 6. 3,030 cm^2

7. 1,636 cm^3

Exercise 17.3.4 page 166

1. a) 2.09   b) 12.4   c) 23.6

2. a) 2.09   b) 4.97   c) 15.7

3. 50 cm^2, 28.5 cm^2 4. 32.6 cm

5. 3.36, 6.71   6. 60.6 cm^2, 30.3 cm

7. 64.9 cm^3, 132 cm^2 8. 11.8 cm^3, 1.05 cm^3

9. 6.28 cu. in.   10. 4 cm

11. 96 litres, 11,300 cm^2 12. 2:3

13. 364 cm   14. 47.1 cm

15. 216°   16. 302 cu ft, 188 sq ft

17. a) 8.38 cm b) $1\frac{1}{3}$ cm   c) 7.89 cm d) 14.7 cm^3

18. 4:1, 8:1   19. $\frac{3}{4}$ 20. 8

21. 101:100, 102:100, 2%, 3%

22. a) 8:1   b) $\frac{7}{8}$

23. a) 1,230 cm^3, 550 cm^2

   b) 1:6   c) 1:36   d) 535 cm^2

Chapter 18

Exercise 18.1.2 page 171

1. a) 0.391   b) 0.682   c) 3.732

   d) 0.901   e) 1.000

2. a) 53.1° b) 76.1° c) 32.2° d) 78.5°

3. a) 7.71   b) 8.66   c) 7.46

4. a) 5.14   b) 4.77   c) 5.31

5. a) 26.6°   b) 36.9°   c) 55.2°

6. a) 73.4°   b) 53.1°   c) 22.0°

7. 7.13 cm, 3.63 cm   8. 66.2°, 23.8°

9. 35.2°, 54.8°   10. 60.7 m, 56.6 m

Exercise 18.1.4 page 173

1. a) 7   b) 7.15   c) 5.85   d) 7.71

2. a) 77.6°   b) 47.2°

3. 7.66, 10   4. 2.75   5. 33.2°

6. 53.1°, 126.9° 7. 2.72 in   8. 2.66 m

9. 28°   10. 11.3°   11. 63.4°, 36.9°

12. 33.9°

13. a) 72°   b) 4.85 cm c) 7.05 cm d) 85.6 cm^2

Exercise 18.2.2 page 175

1. 89 m   2. 1.4°   3. 1.64 m   4. 31°

5. 26 ft   6. 26.6°   7. 7.66°   8. 62°

9. 153 m   10. 163 m   11. 21.2 m

12. a) 5.7°   b) 9.95 m

13. 195 m/sec   14. 21.8°

15. 218 km, 335 km   16. 149°   17. 33.3 km

Exercise 18.2.3 page 176

1. 186.6 m. ,53.8° 2. 100°   3. 570 m

4. 113 m   5. 955 m   6. 30°, 10.4 ft

Exercise 18.3.2 page 176

1. a) 0.5   b) −0.940   c) −0.364

   d) −1   e) 0.5   f) −5.67

2. a) 30°, 150°   b) 23.6°, 156.4°

   c) 197.5°, 342.5°

3. a) 60°, 300°   b) 143.1°, 216.9°

   c) 120°, 240°

4. a) 63.4°, 243.4°   b) 135°, 315°

   c) 145.0°, 325.0°

5. a) 30°<x<150°   b) 90°<x<270°

   c) 0°<x<90° & 180°<x<270°

Exercise 18.4.2 page 178

1. a) 8.38   b) 4.64   c) 5.52

2. a) 46°   b) 18.5°   c) 121°

3. 5.19, 2.19   4. 43°, 75°   5. 18.6 m, 7.27 m

6. 1,010 m   7. 76°, 2.01 m 8. 148 miles

9. a) 9.67   b) 22.7

10. a) 41.4°   b) 134.4°

11. 15.7   12. 31.3°

13. 121 miles   14. 40.7 cm

15. 101°, 79°, 12.7 cm   16. 215°, 331°

18. 6.6 m

19. a) 9.98   b) 7.42   c) 10.3

20. a) 36°   b) 102°   c) 135°
   d) 62°

21. 11.3, 48°   22. 83°

Chapter 19

Exercise 19.1.2 page 182

1. a) 12.04   b) 13   c) 15   d) 7.14

2. 5.83 miles   3. 10.63 cm

4. 93 ft   5. 8.2 cm

6. 5.66 cm   7. 19.84

8. 3 in, 5.29 in   9. 50 m, 50.06 m

10. 4 ft   11. 199.75 cm, 0.25 cm

14. 6.32   15. 3.46 cm, 6.93 cm

16. a) 1.414 cm   b) 2.236 cm   c) 3.606 cm

17. a) 2.236 cm   b) 5.831 cm   c) 7.211 cm
   d) 5.385 cm

Exercise 19.2.2 page 185

1. 10.49, 28.5°, 55.1°, 48.1°

2. 8.60, 8.66, 5.10, 35°, 54°

3. 5.66 cm, 5.29 cm, 61.9°, 70.5°

4. 10.44 cm, 73.3°

5. 3.54 ft, 34.4°, 45°, 25.1°

6. 141.4 m, 86.6 m, 70.7 m, 35.3°, 45°

7. a) both 6.93 cm   b) 70.5°
   c) 2.31, 4.62   d) 54.7°

Exercise 19.3.2 page 187

1. a) 4.60   b) 3.86   c) 8.45

2. 8.70   3. 166 cm^2

4. 10.2 cm, 248 cm^2 5. 8.45 cm, 549 cm^2

6. 9 cm, 5 cm^2 7. 8.19 cm^2, 98.3 cm^3

8. 23 cm^2, 230 cm^3

Chapter 20

Exercise 20.1.2 page 190

10. a) 2   b) 4   c) 3

11. 8008   12. 6009

13. a) translation, 3 right, 1 down
   b) trans. 3 right, 2 up
   c) D, K
   d) reflection in CIOU
   e) rotation of 180°
   f) 90° clockwise rotation
   g) enlargement, factor 2

15. a) translation   b) reflection   c) rotation

Exercise 20.1.4 page 192

1. a) (1,−1) (3,−1) (2,−5)
   b) (−1,1) (−3,1) (−2,5)
   c) (1,1) (1,3) (5,2)   d) (−1,−1) (−1,−3) (−5,−2)

2. a) (1,2) (1,1) (−3,2)   b) (1,4) (1,3) (−3,4)
   c) (−1,−2) (−1,−3) (−5,−2)

     d) (5,6) (6,6) (5,2)   e) (0,1) (−1,1) (0,5)

3. reflection in y−axis   4. reflection in y=x

5. rotation of 90° clockwise about (0,0)

6. rot. of 180° about (0,0)   7. (2,−1), (2,1) (5,1)

8. (4,−1) (5,1) (3,2)   9. enlargement, factor 2

10. rot. 90° anti cw about B   11. rot 180° about A

12. rot 90° cw about (0,−1)   13. rot. 180° about (−1,−1)

14. reflection in y=1   15. refl. in x=−2

16. enlarg. factor 4 about B

17. enlarg. factor −2 about P

18. (6,2) (4,3) (4,2)   19. (0,−4) (2,−5) (2,−4)

20. (0,0) (−1.3,1.5) (0.8,0.6) (−0.5,2.1)

21. (0,2) (1.9,0.8) (1,0.3)   22. (−1,3) (3,5) (3,3)

23. (1,1) (2,1$\frac{1}{2}$) (2,1)

24. (8,4) (8,−2) (5,−2) (5,4)

Exercise 20.2.2 page 194

1. trans of 2 to right   2. trans of 4 down

3. rot of 180° about A   4. rot of 180° about (2,−1)

5. trans 2 to left, 2 down   6. rot of 180° about L

7. trans 2 to left 6 down

8. a) rot 180° about (0,0)
   b) rot 90° cw about (0,0)
   c) trans 2 left   d) trans 2 up
   e) trans 2 right   f) trans 2 down
   g) trans 2 down. $(PQ)^{-1}=Q^{-1}P^{-1}$

9. a) refl in x=y   b) nothing
   c) rot 90° acw about (0,0)
   d) rot 90° cw about (0,0)

Exercise 20.3.2 page 196

1. A (1,−2) (3,−2) (2,−1)   B (−1,2) (−3,2) (−2,1)
   C (2,−1) (2,−3) (1,−2)   D (−2,1) (−2,3) (−1,2)
   E (−2,−1) (−2,−3) (−1,−2)

2. A (1,−1) (1,−3) (2,−3) (2,−1)
   B (−1,1) (−1,3) (−2,3) (−2,1)
   C (1,−1) (3,−1) (3,−2) (1,−2)
   D (−1,1) (−3,1) (−3,2) (−1,2)
   E (−1,−1) (−3,−1) (−3,−2) (−1,−2)

3. A. reflection in x−axis   B. refl. in y−axis
   C. 90° cw rotation   D. 90° acw rotation
   E. refl. in x=−y

4. F (0,0) (3,0) (3,3) (0,3)   G (0,0) ($\frac{1}{2}$,0) ($\frac{1}{2}$,$\frac{1}{2}$) (0,$\frac{1}{2}$)
   H (0,0) ($\frac{1}{2}$,0) ($\frac{1}{2}$,1) (0,1)   J (0,0) (1,0) (1,2) (0,2)
   K (0,0) (−2,0) (−2,−2) (0,−2)

5. F. enlargement factor 3   G. enlargement factor $\frac{1}{2}$
   H. compression along x−axis
   J. stretch parallel to y−axis
   K. enlargement factor −2

6. BA = $\begin{pmatrix} 0 & -1 \\ -1 & 0 \end{pmatrix}$ 7. DC = $\begin{pmatrix} 1 & 2 \\ 0 & -1 \end{pmatrix}$

8. A^{17} = A   10. $\begin{pmatrix} 4 & 2 \\ -2 & -1 \end{pmatrix}$

11. $\begin{pmatrix} 2 & 3 \\ 3 & 2 \end{pmatrix}$ 12. $\begin{pmatrix} \frac{2}{3} & \frac{4}{3} \\ \frac{5}{3} & \frac{2}{3} \end{pmatrix}$

Chapter 21

Exercise 21.1.2 page 200

1. a) $\begin{pmatrix} 7 \\ -6 \end{pmatrix}$ b) $\begin{pmatrix} 29 \\ -9 \end{pmatrix}$ c) $\begin{pmatrix} -11 \\ -17 \end{pmatrix}$ d) $\begin{pmatrix} -3 \\ -8 \end{pmatrix}$

2. a) 7.28 b) 5.10 c) 9.22
   d) 21.95 e) 8.54

3. $x = 4 \ y = 7$ 4. $x = 1 \ y = 2$

5. a) $\begin{pmatrix} 3 \\ 8 \end{pmatrix}$ b) $\begin{pmatrix} 9 \\ -13 \end{pmatrix}$
   c) $x = 1 \ y = 4$ d) $\lambda = 2 \ \mu = 1$

8. $\begin{pmatrix} 1 \\ 1 \end{pmatrix}$ $\begin{pmatrix} 1 \\ -2 \end{pmatrix}$ $\begin{pmatrix} 2 \\ 0 \end{pmatrix}$

9. a) $(3, -6)$ b) $(6, 2)$ c) $(8, -5)$

10. $\begin{pmatrix} 1 \\ -2 \end{pmatrix}$ $\begin{pmatrix} -1 \\ 2 \end{pmatrix}$

Exercise 21.2.2 page 202

1. $\underline{AB} = \begin{pmatrix} 1 \\ 2 \end{pmatrix} \underline{AC} = \begin{pmatrix} 1 \\ -6 \end{pmatrix} \underline{CD} = \begin{pmatrix} 2 \\ 4 \end{pmatrix} \underline{BD} = \begin{pmatrix} 2 \\ -4 \end{pmatrix} \underline{DA} = \begin{pmatrix} -3 \\ 2 \end{pmatrix}$
   $\underline{AB}$ & $\underline{CD}$

2. $\underline{JL} = \begin{pmatrix} 2 \\ 5 \end{pmatrix} \underline{JK} = \begin{pmatrix} 1 \\ -3 \end{pmatrix} \underline{ML} = \begin{pmatrix} -3 \\ 9 \end{pmatrix} \underline{MK} = \begin{pmatrix} -4 \\ 1 \end{pmatrix} \underline{MJ} = \begin{pmatrix} -5 \\ 4 \end{pmatrix}$
   $\underline{JK}$ & $\underline{ML}$

3. 3:1, no   4. no

5. PQ=SR= $\sqrt{13}$, QR=PS= $\sqrt{2}$, no

8. no, no   9. ABD in straight line

10. L, M, K   11. (0,0)

12. (5,−1)   13. 3

14. $\mathbf{c-b}$, $\frac{1}{4}\mathbf{b}$, $\frac{1}{4}\mathbf{c}$, $\frac{1}{4}(\mathbf{c-b})$, $\frac{3}{4}\mathbf{b}$, $\mathbf{c}-\frac{1}{4}\mathbf{b}$, BC parallel to DE, DB parallel to AD, 30

15. $\frac{1}{2}(\mathbf{a+b})$, $\frac{1}{2}(\mathbf{b+c})$, $\frac{1}{2}(\mathbf{c+d})$, $\frac{1}{2}(\mathbf{d+a})$, $\frac{1}{2}(\mathbf{c-a})$, $\frac{1}{2}(\mathbf{d-b})$,
    $\frac{1}{2}(\mathbf{a-c})$, $\frac{1}{2}(\mathbf{b-d})$, parallelogram

16. $\mathbf{c}$, $\mathbf{b}$, $2\mathbf{c}$, $\mathbf{c-b}$

17. $\mathbf{e}$, $-\mathbf{a}$, $2\mathbf{a}$, $2\mathbf{a+e}$

20. $\mathbf{a}$, $\mathbf{a+b}$, $\mathbf{a-b}$

21. $\mathbf{b+d}$, $\frac{1}{2}\mathbf{b}$, $\mathbf{d}+\frac{1}{2}\mathbf{b}$, $\mathbf{d}$, XY parallel to AD

22. $\mathbf{b=a+c}$, $\mathbf{c}$, $\frac{1}{2}(\mathbf{c-a})$, $\frac{1}{2}(\mathbf{a+c})$, $\frac{1}{2}(\mathbf{c+a})$ X=Y

23. $\frac{1}{2}\mathbf{c}$, $\frac{3}{4}\mathbf{b}$, $\frac{3}{4}\mathbf{b-c}$, $\frac{3}{8}\mathbf{b}-\frac{1}{2}\mathbf{c}$, $\frac{1}{2}\mathbf{c}+\frac{3}{8}\mathbf{b}$, $\frac{3}{8}\mathbf{b}$, ZY parallel to AB

24. $\frac{1}{2}(\mathbf{a+b})$, $\frac{2}{3}\mathbf{a}$, $2\mathbf{b}$, $\underline{PQ}=\frac{1}{6}\mathbf{a}-\frac{1}{2}\mathbf{b}$, $\underline{PR}=1\frac{1}{2}\mathbf{b}-\frac{1}{2}\mathbf{a}$, PQR in a straight line

Exercise 21.3.2 page 204

2. 5.83 km.p.h. 31° to line across river.

3. 37° to line across river. 3 minutes.

4. 097°. 403 m.p.h.   5. 83°. 397 m.p.h.

6. 8.60 N. 54°.7.   9. 22 N. 184°

8. $\begin{pmatrix} 5 \\ 12 \end{pmatrix}$ 8.25, 5, 13.   9. $\begin{pmatrix} -12 \\ -32 \end{pmatrix}$

Chapter 22

Exercise 22.1.2 page 207

1. a) 1   b) 5   c) 29
2. a) 4   b) 12   c) 33
3. a) 23   b) 129   c) 421   d) 194
4. a) 15   b) 150°
5. a) 6   b) 15°   c) 16
6. a) £20   b) £7.50
9. a) 24   b) vanilla   c) 56

Exercise 22.2.2 page 210

1.

Number of letters	1	2	3	4	5	6	7	8	9+
Frequency	12	13	11	13	14	10	7	3	8

2.

Score	1	2	3	4	5	6
Frequency	13	4	10	12	6	15

Exercise 22.3.2 page 213

1. b) 30–35   c) 0.382
2. b) 20–30   c) 0.6

Exercise 22.4.2 page 215

1. a) 1.4   b) 0.75   c) −5
2. 112 lb   3. 43

Chapter 23

Exercise 23.1.2 page 217

1. 6, 5, 7   2. 27, $26\frac{1}{2}$, 19
3. $118\frac{1}{3}$ m, 110 m   4. 6 cm, 5.9 cm
5. 103, 103, 21   6. 2,0
7. 48   8. 47.2 g
9. £9,050, £7,450   10. 2.1, 1
11. 82.5 p, 82 p   12. $249\frac{1}{8}$ g, 249 g
13. 7 hrs, 7.3 hrs, 6.9 hrs   14. 4.44 mill, 4.3 mill
15. $42\frac{7}{9}$ hrs, 45 hrs, 34 hrs   16. 3, 3
17. $25\frac{3}{4}$, $25\frac{1}{2}$

Exercise 23.2.2 page 220

1. 5.184, 5, 5   2. 1.375
3. 2.758, 2   4. 2.076
5. 1, 1.7475   6. 0.833, 1
7. 1, 1.25   8. 3.035
9. 20–25, 23.8   10. 60–70 cm, 67.2 cm
11. £10,600   12. 2–4 hrs, 3.6 hrs
13. 1 min, 2.7 min   14. 2,510 m
15. 53

Exercise 23.3.2 page 223

1. a) 10, 33, 51, 60   c) 169, 163, 177, 14
   d) $\frac{2}{3}$

2. a) 7, 22, 50, 69, 80   c)   77, 69, 89, 20
    d) 0.7
3. a) 74, 147, 251, 367, 434, 500   c)   30, 24, 36, 12
4. a) 22, 40, 56, 68, 80 c) 30, 19, 43, 24
5. a) 24, 39, 56, 75, 85, 100
    c) £10,600, £9,100, £12,000, £2,900
6. a) 20, 29, 44, 54, 60 c) 3.5, 2.7, 4.1, 1.4
7. a) 4, 17, 26, 37, 41, 50   c)   15.4, 14.7, 16.1, 1.4
8. For A, 1, 4, 14, 28, 30, 30: for B, 2, 7, 14, 23, 27, 30.
    IQ range for A=12, for B=19.
9. For Lucy, 2, 5, 28, 35, 40, 40 for Liz, 5, 13, 27, 32, 36, 40.
    IQ range for Lucy=4, for Liz=10

Exercise 23.4.2 page 225
1. 8.34   2. 2.35
3. 4.85   4. 1.45
5. C:0.290, D:17.4   6. B: 13, 4,45 G:14, 1.73
7. A:23, 2.55 B:23.5, 8.65 8.   25, 8.   2, 0.6

Chapter 24

Exercise 24.1.2 page 229
3. $\frac{1}{6}$ 4. $\frac{1}{2}$ 5. $\frac{1}{4}$ 6. $\frac{17}{25}$
7. $\frac{4}{11}$ 8. $\frac{1}{5}$ 9. $\frac{93}{100}$ 10. $\frac{1}{37}$
11. $\frac{2}{5}$ 12. $\frac{1}{99}$ 13. $\frac{1}{13}$
14. a) $\frac{1}{11}$ b) $\frac{2}{11}$
15. a) $\frac{1}{8}$ b) $\frac{1}{2}$
16. a) $\frac{1}{24}$ b) $\frac{1}{6}$ c) $\frac{1}{8}$
17. a) $\frac{1}{36}$ b) $\frac{1}{6}$
18. $\frac{5}{6}$ 19. $\frac{36}{37}$ 20. $\frac{3}{4}$

Exercise 24.2.2 page 231
1. a) $\frac{1}{25}$ b) $\frac{16}{25}$
2. a) $\frac{1}{4}$ b) $\frac{1}{4}$ c) $\frac{1}{2}$
3. $\frac{2}{3}$ 4. $\frac{1}{6}$
5. a) $\frac{1}{36}$ b) $\frac{1}{6}$ c) $\frac{8}{9}$
6. a) $\frac{1}{36}$ b) $\frac{1}{6}$ c) $\frac{5}{12}$
7. $\frac{1}{21}$ 8. $\frac{1}{6}$ 9. $\frac{22}{57}$

Exercise 24.3.2 page 233
1. a) $\frac{81}{100}$ b) $\frac{9}{50}$
2. a) $\frac{4}{7}$ b) $\frac{3}{7}$
3. a) $\frac{28}{171}$ b) $\frac{83}{171}$ c) $\frac{88}{171}$
4. a) $\frac{9}{16}$ b) $\frac{15}{16}$

5. a) $\frac{1}{36}$ b) $\frac{25}{36}$ c) $\frac{11}{36}$
6. a) $\frac{66}{325}$ b) $\frac{168}{325}$
7. a) $\frac{1}{15}$ b) $\frac{9}{20}$ c) $\frac{7}{60}$
    d) $\frac{3}{25}$
8. a) $\frac{1}{4}$ b) $\frac{1}{8}$ c) $\frac{1}{4}$
9. $\frac{9}{14}$
10. a) $\frac{2}{5}$ b) $\frac{4}{15}$
11. a) $\frac{1}{17}$ b) $\frac{4}{17}$ c) $\frac{1}{52}$
12. a) $\frac{1}{6}$ b) $\frac{5}{36}$ c) $\frac{25}{216}$
13. a) $\frac{1}{2}$ b) $\frac{3}{20}$ c) $\frac{1}{20}$ d) $\frac{1}{5}$
14. a) $\frac{3}{16}$ b) $\frac{1}{6}$ c) $\frac{1}{24}$ d) $\frac{5}{16}$
15. a) $\frac{1}{3}$ b) $\frac{4}{15}$
16. a) $\frac{7}{15}$ b) $\frac{17}{60}$

Chapter 25

Exercise 25.1.2 page 236
3. a) A–B–C. 8 miles   b) A–D–C 8 miles
    c) A–B–C 14 miles
4. a) 18 miles   b) 18 miles   c) 29 miles
5. a) 18 miles   b) 21 miles   c) 29 miles

Exercise 25.2.2 page 237
1. 18, 24, 45   2. £18.50, £22.50 3. 10, 26, 186
7. 1.732   8. 3.162   9. 24, 720
11. primes 5, 17 in A, rest in B

Exercise 25.3.2 page 240
7. a) 18, ABD   b) 21, CDE   c) 23, CEF
8. a) C, 2   b) A, 1. B, 1 c) A, 9. B, 2. D, 6
9. 39 minutes. BADEFGIH
10. 41, EFABIH

Solutions to test papers

Exam 1 *page 243*

		AT's 2	3	4	5
1.	11 52	1			
	18 minutes	2			
2.	shading	2			
3.	9 ounces	2			
4.	a) 5	1			
	b) 4	1			
	4+6+8	2			
5.	6, 3, 2		3		
6.	a) 8	1			
	b) 2.57	2			
7.	a) £12.40	2			
	b) £13.64	2			
8.	Tessellation			3	
	60°			2	
9.	$x = 1.3$		5		
10.	35 km			2	
	020°			2	
11.	a) $\frac{1}{3}$				2
	b) 0				1
12.	15	4			
	Table	3			
	34	4			
13.	21, 28	3			
	1, 8, 27, 64, 125,	2			
	1, 9, 36, 100, 225	2			
	Square of terms	2			
14.	Table				4
	5, 0.25				3
15.	lines			4	
	a, 2			3	
	shape			3	
16.	35 m³			3	
	$3\frac{1}{2}$ litres			2	
	£19.25	2			
17.	net			4	
	4 cm²			1	
18.	Frequencies: 4 9 8 10 16 9 4				4
	histogram				4
19.	Cost: 2 4 7.5 11 14.5 17.5 20.5		4		
	32.5				
	graph		4		
	30 miles		1		
	a) 40p, b) 32.5p.		4		
20.	Totals: 13 8 6 9				2
	A most popular				1
	bar chart				3
	pie-chart. 130°, 80°, 60°, 90°				4
	$\frac{13}{36}$				2
Total		**40**	**21**	**29**	**30**

Exam 2 *page 246*

		AT's 2	3	4	5
1.	a) 23	1			
	b) e.g. 36	1			
	c) e.g. 27	1			
2.	£7	2			
3.	a) $\frac{3}{5}$	2			
	b) $\frac{4}{9}$	2			
4.	£36.9	2			
5.	a) 88 minutes	2			
	b) 8 26	2			
6.	167.64 cm	3			
7.	a) £1.80	1			
	b) £5.20	1			
8.	A(1,1) B(-2,-1) C(2,-1)		3		
	$y = 0$		1		
	x is negative, y is positive		2		
9.	1, 4, 9, 16		2		
	squares		2		
	25, 36		2		
10.	a) 11, 8		3		
	b) 57, 2		3		
	c) leads to 3 = 5		2		
11.	diagram			5	
	78°			2	
12.	a) 5			2	
	b) 20			2	
13.	net			3	
	3 cm²			3	
	Height less than 1 cm			3	
14.	Frequencies: 4, 6, 7, 2, 1				2
	bar-chart				3
	1.5				2
	2				1
15.	a) 11 lb				1
	b) 3 kg				1
	c) 66 lb				1
	conversion chart				6
16.	12,000				2
	120°				1
Total		**20**	**20**	**20**	**20**

Exam 3 *Page 248*

		AT's 2	3	4	5
1.	a) 28 miles	2			
	b) 49 miles	2			
	c) 12 minutes.	2			
2.	a) 0.125	2			
	b) 22.15	2			
	c) sequence	3			
3.	16		2		
	2		3		
4.	a) (1,1)		1		
	b) plotting			1	
	c) quadrilateral			2	

d) A'(1,3) B'(2,1) C'(4,1) D'(3,3)			4	
5. a) $\frac{4}{7}$				2
b) $\frac{3}{7}$				3
6. table		7		
71		3		
$x + y + xy$		3	3	
7. a) £10.50	2			
b) 50p	2			
c) 1/21	3			
d) £11.55	2			
8. bar chart				2
frequencies 3, 2, 4, 1				2
£65				4
£10.67				3
9. a) 48	2			
b) 768	2			
c) 576	4			
10. a) 1 mile		1		
b) 30 minutes		1		
c) 2 m.p.h., 4 m.p.h.		4		
d) 3 m.p.h.		2		
e) 1 hour		3		
11. a) 023°			3	
b) 3.5 cm			2	
c) 1 cm to 5 km			2	
12. net			4	
1 cm^3			1	
160 (if 8 by 8 by 10)			4	
13. Diagram			1	
2.05 km			3	
103°			3	
14. a) Frequencies 10, 6, 15, 13, 6, 10				2
b) bar chart				2
c) 3.483				3
d) Not strong evidence				4
e) $\frac{1}{10}$				3
Total	30	30	30	30

15. POS POS OSP OPS SOP SPO ... (3)
6 ways. ... (4)
6 ways. ... (4)
a) and c) affected. ... (4) [15]

16. Measurement and calculation for area. ... (5)
Finding prices and finding cost. ... (5)
Estimating time. ... (5) [15]

Exam 4 *page 251*

1. a) e.g. A – B – C – B – A ... (3)
b) All the distances even. ... (3)
... [6]
2. ... [6]
3. a) diagram ... (2)
b) two diagrams ... (3)

c) 5 shapes. 4 in a row, two of 3 in a row, two of 2 in a row. ... (5) [10]
4. ... [6]
5. Plan and elevations. ... (5)
Sees front wall and roof. ... (5) [10]
6. 2, 3, 3, 4, 4 ... (4)
51, 51 ... (4)
If even, add 2 and halve. If odd, add one and halve. ... (4) [12]
Total [50]

Exam 5 *page 252*

		AT's 2	3	4	5
1.	2142	2			
	Last digit 0 or 5	3			
2.	$x = -1$		2		
	$x^2 - 2x - 15$		2		
3.	30°			3	
	$\sqrt{2} = 1.41$ ft			4	
4. a)	$\frac{9}{10}$				1
b)	$\frac{63}{80}$				2
c)	$\frac{8}{80}$				3
5. a)	£200	3			
b)	58%	3			
6.	$\pi r^2, 2\pi r$			1	
	equation		3		
	(graph)		4		
	3.6		2		
7.	T			3	
	U			2	
	Translation, 2 to left			3	
8. a)	e.g. BDE			1	
b)	ADF			2	
c)	ABE ACE ABDE ACFE ABCE etc				4
d)	$\frac{1}{18}$				3
9. a)	£7.87	2			
b)	£15.5	1			
c)	£7.20	2			
d)	£48.32	1			
10. a)	i) 30 minutes		1		
	ii) 60 m.p.h., 70 m.p.h.		2		
b)	travel graph		5		
11.	26.5				1
	28.5				2
12.	303 pages				2
	296 pages. A bit under 299.				3
	Frequency polygon				4
13.	4 hours.	1			
	1100	2			
	5 hours	1			
	3 hours	3			
14.	Reflection in x-axis			2	
	Enlargement, scale factor 2			2	
15.	6.24 cm			4	

	2	3	4	5
16. $x \geq 0,\ y > 1,\ y + x \leq 2$		3		
17. $1^3 < 1 + 1,\ 2^3 > 2 + 1$	2			
$\quad x = 1.3\ 3$	3			
18. £15,000				1
$\quad$ £7,000				3
$\quad \dfrac{1}{4}$				2
19. a) rectangle			1	
$\quad$ b) square			1	
$\quad$ c) rhombus			1	
20. $3 \times \sqrt{(4 + 1)} = 6$	3			
21. $Y = X/(a + b)$		3		
Total	**32**	**27**	**30**	**31**

Exam 6 *page 255*

	AT's 2	3	4	5
1. a) $\dfrac{5}{8}$	2			
$\quad$ b) $\dfrac{3}{28}$	2			
$\quad$ c) $\dfrac{7}{20}$	2			
2. A 26, B 25	1			
$\quad$ Win	2			
$\quad$ A wins or second: A third B second or unplaced: A unplaced, B third or unplaced	4			
3. 99,999	1			
$\quad$ –999,999	1			
$\quad$ -9.9999×10^{10}	2			
4. a) 9 $\quad$ b) 1024 $\quad$ c) 4	3			
$\quad 3^4$	2			
5. a) 7		1		
$\quad$ b) 2		2		
6. $4x + 16 = 2(x + 16).\ x = 8$		6		
7. a) $2x + 3y$		1		
$\quad$ b) $15x^3y^8$		1		
$\quad$ c) $3p^2s^{-2}$		1		
8. $y = 5, 2, 1.25, 1, 1.25, 2, 5$		3		
$\quad$ Graph		5		
$\quad$ 0 or 2		1		
$\quad$ 1.7 or 0.3		2		
9. (2,1), (1,3)			2	
$\quad$ (3,4). Illustration			3	
$\quad$ 3 sides of triangle, or of parallelogram			3	
10. a) and c)			3	
11. diagram			3	
$\quad$ 11.2 cm			2	
$\quad$ 5.3 km			3	
12. 36.1 m			2	
$\quad$ 33.7°			2	
$\quad$ 24.9 m			3	
13. (1,0,3) (1,3,3) (4,3,3) (4,0,3)			2	
$\quad$ (2.5,1.5,1.5)			2	
14. HTH HTT THH THT TTH				2
				2
$\quad$ a) $\dfrac{3}{8}$				2
$\quad$ b) $\dfrac{7}{8}$				2

				5
15. Frequencies: 3 5 8 11 11 6 6 6 3 1				3
$\quad$ 45,000 miles				3
$\quad$ Polygon				3
16. Graph				3
$\quad$ 4,400 per year				4
$\quad$ 1979				3
Total	**22**	**23**	**30**	**25**

Exam 7 *page 257*

	AT's 2	3	4	5
1. 1.17×10^{13} miles	5			
2. a) $x > 2$		2		
$\quad$ b) number line		2		
$\quad$ c) $-2, -1, 0, 1, 2$		2		
3. a) $10x = 13(x - 3).\ x = 13$		3		
$\quad$ b) $x = -1,\ y = 3$		3		
4. $\dfrac{1}{6}$				3
$\quad$ 10				3
$\quad$ Azaeez, by £1				4
5. a) 14.9376	2			
$\quad$ b) 0.332	2			
$\quad$ c) (i) 39 $\quad$ (ii) $-2\dfrac{5}{9}$	3			
$\quad$ d) (i) $(3.2+4.1)\surd$ (ii) $3 \div (4.2+5.7) =$	4			
6. $60 - x$. Formula for A.		5		
$\quad$ A values: 500 800 900 800 500 0. graph		7		
$\quad$ 17, 43 m		3		
$\quad$ 900 m²		3		
7. diagram			7	
$\quad$ 23.3 m			4	
8. Line				4
$\quad$ 41 p. Year 11				4
9. $a^2 + b^2 = c^2$			2	
$\quad$ 25	1			
$\quad$ 5.74	2			
$\quad$ 52, 65 or 760, 761	3			
10. a) diagram			6	
$\quad$ b) clockwise, 10 r.p.m.			2	
11. a) (0,4), (1,4), (1,2)			4	
$\quad$ b) (0,0), (0,1), (2,1)			3	
$\quad$ c) (1,0), (-1,0), (1,4)			4	
12. CF's: 24 66 94 118 136 146 154 160				5
$\quad$ 27. 14 years				4
$\quad$ 0.17				3
13. 0.0552 cu. in.			2	
$\quad$ £2,390	4			
Total	**26**	**30**	**34**	**30**

14. (2) for each of the four pairs of lines. All angles of the form $\tan^{-1}p + \tan^{-1}q$, where p and q are rational, can be constructed from the dots. (7) for this. (4) for giving all angles of the form $\tan^{-1}p$, (without explanation that this includes the above.) (2) for all angles of the form $\tan^{-1}n$, where n is an integer. [15]

15. Ensure sum is multiple of 10 (5)
    First player. Pick 3, then proceed as above. (5)
    Different total, different numbers allowed, repetition not allowed etc. (5) [15]

Exam 8 *page 260*

1. 2,000,000 grains. Taking 1 grain = 0.2 mm in width, and matchbox to contain 16 cm^3 (1) for guess. (2) for calculation. [3]
2. Only 3 doctors were asked. (1) for answer. (2) for suspicion about accuracy of percentage. [3]
3. With 1, 2, 3, 4 lines, there are 0, 1, 3, 6 crossing points and 2, 4, 7, 11 regions. (4) for this. These sequences can be continued: 10, 15 etc and 16, 22 etc. (4) for this. [8]
4. a) and b). Pictures or words. (3)
   c) 7 are possible. A minimum of (4) for any systematic approach, e.g. by taking 4 in a row, then 3 in a row etc. (7) [10]
5. Move 120 Turn 90 Move 120 Turn 90 Move 120 Turn 90 Move 120 (3)
   e.g. Move 2.1 Turn 1, repeated 360 times. (5) [8]
6. No. Marcia has the advantage. (2)
   P(Marcia wins) = $\frac{6}{11}$. (3) for any sensible guess. Repeat the game many times, find the proportion of times that Marcia wins. (3) [8]
7. Questionnaire with sensible questions, e.g. about the effect to which computers have affected the respondent's life. (3)
   Varied sample of people asked, e.g. different ages, occupations etc. (3)
   Presentation of answers in bar-charts, pie-charts, relevant to the questions. (4) [10]

Total [50]

Continuation (Exam 8 AT's):

	2	3	4	5
$n + \frac{1}{2}$				2
9. histogram				4
Modal time 34-35				2
histogram				4
10. a) 6.56 miles			4	
b) 129°			4	
c) 3.42 miles			3	
11. a) Formula		3	3	
b) Area		2		
c) Graph		4		
d) 8 cm^2		2		
e) 1.8 cm or 0.2cm		2		
12. a) 13.5°	3			
b) £43,000	3			
13. a) $F = 100/d^2$		3		
b) $\frac{1}{2}$		2		
14. a) not a ratio of integers	2			
b) (i) true	2			
(ii) false	2			
(iii) true	2			
15. 31, 4.82				3
37.7 (Taking midpoint of each group)		2		3
16. a) i) $(2p+3)/p^2$		2		
ii) $q^3/(2r^3)$		2		
b) $(x + 15)(x - 4)$		2		
17. $(x+y) \times 10^5 \, xy \times 10^{10}$	3			
Total	**30**	**30**	**30**	**30**

Exam 9 *page 260*

	AT's			
	2	3	4	5
1. $a = 2,\ b = 2,\ c = -5\frac{1}{2}$		2		
2. £15,400	3			
£4,720	3			
3. $\frac{1}{2}x$, 180° - 2x. x = 78°			4	
4. a) 156°			2	
b) 323°			2	
5. a) £2,800	1			
b) £4,000	3			
c) 12%	3			
6. a) $\frac{1}{2}\mathbf{b}$, $\mathbf{c} - \mathbf{b}$, $\frac{1}{2}(\mathbf{c} - \mathbf{b})$, $\frac{1}{2}(\mathbf{b} + \mathbf{c})$			4	
b) Similar			2	
c) 6			2	
7. Explanation, by Tree diagram or otherwise				5
$2x - x^2 = 1.75x.$ $x = \frac{1}{4}$ or 0			4	3
8. 5, 5				2
$n + 1$				2

Exam 10 *page 263*

	AT's			
	2	3	4	5
1. a) $((5 + 1)/3)^2 = 4$	3			
b) 3.75	1			
2. D beat B, E beat B	2			
Points 7, 2, 1, 4, 6	3			
15	2			
3. 5.1	2			
5.02 to 5.24	4			
4. a) 2 b) 6.32 c) 2×10^9 d) 0.000632	2			
(a) and (c)	2			
even values	2			
64 (any sixth power of a rational)	2			
5. a) −4 or −6		3		
b) 0.562, -3.562		3		
6. a) 2 o'clock		2		
b) 4 cm hr^{-1} (approx)		3		
c) extension		2		
7. a) $y = 4x - 3$		3		
b) (0,3), (2,0). −1.5. 3		3		
8. (b) shift 2 to left		2		
(c) squash horizontally by factor of 2		2		
(d) shift up by 2		2		
9. a) 226 m			2	
b) (−200,100,30)			2	

10. a) enlargement		2	
b) $\begin{pmatrix}3 & 0\\0 & 3\end{pmatrix}$		2	
11. a) 6, 120			2
b) $4\frac{2}{3}$			4
finds mean			3
12. 179 cm^3		2	
47.1 cm^3		2	
13. plot		1	
(−1,1) (−3,1) (−2,5)		3	
reflection in y-axis		2	
$\begin{pmatrix}1 & 0\\0 & -1\end{pmatrix}$		2	
14. a) 12, 45, 83, 100			1
b) graph			4
c) 13 cm			2
d) 0.35			2
15. $x \geq 1,\ 6x + 10y \geq 15,\ 5x + 2y \leq 10$			4
graph			4
100 tons coal, 90 tons oil			4
Total 25	25	20	30

Exam 11 *page 265*

AT's

	2	3	4	5
1. a) $10x^2 - 29xy - 21y^2$		2		
b) $a^2 - 2$		2		
2. diagram			4	
3. A: 30, 3.74 B: 28, 1.18			4	
B cheaper, more uniform				3
4. a) $\frac{2}{3}$				1
b) $\frac{19}{27}$				3
c) $\frac{4}{27}$				3
5. 1.192	7			
6. 32.6°			4	
114°. 4.58 km hr^{-1}			4	
7. a) iii) Curved, maximum		2		
b) i) Straight, positive gradient		2		
c) ii) Straight, negative gradient		2		
d) iv) Curved, minimum		2		
8. FFF FT TF TT FFT				4
$\frac{3}{8}$				3
9. £2998.22	2			
a) £508.48 b) 17%	4			
£5,152.49	4			
10. a) 30 m s^{-1}			1	
b) graph			4	
c) 85 seconds, 2175 m			3	
11. a) $x = 2$			3	
b) 5^{13}			3	
c) $T = 1.5S^2$			4	
$\frac{1}{3}$			4	
12. a) plotting			3	
b) $y = 7.8 - 0.8x$			4	

13. c) after 10 minutes		2		
a) 1:3			1	
b) 2 cm			2	
c) $5\frac{1}{3}, 138\frac{2}{3}$				3
d) 1:27. Cube				3
e) 1:8				3
14. $-3\frac{5}{8}, 0, 1\frac{5}{8}, 2, 1\frac{7}{8}, 2, 3\frac{1}{8}$		3		
plotting		3		
$x = -1.1$		3		
0.5			4	
Total	19	56	24	21

Section B

15. e.g. At noon, measure the angle of elevation of the sun at two places d miles on same Longitude. (7)
If difference is α, then radius of Earth is $360d \div 2\pi\alpha$ (8) [15]

16. 8 ways if writing 5 as 1's and 2's (3)
For n = 1, 2, 3, 4, 5, 6, number of ways 1, 2, 3, 5, 8, 13. For systematic investigation (8)
$f(n) = f(n - 1) + f(n - 2)$. (4) [15]

Exam 12 *page 268*

1. a) 7 b) 15. (1)
a = 1.5, b = 0.5. (3)
Check: 26 cards for 4 storeys (2) [6]

2. If proportion of drug-takers is p, proportion of 'A' replies $= \frac{2}{3} - \frac{1}{3}p$. (4)
Finding p from results, (2) [6]

3. Starting with (say) $\frac{1}{1}$, next terms are
$\frac{1}{2}, \frac{2}{3}, \frac{3}{5}$ etc. (2)
Sequence tends to 0.618 (3)
Sequence tends to root of $\alpha^2 + \alpha = 1$ (3) [8]

4. Showing diagram of circles and rectangle touching sides of square. (4)
Calculation: $r = 1.27$, $h = 2.91$ (4) [8]

5. Cube. $F = \frac{3}{8}v^3$ (4)
$2a + 4b = 3$, $4a + 16b = 24$. $a = -3$, $b = 2.5$ (4)
By graph or table, second formula is closer (4) [12]

6. Sensible list of tasks. (2)
Sensible estimation of times. (2)
Diagram. (4)
Shortest time. (2) [10]

Total [50]

Solutions to mental tests

Mental Test 1 *page 270*

1) $\frac{33}{77}$

2) 35 m^2

3) 8 cm

4) $\frac{3}{8}$

5) 8

6) 60°

7) £48

8) £3.60

9) $15x$

10) 5

11) 75

12) 26

13) Sept 6

14) 21 Dec

15) 5217

17) £50

18) £58.75

19) 20

20) 72°

Mental Test 2 *page 272*

1) 10,243

2) 32 m

3) £218

4) 6

5) equilateral

6) 0.2

7) 3:2

8) 1.92 m

9) 8, 17

10) 130 mins

11) 22

12) £72,000

13) 259

14) 16

15) $12x$

16) 19°

17) 6

18) £1550

19) $\frac{2}{3}$

20) £3.36

Mental Test 3 *page 273*

1) 8 cm

2) 4

3) 240

4) 132 kg

5) 3, 4, 5

6) 4 am

7) 150 miles

8) 0.4 ft

9) 1×10^6

10) 2, 3, 5

11) −3

12) x

13) 3

14) $1 - x$

15) 20 hrs

16) 500 mins

17) £235

18) −73°

19) $4\frac{1}{2}$ secs

20) £182

Index

Mathematics Attainment Tests

Key Stage 1
Stafford Burndred
ISBN 1 85805 037 5 Date 1993 Edition 1st Extent 96 pp

Key Stage 3
Stafford Burndred
ISBN 1 85805 034 0 Date 1993 Edition 2nd Extent 192 pp

Each of these books enables pupils, parents or teachers to know what level a child has reached in mathematics in relation to the standards laid down by the National Curriculum. The books are ideal for self-testing (or testing by parents at Key Stage 1), for identifying any areas of weakness and for giving practice of Standard Attainment Test-style questions where appropriate.

GCSE Mathematics Revision Course – Intermediate Tier Incorporating One-Hour Practice Papers

GCSE Mathematics
Revision Course – Higher Tier Incorporating One-Hour Practice Papers
JJ McCarthy

Intermediate – ISBN 1 85805 039 1 Date 1993 Edition 1st Extent 96 pp (approx)

Higher – ISBN 1 8505 040 5 Date 1993 Edition 1st
Extent 96 pp (approx)

These books provide all the practice required to meet the requirements of all the GCSE Mathematics examination board syllabuses at either the Intermediate or Higher Tiers and the National Curriculum at Key Stage 4. Pupils can use them for pre-exam revision to identify topics they need to practise further, or they can be used in the classroom.

Section 1 – One-Hour Practice Papers made up of

❐ 15 one-hour cross-topic practice papers that together will cover the whole of the GCSE syllabus and test the Attainment Targets 2 to 4. Each question in the papers has a 'helpline' reference to a page of Section 2 where you can find help assistance with the techniques behind that specific question and an indication of which level it is (ie Level 7, 8, 9 or 10).

❐ 15 coursework practice assignments

❐ 15 aural tests (each a set of 20 mental arithmetic questions),

Section 2 – Mathematical principles and formulae

A brief summary of the essential mathematical principles and formulae of the GCSE syllabus, organised by topic, with examples where necessary

Mathematics Key Stage 4 Incorporating GCSE
RC Solomon *Date 1992 Edition 1st*

Volume 1 – Levels 4, 5 and 6
ISBN 1 873981 21 X Extent 272 pp

Volume 2 – Levels 7 and 8
ISBN 1 873981 22 8 Extent 256 pp

Volume 3 – Levels 9 and 10
ISBN 1 873981 23 6 Extent 288 pp

This is a totally new series of textbooks written to meet all the requirements of Key Stage 4 and the revised GCSE in 1994. Many schools have already adopted the series after its publication in June 1992 appreciating the comprehensive coverage, competitive price and clear links with the National Curriculum.

Free Teachers' Guide

Advanced Level Mathematics Revision Course
Incorporating One-Hour Practice Papers

RC Solomon

ISBN 1 85805 041 3 Date 1993 Edition 1st Extent 200 pp (approx)

This book provides practice and reinforcement of the common core, mechanics and statistics papers of A Level Maths and helps pupils identify topics that they need to practise further. The book has three main sections:

Section 1 – 30-One-hour Practice Papers on the common core, mechanics and statistics

Each question has a 'helpline' reference to a page of Section 2 where you can find help with the techniques behind that specific question.

Section 2 – Statements of principles, formulas, and techniques

A brief summary of the essential mathematical principles, techniques and formulae of the syllabus, organised by topic, with examples where necessary.

Section 3 – Answers and marking guide

GCSE Mathematics Revision Course – Intermediate Tier Incorporating One-Hour Practice Papers

GCSE Mathematics
Revision Course – Higher Tier Incorporating One-Hour Practice Papers

JJ McCarthy

Intermediate – ISBN 1 85805 039 1 Date 1993 Edition 1st Extent 96 pp (approx)

Higher – ISBN 1 8505 040 5 Date July 1993 Edition 1st
Extent 96 pp (approx)

These books provide all the practice required to meet the requirements of all the GCSE Mathematics examination board syllabuses at either the Intermediate or Higher Tiers and the National Curriculum at Key Stage 4. Pupils can use them for pre-exam revision to identify topics they need to practise further, or they can be used in the classroom.

Section 1 – One-Hour Practice Papers made up of

❐ 15 one-hour cross-topic practice papers that together will cover the whole of the GCSE syllabus and test the Attainment Targets 2 to 4. Each question in the papers has a 'helpline' reference to a page of Section 2 where you can find help assistance with the techniques behind that specific question and an indication of which level it is (ie Level 7, 8, 9 or 10).

❐ 15 coursework practice assignments

❐ 15 aural tests (each a set of 20 mental arithmetic questions),

Section 2 – Mathematical principles and formulae

A brief summary of the essential mathematical principles and formulae of the GCSE syllabus, organised by topic, with examples where necessary